institut français du pétrole publications

G. MONNOT

A. FEUGIER, F. LE BOUC
F. MAUSS, C. MEYER
E. PERTHUIS
G. DE SOETE

Preface by **BERNARD SALÉ**
Director, Energy
Institut Français du Pétrole

PRINCIPLES OF TURBULENT FIRED HEAT

Translated from the French by
ETHEL B. MILLER & RYLE L. MILLER

1985

GULF PUBLISHING COMPANY • BOOK DIVISION • HOUSTON, LONDON, PARIS, TOKYO

ÉDITIONS TECHNIP 27 RUE GINOUX 75737 PARIS CEDEX 15 

This Edition, 1985
Gulf Publishing Company
Book Division
Houston, Texas

ISBN 0-87201-724-9
Library of Congress Catalog Card No. 84-073311

Printed in France
by Imprimerie Louis-Jean, 05002 Gap

The several authors have contributed
to this book as indicated below

	CHAPTER No.
Georges MONNOT Docteur ès Sciences	1 , 2 , 3 , 4 5 , 6 , 7 , 8
Alain FEUGIER Docteur ès Sciences	7
François LE BOUC Ingénieur principal	3
François MAUSS Ingénieur de l'Institut Chimique de Nancy	1 , 6 , 7
Colette MEYER Docteur ès Sciences	7
Edmond PERTHUIS Ingénieur Ecole Nationale Supérieure du Pétrole et des Moteurs	7 , 8
Gérard de SOETE Docteur ès Sciences	2 , 7

All of these authors are on the staff of Institut Français du Pétrole

preface

Procedures for transforming the energy of chemical bonds into heat all share the same fundamental characteristics, whether for domestic space heating or for industrial heating. Thus combustion in furnaces and fired heaters is an aspect of technology with considerable practical value. This value is reflected by the economic position occupied by energy demands for domestic and industrial heating: worldwide, heating accounts for 80% of the fossil energy consumed, amounting to almost all the gas and coal and 60% of the oil.

Since humanity is faced with an inescapable necessity for reducing energy consumption to the point where supply and demand reach an equilibrium without economic recession, research toward reducing the specific consumptions of the various heating applications holds vital importance. Indeed, heating operations offer a prime area for fossil fuels to be replaced by electricity, to open the way for an expansion of nuclear energy, and thus to strengthen the position of nuclear energy with respect to satisfying energy needs. Although the present trend in this direction may be expected to continue, the actual substitutions will proceed slowly and the consumption of fossil fuel for domestic and industrial heating will surely remain very important for several decades. The Commission of the *Worldwide Conference of Energy* estimates that heating will still account for half of the energy consumption by the years 2000-2020 based on forecasts of the world energy balance. In France about half of the petroleum consumption will still be for industrial and domestic fuels in the years 1985-1990. Actions aimed at economizing these fuels are of considerable interest and will be for a long time.

Although the margin for improvement to be achieved outside the fireboxes of furnaces and heaters —improvements in insulation, control, process design, etc.— is important, improvement in the combustion process remains indispensable to energy conservation. Improved combustion permits a reduction in excess air, which in turn permits improving the productivity of installations as well as the reliability and longevity of construction materials through optimizing heat flows in the combustion chambers. Since the combustion process in large measure determines the emissions of pollutants, the emission of solid particles is the first criterion for evaluating combustion; acid soot (which is eliminated or reduced as low excess-air limits the formation of sulfuric acid and nitrogen oxides) is directly tied to the structure of the flames. At the level of combustion, therefore, a profound interaction exists between the objectives of saving energy and reducing pollution, which retains all its weight in the search for a more rational use of energy better adapted to the needs of mankind. The balance between maximum combustion and minimum air thus becomes the exciting and difficult task incumbent upon the combustion specialists.

In addition to this search for efficiency, the economics of energy supply impose changes in the nature of the fuel; and these in turn pose new problems in combustion. This is typified by the increase in the carbon-to-hydrogen ratio of fuels resulting from distilling off lighter hydrocarbon fractions for use as petrochemicals, gasoline and diesel fuel.

In confronting these problems, the combustion engineer finds himself in the grips of an extremely complicated collection of closely related phenomena that are subjects of a variety of disciplines, including fluid dynamics, mechanics of suspensions, chemical kinetics, convection, radiation, etc., all interacting on chemical factors (nature of the fuel, additives) and mechanical factors (design of the burner and the firebox). The consequent complexity gives rise to the empiricism that has marked this field for a long time.

Given the increasing difficulty of the interrelated problems, it becomes obvious that progress depends on a synthesis that can relate the divers phenomena in a logical structure. Accordingly, the value of this book by G. Monnot and his colleagues rests above all on the synthesis that they have been able to make of contemporary knowledge without neglecting any of the essential phenomena that relate to the quality of combustion and that must be considered as part of a contiguous whole.

For reasons of scientific and technical unity, as well as of personal experience, the authors have limited themselves to liquid and gaseous fuels. The first chapter gives the principal physical-chemical characteristics of these fuels when put to work in combustion. Subsequently the reader will find a thorough and practical discussion of the two types of fundamental phenomena competing to determine the characteristics of flames: the physical chemistry of combustion (Chapter 2) and the dynamics of gas flow in the firebox (Chapter 3).

The following chapters are devoted to overall events that are actually experienced in furnaces and heaters; and they correspond to similar sub-groups:

(a) Different ways of stabilizing a flame, which is the basic function of combustion systems, in Chapter 4.

(b) Characteristics of premixed flames (as distinguished from diffusion flames typical of contemporary practice and covered in the remaining chapters) is the subject of Chapter 6.

(c) Effects of preheating the combustion air (which has taken on much importance as a way of saving energy) as well as of oxygen enrichment (which has proven particularly useful in steel making) are compared in Chapter 5.

(d) Pollutant formation and the foreseeable ways for reducing emissions present an area where the scientific approach is particularly useful in arriving at a compromise between reducing either energy consumption or pollution, in Chapter 7.

(e) Industrial combustion is represented through the relations between the flames and radiant heat transfer in Chapter 8.

The team led by G. Monnot has been well armed to undertake this book. All the authors have been collaborating for several years on a project of the *Institut Français du Pétrole* that involved methods for improving the use made of petroleum products with respect to energy conservation and pollution reduction. Basic researchers, as well as furnace and heater combustion engineers, these are people who have worked together a long time in applying available knowledge to find practical solutions to the problems of using fuels. Their names are well known among combustion and heating specialists, and their numerous and important achievements demonstrate their competence. Most of them have participated actively in the work of *Fondation de Recherches Internationales sur les Flammes (FRIF) (International Flame Research Foundation-IFRF)* and the *Groupe d'Etude des Flammes de Gaz Naturel (GEFGN) (The Study Group on Natural Gas Flames)*. In addition, the principal author, G. Monnot, has been President of the Technical Committee of *GEFGN* and President of the French Committee of the *IFRF*. Besides extensive contacts with their colleagues and electricians and users in the petroleum, gas and coal industries, the authors have here found an opportunity for widening their knowledge and testing their ideas with the most eminent French and foreign combustion specialists. The reader can thus be assured that, with "Principles of Turbulent Fired Heat", he has at his disposal a working tool that is effective and reliable. I am convinced that this book will render great service to those engineers and technicians handling heating problems, and in a more general way all those concerned with the better use of energy.

The authors have dedicated their work to the memory of Professor Ribaud. Allow me to associate myself with their homage and add my homage to the man whose scientific work and personality has affected the heating field so profoundly and who inspired an Ecole Française whose vitality is attested by this book.

B. SALÉ
Director, Energy
Institut Français du Pétrole

foreword

This book is the fruit of efforts by a group of engineers from the *Institut Français du Pétrole*, who have each contributed to one or more of the chapters according to his or her field of interest.

A portion of the information presented here results from research done in the laboratories of *IFP*, specifically in Chapter 2, "Physical-chemical aspects of combustion," and Chapter 7, "Pollution due to combustion". The contents of Chapter 6, "Premixed combustion," owe much to various publications by engineers of *Gaz de France*. And the last chapter on heat transfer in fireboxes discusses experience acquired at *IFP* through application of the Hottel-Sarofim calculations.

Those chapters on the stability of flames (Chapter 4), on the influence of the fuel (Chapter 6) and on the kind of oxidizer (Chapter 5) are directly inspired by research carried out at the *International Flame Research Foundation (IFRF)* and the *Groupe d'Etude des Flammes de Gaz Naturel (GEFGN)*. IFRF has been working in Ymuiden, The Netherlands, since 1948: it has an international character and has benefited from the support of the *Communauté Européenne du Charbon et l'Acier (CECA)*. GEFGN comprises the French members of *IFRF* and the Italian *SNAM (Società Nazionale Metanodotti)*. The leader of the group and the principal source of funds is *Gaz de France*. Research has been carried out at Toulouse since 1959.

IFRF and *GEFGN* publish their work first for the benefit of their members and then distribute it for all users of thermal energy.

The first president of *IFRF* was Professor G. Ribaud, member of the *Académie des Sciences*, whose scientific stature and personality imbued the cooperative efforts toward this applied research with a common good will. This book is dedicated to his memory, as a homage to his work in the field of heating.

contents

Chapter 3
FLUID DYNAMICS IN THE FIREBOX

Chapter 4

THE STABILIZATION OF FLAMES

Chapter 5

HIGH-TEMPERATURE FLAMES

Chapter 6

PREMIXED COMBUSTION

Chapter 7

POLLUTION CAUSED BY COMBUSTION

Chapter 8

HEAT EXCHANGE IN THE FIREBOX

symbols for units of measurement

Symbol	Denomination	Symbol	Denomination
a	annum	l	liter
a	are	lm	lumen
a	atto	lx	lux
A	ampere		
		m	meter
bar	bar	m	milli
Bq	becquerel	M	mega
		min	minute
c	centi	mol	mole
C	coulomb		
cd	candela		
		n	nano
		N	newton
d	day		
d	deci		
D	darcy	p	pico
da	deca	P	peta
		P	poise
E	exa	Pa	pascal
f	femto		
F	farad	q	quintal
g	gram	rad	radian
G	giga		
Gy	gray	s	second
		S	siemens
h	hecto	sr	steradian
hr	hour	St	stokes
H	henry		
ha	hectare	t	ton
Hz	hertz	T	tera
		T	tesla
j	jour		
J	joule	V	volt
k	kilo		
K	kelvin	W	watt
kg	kilogram	Wb	weber

1

introduction
petroleum fuels

Before considering the use of petroleum fuels as a source of thermal energy, it seems worthwhile to indicate briefly the origins of those fuels that a user might have at his or her disposal, as well as specifications typical of the different types of fuels, and some numerical data related to the physical and chemical properties of those fuels.

Some references at the end of this chapter give sources of more extensive information on the manufacture and properties of fuels. The following discussion should suffice for an understanding of the subsequent chapters.

1.1. GASEOUS FUELS DERIVED FROM PETROLEUM

The petroleum industry is based on the exploitation of reserves of hydrocarbons trapped at considerable depth inside reservoir-rock in the earth — that is, trapped in porous rock to which the hydrocarbons have penetrated through the combined effects of pressure and capillary action.

These reservoirs generally enclose very complex mixtures of hydrocarbons along with salt water. When these mixtures are brought through drilled wells to the surface, where the pressure is generally lower than in the deposit, and the temperature is ambient, the separation of a gaseous phase and a liquid phase will occur.

The gases collected this way are "wet", that is, they contain vapors of higher-boiling hydrocarbons that have been carried out at the reduced partial pressures caused by association with the gases. Reciprocally, the liquid phase contains light hydrocarbon gases that have been retained at the reduced mole fractions of those gases in association with the bulk of the liquid.

The possibilities in the reservoir include natural gas in which gaseous hydro-carbons predominate over a small quantity of liquid products enclosed in the rock. Even so, the quantities of liquid or so-called condensate carried by the gas can be considerable, as for example 615,000 tons of condensate in 2,860 million cubic meters of gas from the gas reservoir at Hassi-R'Mel in the Sahara. Conversely, in a number of fields, the production of oil is accompanied by a considerable production of so-called flash gas, as for exemple, 200 m³ of gas at atmospheric pressure flashed off for each cubic meter of oil extracted at Hassi-Messaoud.

The crude products extracted from these reservoirs are refined before being used, except for unusual cases of direct use in the immediate vicinity of the wells. The purpose of the refining operation is to separate the heavy ends and to draw off well-defined by-products with specifications for particular uses.

1.1.1. Natural gas

The term "natural gas" designates the gas that does not liquefy at ambient temperatures and a few bars of pressure, at which it is generally transported and handled.

The principal constituent of natural gas is methane, the simplest and lightest of all the hydrocarbons. Methane comprises 69% of the crude gas and 97.6% of the refined gas from Lacq (France); it comprises 81% of the crude gas from Hassi-R'Mel (Algeria), 96% of the gas from Cortemaggiore (Italy), and 81% of the gas from Groningue (Netherlands). Along with methane, natural gas also contains ethane, propane, butane and even small quantities of C_5 or C_6 hydro-carbons. Also, natural gas can carry non-combustible gases, such as nitrogen

TABLE 1.1

PROPERTIES OF METHANE

Density at standard conditions (g/dm³)	0.7168
Specific gravity with respect to air	0.554
High heating value (kcal/m³)	9,510
Low heating value (kcal/m³)	8,570
Stoichiometric volume of air required for combustion (m³/m³)	9.52
Temperature achieved from adiabatic combustion with a stoichiometric volume of air at 25(°C)	1,952
Concentration limits of flammability in air (%):	
lower	5
upper	15
Temperature of autoignition (°C)	580
Speed of flame front in a stoichiometric mixture at 25° C (cm/s)	43.4

TABLE 1.2

SPECIFICATIONS FOR COMMERCIAL PROPANE AND BUTANE

	Commercial propane	Commercial butane
Definition	Mixture of hydrocarbons made of about 90% propane and propylene and the rest made of ethane, ethene, butanes and butenes	Mixture of hydrocarbons consisting mainly of butanes and butenes, with less than 19% vol. propane and propylene
Odor	Characteristic	Characteristic
Vapor pressure at 50°C, bars gauge	$11.5 \leqslant P \leqslant 19.3$	$P \leqslant 6.9$
Evaporation: Temperature °C for 95% vol. evaporated	$T \leqslant -15°C$	$T \leqslant +1°C$
Corrosion Copper strip	max. 1 b	max. 1 b
Sulfur compounds	$\leqslant 0.005\%$ max.	"Doctor test" negative
Water content	No moisture detected by the "cobalt bromide" test	No free water observed by decantation

TABLE 1.3

TYPICAL YIELDS OF STRAIGHT-RUN REFINERY
PRODUCTS BASED ON CRUDE

Product	Typical yield on crude (%)
Liquefied petroleum gas	1-2
Light naphtha	15-30
Heavy naphtha	5-10
Kerosene	10-15
Gas oil	15-20
Residual oil	40-50

(14% in the gas from Groningue) or carbon dioxide (9.6% in the Lacq crude). Because of the high content of methane in natural gas, it generally has physical and chemical properties that are close to those of methane. Table 1.1 shows some of the properties of methane.

1.1.2. Liquefied petroleum gas (LPG)

Two different gases, propane and butane, are typically offered for consumption as liquefied petroleum gas (LPG). Typical commercial products correspond to the specifications shown in Table 1.2. These gases are furnished for the most part by refineries that recover them from the distillation of crude oil (Table 1.3).

1.1.2.1. Preparation of liquefied petroleum gas

In the gas fields a portion of the propane and butane is found among the products condensed at the wellhead, although some propane and butane is also carried with the methane in the natural gas product. No matter whether it is a question of treating condensates at the gas wellhead or of treating crude oil, propane and butane are separated through fractional distillation, i.e. through passing vapors at their condensing point upward through downward flowing liquid at its boiling point by means of trays spaced one above another in a vertical tower, so that there is an equilibrium exchange to concentrate high boilers at the bottom and low boilers at the top. The relative concentrations depend on the number of trays and the amount of vapor-liquid recirculation. The trays in these distillation towers create close contact between the liquid and vapor phases. The ascending vapors are produced by boiling the bottoms with heat supplied in a reboiler. The descending liquid is produced by condensing a corresponding amount of vapors in an overhead condenser. Because low boilers are concentrated at the top, with high boilers at the bottom, the reboiler operates at a higher temperature than the condenser, with a temperature gradient occuring across the trays in between (Fig. 1.1). In special columns furnished with numerous trays, propane and butane are separated, and these columns are used as well for the separation of saturated and unsaturated products such as propylene and butene. Such separations are incomplete, however, and commercial products are mixtures of different close-boiling hydrocarbons.

Some of the properties of the hydrocarbons making up LPG are shown in Table 1.4.

For handling commercial products, it is important to know the variation with pressure of the so-called "dew point" at which the gas begins to condense (Table 1.5). In order that users will detect a leak before the air-borne concentration of the gas reaches its lower limit of flammability, a product with an unpleasant smell is added, usually ethyl mercaptan, tetrahydrothiophane, or

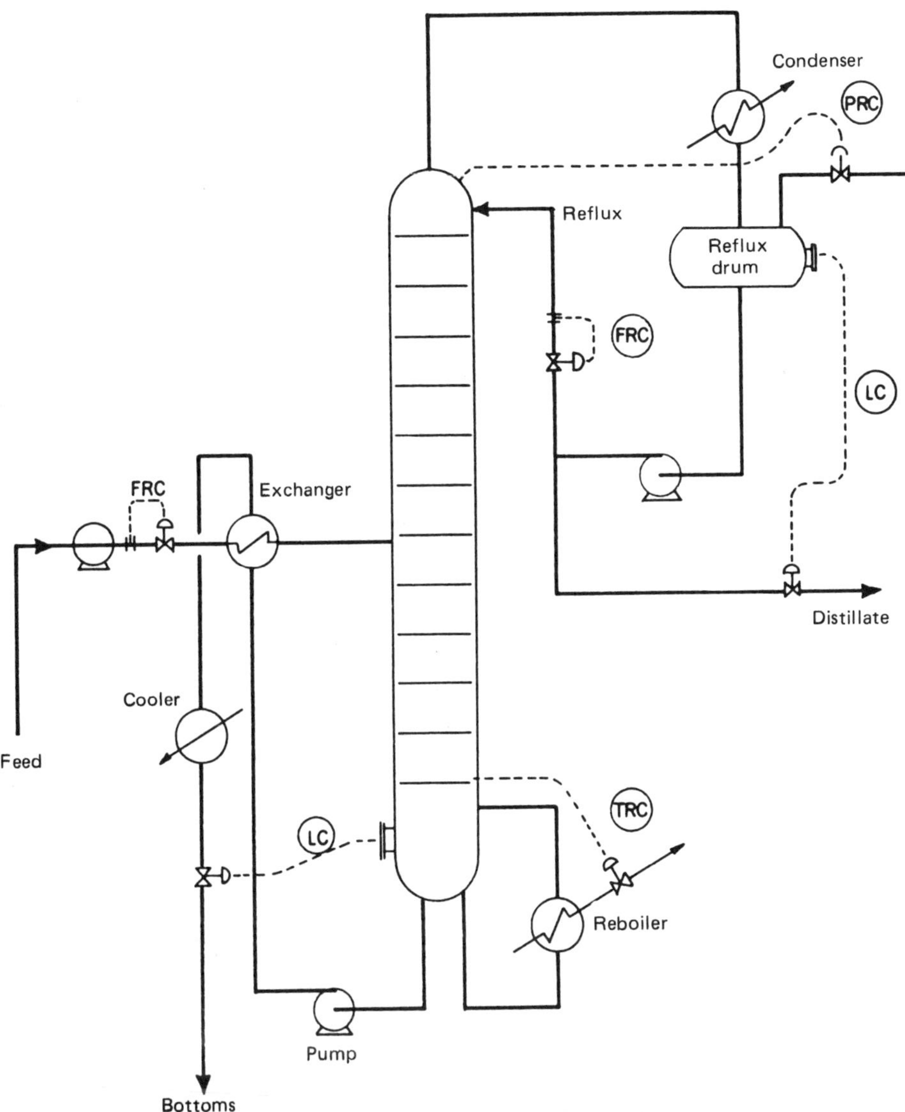

Fig. 1.1. Flow diagram for a distillation column and accessories.

amyl mercaptan, at concentrations of 0.5 to 0.7 kg for 40 m³ of LPG in liquid phase.

Finally, 2‰ by weight of methanol is added during winter to prevent stable hydrates that form in propane up to around 4°C in the presence of very low quantities of water that can be in the gas. For domestic use propane LPG is stored outside of dwellings and consequently allowed to take on the ambient temperature which in winter can be relatively low.

TABLE 1.4

PROPERTIES RELATING TO THE COMBUSTION OF SOME LIGHT HYDROCARBONS

Properties \ Hydrocarbon	Ethane	Ethylene	Propane	Propylene	n-butane	Isobutane	Butene-1	Isobutene	n-pentane	Pentene-1	Isopentane
High heating value { kcal/kg	12,399	12,022	12,034	11,692	11,832	11,797	11,577	11,505	11,715	11,545	11,688
High heating value { kcal/m^3	16,820	15,150	24,290	22,390	31,990	31,480			40,520		
Low heating value { kcal/kg	11,350	11,272	11,079	10,942	10,926	10,892	10,826	10,755	10,840	10,755	10,813
Low heating value { kcal/m^3	15,390	14,200	22,370	20,960	29,540	29,070			37,490		
kg air/kg	16.05	14.74	15.63	14.74	15.42	15.42	14.74	14.74	15.58	14.74	15.58
(NTP) m^3 air/kg	12.41	11.40	12.09	11.40	11.93	11.93	11.40	11.40	12.05	11.40	12.05
(NTP) m^3 air/m^3	15.65	14.37	24.41	21.83	32.25	31.83					
Maximum CO_2 in { wet	10.92	13.02	11.56	13.02	11.90	11.90	13.02	13.02	12.10	13.02	12.10
flue gas (%) { dry	13.12	14.98	13.69	14.98	13.99	13.99	14.98	14.98	14.17	14.98	14.17
Temperature (°C) of adiabatic combustion in stoichiometric air at 25°C and 1 bar	1,986	2,096	1,994	2,061	1,996	1,992	2,048	2,040	1,999	2,040	1,996
Flammability limits (vol %) of combustible mixture upper	3	3.1	2.25	2.2	1.85	1.8	1.6	—	1.45	—	1.4
lower	12.5	3.32	9.4	10.5	8.4	8.4	9.35	—	7.8	—	7.5
Temperature of autoignition (°C) in stoichiometric air	490	520	480	460	420	480	385	—	290	—	450

TABLE 1.5

DEW POINT TEMPERATURES ($^\circ$C) FOR COMMERCIAL PROPANE AND BUTANE

Gage pressure (bar)	Propane dew point	Butane dew point
0	− 42	− 0.5
1	− 26.5	+ 19
2	− 14	+ 31
3	− 5.5	+ 40.5

1.2. LIQUID FUELS

1.2.1. Different types of liquid petroleum fuels

Two classes of liquid fuels are derived from crude petroleum:

(a) Distilled fuels, which are drawn off from the sides or overheads of distillation columns in refineries.

(b) Residual fuels, which are the products remaining at the bottoms of the distillation columns.

Because the boiling points of the residual products are often higher than the temperatures at which thermal cracking occurs, the distilled fuels are usually carried overhead by upward flowing steam, called stripping steam. These distilled fuels, which are relatively light, can then be condensed at temperatures well below their cracking temperature.

The residual fuels contain the heaviest molecules that are in the crude oil, or that are formed during certain refinery operations. At atmospheric pressure, these molecules cannot be taken to their boiling point without breaking down, or cracking, due to the high temperature. Besides, a large part of the impurities present in the crude are also concentrated in these fractions, impurities the principal drawbacks of which are described below.

1.2.1.1. Kerosene

Although kerosene was originally produced for burning in lamps with a wick, products of this distillation range have become the principal fuel for gas turbines such as are used in jet planes; and as such, it has come to be known as jet fuel. Its distillation range is roughly from 200 to 300°C. It is practically free of sulfur.

Kerosene is identified by a specification for smoke point, that is, the flame height that can be obtained without smoke in a wick-lamp. This limits its aromatic hydrocarbons content.

Application for the use of kerosene is essentially domestic. Consumption is low. In 1975, in France it came to around 43,000 t, being 0.45% of the total consumption of petroleum products. Thus, it is a product of little importance and will not be discussed in the book. Nevertheless, we note that there is also an oil for lighting that is de-aromatized with a smoking point of 35 mm instead of 21 mm for the usual product (Table 1.6).

TABLE 1.6

TYPICAL SPECIFICATIONS FOR LIGHTING OIL

Property	Requirements
Definition	Either straight run distillate product or synthesized mixture of hydrocarbons.
Appearance and color	Clear. Saybolt color equal to or better than + 21.
Distillation profile	Less than 90% at 210°C. 65% or more at 250°C. 80% or more at 285°C.
Sulfur content	Not more than 0.13 wt.%.
Corrosion of copper cupon at 3 hr and 50°C	Less than 1 b.
Total acidity	Not more than 3 mg/100 cm^3 expressed as potassium hydroxide.
Cloud point	38°C maximum.
Smoke point	21 m/m maximum.

1.2.1.2.   Home heating oil

Typical specifications for domestic fuel oil are shown in Table 1.7. It has a cetane number higher than or equal to 40, which suits it for diesel motors that are less demanding so far as autoignition is concerned.

The boiling specification is for a fraction that is 65% distilled at 250° C and 85% at 350° C. Actually, the residuals in domestic oil will boil off below 400° C. This fuel remains fluid down to temperatures slightly below 0° C; and its maximum viscosity, 9.5 cSt at 20°C, makes handling easy. Because of its derivation, its water and sediment contents are low, so that it will pass through fine filters without risk of warping them. It has a sulfur content less than or equal to 0.50%; and this gives it important advantages for certain uses. Small amounts of colorants and tracing agents are added to this fuel to avoid unethical marketing as fuel for ordinary diesel engines.

TABLE 1.7

TYPICAL SPECIFICATIONS FOR DOMESTIC FUEL OIL

Property	Specification
Definition	A mixture of hydrocarbons derived either from mineral sources or synthesized, and intended for either heat production in combustion facilities or in certain cases power generation in internal combustion engines.
Color	Red.
Viscosity at 20° C	9.5 cSt maximum.
Sulfur content	0.3 wt. % maximum.
Distillation	Less than 65% distilled at 250° C.
Cloud point	55° C maximum.
Content of water and sediment	0.10 wt. % maximum.
Pour point	6° C maximum.
Conradson carbon on residue	0.35% maximum.
Cetane number	40 minimum.
Coloring agent	Up to 1 g/hl of clear colorant identified chemically.
Identifying agents	5 g/hl of diphenylamine; 1 g/hl of furfural.

The consumption of home heating fuel oil in the United States came to 1,040 million barrels in 1975 making it 17% of the total tonnage of petroleum products consumed in that year. Fuel oil for space heating is thus a large-volume product.

1.2.1.3. Special low-sulfur fuels

Typical specifications for special low-sulfur fuel oil are shown in Table 1.8.

This fuel is made up of mixtures of distillates and residual products. Its viscosity, as well as some other properties, indicates some precautions are necessary to its use. It remains liquid above 0° C. Its water and sediment content, as well as its sulfur content, are limited. It finds applications where combustion products should not contain large quantities of corrosive products.

Nevertheless, the presence of residual products causes this type of fuel to produce black smoke and create soot in fireboxes that have poor flues. This

TABLE 1.8

TYPICAL SPECIFICATIONS FOR SPECIAL LOW-SULFUR LIGHT FUEL OIL

Property	Specification
Definition	Hydrocarbon mixtures, either derived from mineral oil or synthesized, which are destined for heat production in combustion facilities.
Color	Black.
Viscosity	15 cSt maximum at 50° C.
Sulfur content	1 wt. % maximum.
Distillation	Less than 65% off at 250°C. Less than 85% off at 350°C.
Cloud point	70° C maximum.
Water content	0.50 wt. % maximum.
Solids content	0.15 wt. % maximum (sediments only).
Pour point	0° C maximum.

tendency has brought on anti-pollution restrictions replacing it with domestic fuel oil in most of the furnaces of buildings in cities.

The consumption of this type of oil was 70 million barrels in the United States of America in 1975 making it 1.2% of the consumption of petroleum products. This proportion has varied little in the petroleum market over the past few years.

1.2.1.4.   No. 1 heavy fuel oil

Heavy fuel oil No. 1 is essentially a residual product. Typical specifications are shown in Table 1.9.

This oil is characterized mostly by its viscosity and its fractions distilled at 250° C and 350° C. Its sulfur content, although limited to 2% by weight, makes it unusable for certain purposes. This fuel is an industrial product, needing reheating for handling and utilization.

1.2.1.5.   No. 2 heavy fuel oil

Heavy fuel oil No. 2 also is essentially residual products. Typical specifications are shown in Table 1.10. This oil is more viscous than fuel oil No. 1, which means this oil needs to be heated more for handling and use. Because the sulfur

TABLE 1.9

TYPICAL SPECIFICATIONS FOR No.1 HEAVY FUEL OIL

Property	Specification
Definition	Hydrocarbon mixtures, either derived from mineral oil or synthesized, which are destined for heat production in combustion facilities.
Viscosity	15 cSt maximum at 50° C.
Sulfur content	2 wt.% maximum.
Distillation	Less than 65% off at 250°C. Less than 85% off at 350°C.
Cloud point	70° C maximum.
Water content	0.75 wt.% maximum.
Solids content	0.25 wt.% maximum (sediments only).

TABLE 1.10

TYPICAL SPECIFICATIONS FOR No.2 HEAVY FUEL OIL

Property	Specification
Definition	Hydrocarbon mixtures, either derived from mineral oil or synthesized, which are destined for heat production in combustion facilities.
Viscosity	15 cSt maximum at 50° C.
Sulfur content	4 wt.% maximum.
Distillation	Less than 65% off at 250°C. Less than 85% off at 350°C.
Cloud point	70° C minimum.
Water content	1.5 wt.% maximum.
Sediment content	0.25 wt.% maximum.

content of this fuel can go to 4% by weight, the flue gas resulting from its combustion contains sulfur oxides.

The *US* consumption of heavy fuel oils (No. 1 and No. 2) came to 883 million barrels in 1975 with the greatest portion being fuel oil No. 2, on the order of 75%.

1.2.1.6.   No. 2 heavy fuel oil with low sulfur content

Some applications require that there be no oxides of sulfur (SO_2 and SO_3) in the combustion products. For these needs, refineries can deliver fuels with the sulfur reduced to less than 2% or less than 1%.

In 1975, deliveries of low sulfur fuel oil came to around 18% of the total deliveries of heavy fuels from United States refineries. Here then is a product of large consumption, whose importance is abetted by certain oils which generally have low sulfur (0.13% for Hassi-Messaoud and Nigerian crudes; 0.45% for crude from Libya). The crudes from the North Sea are also low in sulfur.

1.2.2.   Preparation of liquid petroleum fuels

We will give only some very general points on the preparation of petroleum liquid fuels, sending the reader who would like more information to specialized books.

1.2.2.1.   Distilled liquid fuels

Without going into a detailed description of refinery operations, it is possible to say that a refiner's choice of products depends on:

(a)   The process units represented by the installed equipment, including units for: atmospheric distillation, vacuum distillation, visbreaking, steam-cracking, catalytic cracking, hydrocracking, catalytic reforming, hydrorefining, alkylation, polymerization, finishing treatment and so forth. Some of these units have similar functions; and an initial selection was made in the design of the refinery with respect to the processes and consequently the possible products.

(b)   The kind of crude oils available to the refinery, including crudes that are more or less heavy with a predominance of paraffinic, napthenic aromatic compounds and containing more or less of impurities such as sulfur, vanadium, nitrogen and so forth.

(c)   The market for the finished products which is, finally, the deciding factor.

The multiplicity of the possibilities in these variables has led the refiner to use computers for establishing his manufacturing program. A first study consists of translating all the manufacturing operations and the restrictions that are imposed by the specifications of the products into first-degree equations. In a typical case, a refiner might obtain p linear equations with n unknowns, so that with $(n - p)$ degrees of freedom, there is an infinite number of solutions. Only one of these solutions satisfies all the equations by obtaining an optimum for a function that reflects the profit for the refiner. This one solution indicates the use and the mixture of the different base products from the various process units so that there is no universal "recipe" for making gas oil or fuel oil, but rather a different product blend for each refinery.

The commercial products obtained in these product blends can also have their market quality improved by small quantities of special additives. The effect on the manufacturing cost of this operation, if it is necessary, is taken into account in setting up the manufacturing program for the refinery, following the method described above.

1.2.2.2. Residual fuel oils

The components of residual fuel oil include residues from atmospheric and vacuum distillation, as well as from a variety of refinery operations that produce a high-boiling residue.

Heavy fuel oils are prepared by blending these different components and adding, as in the case of special light fuel, a large quantity (around 50%) of gas oil to give the product the viscosity required by the specifications. Special light fuel oil is thus about equal quantities of distilled products and residual products.

Generally, the relative amounts of the different products going into residual fuels depends, within the framework of a given refinery, on the kind of crude oil treated and the market requirements.

The art of the refiner consists in making up, by means of the available processes, a roster of mixtures that meet specifications for the liquid fuels in current demand, and that also are sufficiently stable to be stored without degradation over a long period of time.

Heavy fuel oils must be heated in storage to keep them ready for pumping at the time of use. Storage can cause a precipitation of asphaltic components. The asphalt is actually present in the form of colloidal micelles, peptized in the oil phase by resins. This blend of resins, asphalts, oil and solubilizer can have its equilibrium displaced by the addition of products of different origins, and this causes precipitation of the asphalts. Sediments and water aggravate the problem of such precipitation by forming a sludge.

To avoid such problems, refiners observe the following principles:

(a) Avoid using paraffinic residues in blending special light fuels with low sulfur content.

(b) Avoid contaminating blends with solid impurities.

(c) Reserve a lower zone for decantation in storage tanks.

(d) Avoid mixing products that are known to be incompatible.

There are tests for the stability and compatibility of fuel oils, principal among which are:

(a) The NBTL test (Navy Boiler and Turbine Laboratory).

(b) The SHFT test (Shell Hot Filtration Test).

REFERENCES

1.1   *Activité de l'industrie pétrolière. Eléments statistiques.*
Tome I. Comité professionnel du pétrole, 51, Boulevard de Courcelles, Paris.
Tome II. Ministère de l'Industrie et de la Recherche. Direction des Carburants, 5, rue Barbet de Jouy, Paris.
These publications appear each year.

1.2   NORMAND, X. – *Leçons sommaires sur l'industrie du raffinage du pétrole.* Editions Technip, Paris, 5th ed., 1977.

1.3   WUITHIER, P. – *Le pétrole. Raffinage et génie chimique.* 2nd edition entirely revised. Editions Technip, Paris, 1972.

2

the physical chemistry of combustion

Combustion, especially the combustion of liquid or gaseous hydrocarbons, results from very rapid reactions between the fuels and oxygen that is usually oxygen of the air. This chapter reviews the fundamentals of these reactions: the energy they liberate, and how they are initiated and propagated.

2.1. COMBUSTION REACTIONS

2.1.1. The nature and the temperature of combustion products

This study considers the overall evolution of the system beginning with the fuel and the oxidizer and ending with flue gases.

To illustrate, assume the combustion of methane, which is the principal constituent of natural gas. Methane is gaseous under normal conditions of temperature and pressure. Its formation as CH_4 from carbon and hydrogen molecules occurs with a liberation of energy. By convention, this energy is called the "heat of formation", when it applies to one molecule of the compound formed in its natural state at 25° C from the elements, whose heats of formation are taken to be zero.

The mixture of methane and oxygen under normal conditions is metastable and does not change unless there happens to be a catalyst present. On the other hand, if a suitable proportion of methane and oxygen is heated high enough, a reaction takes place, with the disappearance of the initial molecules and the appearance of new molecules, such as CO_2 and H_2O and CO. These combustion products plus unreacted fuel and air make up the flue gas. Each one of the compounds in flue gas possesses its own heat of formation, and the energy of the combustion reaction is the difference between the sum of the energies of formation of the final products and the sum of the energies of formation of the initial products, the final products having been previously brought to normal conditions of temperature and pressure.

For a complete combustion, this reaction might be represented by the Equation:

$$CH_4 + 2O_2 \rightarrow CO_2 + 2H_2O - \Delta H \qquad (2.1)$$

where the standard heat of the reaction is shown by ΔH, whose sign indicates whether the reaction is exothermic $(-)$ or endothermic $(+)$.

Depending on how much combustion heat is absorbed by the surroundings, this heat appears as sensible heat in the flue gas, as shown by the temperature of the flue gas; when the overall reaction is adiabatic, the flue gas reaches very high temperatures, at which its molecules can show a certain instability by decomposing, e.g. according to the following equilibrated equations:

$$CO_2 \rightleftharpoons CO + \frac{1}{2}O_2 + \Delta H_1 \qquad (2.2)$$

$$H_2O \rightleftharpoons H_2 + \frac{1}{2}O_2 + \Delta H_2 \qquad (2.3)$$

$$H_2O \rightleftharpoons H + OH + \Delta H_3 \qquad (2.4)$$

$$H_2 \rightleftharpoons 2H + \Delta H_4 \qquad (2.5)$$

$$O_2 \rightleftharpoons 2O + \Delta H_5 \qquad (2.6)$$

As indicated by the equations, these different decompositions absorb energy and thus act to limit the temperature of the flue gas.

The extent of the disassociations is a function of temperature, θ, and is expressed by equilibrium constant, K, which at constant pressure depends only on the temperature:

$$K_1(\theta) = \frac{(CO)(O_2)^{\frac{1}{2}}}{(CO_2)} \qquad (2.2')$$

$$K_2(\theta) = \frac{(H_2)(O_2)^{\frac{1}{2}}}{(H_2O)} \qquad (2.3')$$

$$K_3(\theta) = \frac{(H)\,(OH)}{(H_2O)} \qquad (2.4')$$

$$K_4(\theta) = \frac{(H)^2}{(H_2)} \qquad (2.5')$$

$$K_5(\theta) = \frac{(O)^2}{(O_2)} \qquad (2.6')$$

where the concentrations of the different chemical species in the mixture are expressed by (CO), (O_2), (H_2), etc. Ordinarily, these are gases and the concentrations are expressed as partial pressures, which are proportional to their mole fractions.

Knowing the values of ΔH, ΔH_1, ..., ΔH_6 and the equilibrium constants at different temperatures, the adiabatic equilibrium temperature of the combustion products can be calculated. The calculation can easily take into account a possible preheating of the products.

2.1.2.   Kinetics of combustion

As with all kinetics, the kinetics of combustion reactions depend, first, on the mechanism or sequence of the chemical reactions, and then on the rates of the individual reactions.

2.1.2.1.   Reaction mechanism of combustion

We shall illustrate the reaction mechanism by the combustion of methane:

$$CH_4 + 2\,O_2 \rightarrow CO_2 + 2\,H_2O \qquad (2.1\ bis)$$

In order for the mechanism to follow the overall reaction as expressed above, the combustion of methane would require a three-body collision leading to the formation of three molecules. Such a direct reaction is not statistically probable and the actual mechanism is more like the following:

$$\text{initiation}: CH_4 \rightarrow C\dot{H}_3 + \dot{H} \qquad (2.7)$$

where $C\dot{H}_3$ and $\dot{H}$ are free radicals, with a solitary valence electron. Only a very small amount of free radicals are actually formed by the initiation reaction; it needs a supply of energy, such as heat or photons from electromagnetic radiation, and it is followed by a propagation:

$$\begin{cases} p' & \dot{C}H_3 + O_2 \rightarrow CH_2O + O\dot{H} \end{cases} \qquad (2.8)$$

$$\begin{cases} p & O\dot{H} + CH_4 \rightarrow H_2O + \dot{C}H_3 \end{cases} \qquad (2.9)$$

$$p' \qquad \dot{R}_1 + A \rightarrow C + \dot{R}_2 \qquad (2.8\ bis)$$

$$p \qquad \dot{R}_2 + B \rightarrow D + \dot{R}_1 \qquad (2.9\ bis)$$

Adding these two propagation p' and p reactions with a subsequent reaction between formaldehyde (CH_2O) and oxygen results in the total balance of moles given in Eq. (2.1) because the further oxidation of formaldehyde gives one molecule of CO_2 and another molecule of H_2O.

In this reaction mechanism, the $\dot{C}H_3$ radical is regenerated; it can take part in a new set of propagation reactions indefinitely, and the combustion advances in this way. Equations (2.8 bis) and (2.9 bis) represent generalized statements, in which A, B, C and D are molecules and $\dot{R}_1$ and $\dot{R}_2$ free radicals.

This chain of reactions can be interrupted, however, when the free radicals are removed. Such a break can take place in homogeneous phase, when another substance converts the free radical into a stable compound, or in heterogeneous phase, when a radical hits the wall of the reactor:

$$\text{homogeneous:} \ \ \dot{R}_1 + \dot{R}_2 + (M) \rightarrow \text{stable compounds}$$

$$\text{heterogeneous:} \ \ \dot{R} + \text{wall} \rightarrow \text{stable compounds}$$

Conversely, other reactions than (2.7) can lead to an increase in the number of free radical; they are called chain branching reactions:

$$\dot{C}H_3 + O_2 \rightarrow \dot{C}H + 2\dot{O}H \qquad (2.10)$$

$$\dot{R}_1 + A_2 \rightarrow \dot{R}_3 + 2\dot{R}_2 \qquad (2.10\ bis)$$

The new radicals formed in the course of this reaction are capable of entering into propagation chains p and p'. A review of the possibilities shows that the variation of free radicals in a reaction must be equal to an even number or to zero. The overall combustion reaction will be the result of these different elementary reaction steps occurring simultaneously in the reaction mixture.

2.1.2.2. Reaction rate

The reaction rate can be expressed as the change in the concentration of an initial reactant or a reaction product over a period of time. Using as an example the combustion of methane (reaction 2.1), this overall rate can be represented by the equation:

$$V_R = -\frac{d}{dt}\frac{p_{CH_4}}{p_{total}} = \frac{d}{dt}\frac{p_{CO_2}}{p_{total}} = \frac{1}{2}\frac{d}{dt}\frac{p_{H_2O}}{p_{total}} \qquad (2.11)$$

where p is the partial pressure of the substances identified by the subscripts, and $\dfrac{d}{dt}$ is the derivative with respect to time. The factor $\dfrac{1}{2}$ derives from formation of two molecules of water out of a single molecule of methane.

Taking up the reaction mechanism, the rate of each step can be represented as follows: Let Z_2 be the number of biparticle collisions per unit time and per particle (molecule or radical); let α, β and δ stand for the probabilities of propagation, termination or branching, respectively, that might follow one such collision. Then the probability that a collision will cause a chemical reaction is equal to the sum of $\alpha + \beta + \delta$. And the reaction rate, V_p, can be described as:

1. *The event of a propagation reaction*

$$V_p = Z_2 \times \begin{array}{c}\text{probability} \\ \text{that one of} \\ \text{the partners} \\ \text{will be } CH_3 \\ \text{(or } \dot{O}H \text{ for} \\ V_{p'})\end{array} \times \begin{array}{c}\text{probability} \\ \text{that the} \\ \text{other} \\ \text{partner} \\ \text{will be} \\ O_2 \text{ (or } CH_4 \\ \text{for respec-} \\ \text{tively } V_{p'} \\ \text{and } V_p)\end{array} \times \begin{array}{c}\text{probability} \\ \text{that the} \\ \text{collision} \\ \text{energy will} \\ \text{be suffi-} \\ \text{cient for} \\ \text{the stage}\end{array} \times \begin{array}{c}\text{steric} \\ \text{factor, } f, \\ \text{of the} \\ \text{collisions}\end{array}$$

$$V_p = Z_2 \frac{p_{\dot{C}H_3}}{p_t} \times \frac{p_{O_2}}{p_t} \times \exp\left(-\frac{E_p}{R\theta}\right) \times f$$

If $p_{\dot{R}}$ is the partial pressure of the free radicals, then:

$$V_{p'} = Z_2 \frac{p_{\dot{R}}}{p_t} \frac{p_{\dot{C}H_3}}{p_{\dot{R}}} \frac{p_{O_2}}{p_t} \exp\left(-\frac{E_p}{R\theta}\right) f$$

Grouping the last four factors as:

$$\alpha = \frac{p_{\dot{C}H_3}}{p_{\dot{R}}} \frac{p_{O_2}}{p_t} \exp\left(-\frac{E_p}{R\theta}\right) f$$

allows the equation to be written as:

$$V_p = Z_2 \alpha \frac{p_{\dot{R}}}{p_t} \tag{2.12}$$

2. *The event of a branch reaction*

$$V_r = Z_2 \frac{p_{\dot{R}}}{p_t} \frac{p_{\dot{C}H_3}}{p_{\dot{R}}} \frac{p_{O_2}}{p_t} \exp\left(-\frac{E_r}{R\theta}\right) f'$$

and putting the last four factors together:

$$\delta = \frac{p_{\dot{C}H_3}}{p_{\dot{R}}} \frac{p_{O_2}}{p_t} \exp\left(-\frac{E_r}{R\theta}\right) f'$$

we have:

$$V_r = Z_2 \delta \frac{p_{\dot{R}}}{p_t} \qquad (2.13)$$

3. The event of a homogeneous chain termination

$$V_t = Z_3 \left(\frac{p_{\dot{R}}}{p_t}\right)^2 \exp\left(-\frac{E_t}{R\theta}\right) f''$$

where

Z_3 is the frequency of trimolecular collisions, and $Z_3 \simeq Z_2 \dfrac{\sigma}{\lambda}$,

σ = average diameter of a particle,
λ = average free path of a particle.

Substituting this expression for Z_3:

$$V_t = Z_2 \left(\frac{p_{\dot{R}}}{p_t}\right)^2 \frac{\sigma}{\lambda} \exp\left(-\frac{E_t}{R\theta}\right) f''$$

and grouping the last three terms:

$$\beta = \frac{\sigma}{\lambda} \exp\left(-\frac{E_t}{R\theta}\right) f''$$

$$V_t = Z_2 \beta \left(\frac{p_{\dot{R}}}{p_t}\right)^2 \qquad (2.14)$$

Since the activation energies of homogeneous chain reactions, E_t, will be near zero:

$$\exp\left(-\frac{E_t}{R\theta}\right) = 1$$

The exponential factor $\exp\left(-\dfrac{E}{R\theta}\right)$ represents that fraction of the collisions that will involve molecules with an energy level high enough to activate the reaction. This is why exponent E has the dimensions of energy and has come to be called activation energy; it represents the minimum energy necessary at the moment of the collision for a given chemical reaction to take place. If the available energy is less than E, the collision may be simply elastic.

Now, if in the mechanism of a chain reaction, the overall reaction is attained after a propagation reaction, the result will be:

$$V_R = V_p = Z_2 \, \alpha \frac{p_{\dot{R}}}{p_t} \tag{2.15}$$

The overall reaction rate V_R will be proportional to the number of free radicals. If there is an increase in the number of free radicals, as when $\delta > \beta$, the reaction rate will be accelerated, and this in turn usually brings an increase in temperature, θ, for exothermic reactions. These accelerating effects lead to uncontrolled reactions which characterize explosions.

Finally, let us note that the activation energies of reactions progressing through free radicals are low compared to those taking place without free radicals (Table 2.1).

TABLE 2.1

ENERGIES OF ACTIVATION OF SOME COMBUSTION REACTIONS

$$
\begin{array}{lll}
\dot{C}H_3 + O_2 & \rightarrow CH_2O + \dot{O}H & E = 19 \text{ kcal/mol} \\
 & \hookrightarrow CO + H_2 & \\
\dot{O}H + CH_4 & \rightarrow H_2O + \dot{C}H_3 & E = 9 \text{ kcal/mol} \\
CH_4 + 2\,O_2 & \rightarrow CO_2 + 2\,H_2O & E = \text{very high but unknown}
\end{array}
$$

The energies of activation of chain termination are usually equal to zero. The energies of activation of branching reaction are of some tens of kilocalories/mole.

2.2   AUTOIGNITION

Under normal conditions of temperature and pressure, mixtures of fuel and oxidizer generally do not react; that is, the reaction rate of the oxidation is infinitely small. When the fuel and the oxidizer are simply put in contact without being intimately mixed, any combustion reaction that may be started will be controlled by the rate of diffusion, which is usually much slower than the rate of the reaction.

Therefore, a homogeneous mixture of fuel and oxidizer will ignite spontaneously only under certain limited conditions of temperature, pressure and concentration.

The concentrations are usually identified by the equivalence ratio r, as defined by the following dimensionless relationship:

$$r = \frac{(\text{fuel/oxidizer})_{\text{actual}}}{(\text{fuel/oxidizer})_{\text{stoichiometric}}}$$

r is equal to 1 for a stoichiometric mixture. For "lean" mixtures having an excess of oxidizer:

$$0 < r < 1$$

and for "rich" mixtures having an excess of fuel:

$$r > 1$$

In those correlations that employ "excess air," λ, as a parameter, λ is defined as:

$$\lambda = \frac{1}{r}$$

For a number of hydrocarbons, the limits with oxygen or air extend to both sides of the stoichiometric, or above and below $r = 1$. The field of autoignition of a given fuel mixed with an oxidizer is limited by conditions of pressure (pressure limit of autoignition), of temperature (temperature limit of auto-ignition, also called ignition temperature) and of composition.

When an explosive mixture is taken to the temperature and pressure of auto-ignition, fast combustion starts in the whole volume. However, there is a time lapse between the instant that the conditions of autoignition are reached and the instant the autoignition takes place. This time interval is called autoignition delay.

2.2.1. Experimental results

2.2.1.1. The limit of autoignition

The autoignition of hydrocarbon/oxygen/nitrogen mixtures is frequently described by the autoignition diagram shown in Fig. 2.1. In the area of low pressure and high temperature at the lower right in this diagram the data are represented by an S-curve, with three pressure limits for one value of the temperature.

In the area of high pressure and moderate temperature, by contrast, there is a dip and rise in the curve, indicating three temperature limits for one value of pressure. The dip exhibited by the curve at this point corresponds to the appearance of cold flames within certain chemical species. When observed with a spectroscope, these cold flames are seen to emit light waves characteristic of formaldehyde. Actually, in this region, the chemical transformation does not proceed to the limits of oxidation (CO_2 and H_2O) but is interrupted at the level of intermediates characterized by aldehydes.

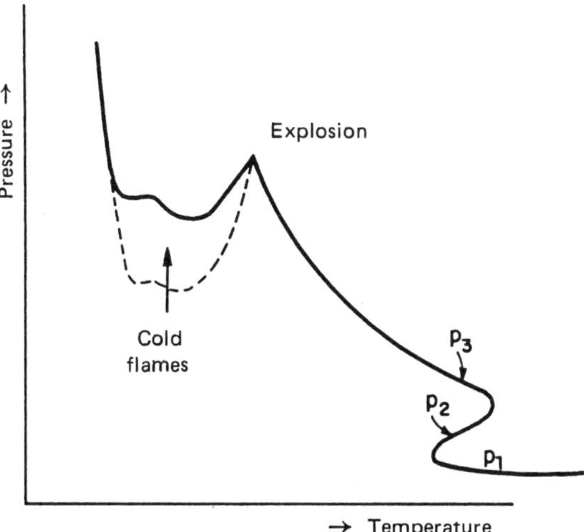

Fig. 2.1. Limits of autoignition for hydrocarbons.

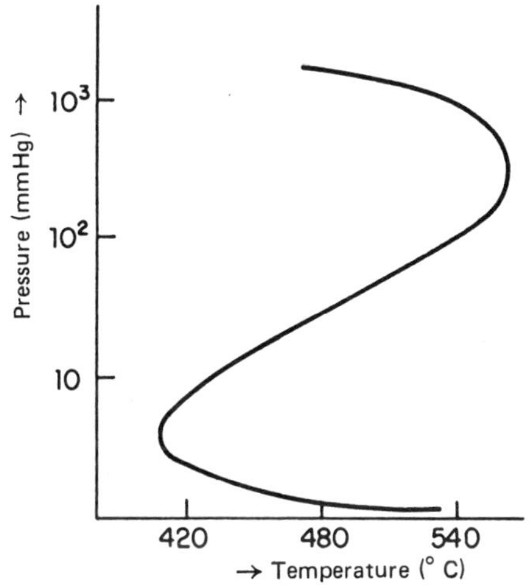

Fig. 2.2. Limits of autoignition for H_2/O_2 mixtures.

Mixtures of hydrogen/oxygen/nitrogen or carbon monoxide/oxygen/nitrogen exhibit S diagrams without a region of cold flames. Fig. 2.2 shows the H_2/O_2 system. These systems thus exhibit an area of three successive pressures of autoignition for a large range of temperatures. Experimental determination of such curves is difficult, all the more so because there is for each point a unique delay of autoignition whose value is sensitive to the values of pressure and temperature.

Two methods, the crucible method and the gaseous current or pyrometer

TABLE 2.2

AUTOIGNITION TEMPERATURES FOR SOME HYDROCARBONS IN AIR

Hydrocarbon		Temperature ($^\circ$C)
H_2	Hydrogen	570
NH_3	Ammoniac	650
N_2H_4	Hydrazine	270
H_2S	Hydrogen sulfide	290
HCN	Hydrogen cyanide	540
$(CN)_2$	Cyanogen	850
CS_2	Carbon disulfide	130
CO	Carbon monoxide	630
CH_4	Methane	580
C_2H_6	Ethane	490
C_3H_8	Propane	480
C_4H_{10}	Butane	420-480
C_6H_{14}	*n*-hexane	260 (?)
C_7H_{16}	*n*-heptane	285 (?)
C_8H_{18}	*n*-octane	220 (?)
$C_{10}H_{22}$	*n*-decane	240 (?)
C_6H_{12}	Cyclohexane	265
CH_2Cl_2	Dichloromethane	650 ⎫ not combustible in
C_2HCl_3	Trichlorethylene	460 ⎭ air
C_2H_4	Ethylene	520
C_3H_6	Propylene	460
C_4H_8	Butene-1	385
	Butene-2	435
C_4H_6	Butadiene	420
C_2H_2	Acetylene	320
C_6H_6	Benzene	620 (?)
$C_6H_5CH_3$	Toluene	585 (?)
$C_6H_4(CH_3)_2$	Xylene	520
$C_{10}H_8$	Naphthalene	575
CH_3OH	Methanol	510 (?)
C_2H_5OH	Ethanol	490 (?)
$(C_2H_5)_2O$	Ethyl ether	190 (?)
$(CH_3)_2CO$	Acetone	560 (?)
CH_3COOH	Acetic acid	575
CH_3NH_2	Methylamine	430
CH_3Cl	Methyl chloride	630

method (attributed to Mallard and Le Chatelier) are suitable for measuring autoignition delays on the order of a second or more. Shorter delays are measured in rapid compression machines or shock tubes, which produce an abrupt heating by means of adiabatic compression. Descriptions of these methods are found in works on chemical kinetics.

Table 2.2 lists temperature limits of autoignition in air at atmospheric pressure for various substances.

2.2.1.2.   The autoignition delay

We have seen that the autoignition delay varies considerably over the range of possible temperatures and pressures. Those delays occurring at the low pressure dip in the curve of Fig. 2.1 are relatively long, on the order of minutes, even hours. Fig. 2.3 shows this region for a typical mixture comprising normal hexane and air with $r = 0.69$ in more detail. On branch $p_2\text{-}p_1$ of this curve, delays are much longer than those measured on branch p_3 where they are barely more than a few seconds. The extrapolation of the branch p_3 along AB actually corresponds to a different physical phenomenon. In the area above AB are observed short delays corresponding to explosions in a single stage. However, below AB, the observed delays correspond to two stages. In the first and longer stage a gradual increase in pressure occurs along with an increase in concentrations of aldehyde, carbon monoxide and hydrogen, as fuel is transformed but little carbon dioxide is formed.

This transformation is illustrated in Fig. 2.4 with a stoichiometric mixture of ethane and oxygen at a temperature of 600° C. After the first stage, a second, much shorter one appears characterized by relatively constant concentrations of chemicals (Fig. 2.4), as well as by constant pressure. Then, at the end of this second stage, the explosion occurs with a sudden decrease in pressure, due to a reduction in the number of molecules with formation of carbon dioxide. At the moment of this explosion almost all of the initial fuel has already disappeared, transformed into intermediates, such as aldehydes, carbon monoxide and possibly peroxides. Formaldehyde is formed near the end of the first stage with the appearance of a blue luminescence characteristic of this molecule when excited.

Curves like $p_2\text{-}p_1$ of Fig. 2.3 occur only with lean hydrocarbon/oxygen mixtures, as for example, $(O_2) \geqslant 63\%$ for CH_4/O_2 mixtures.

The high pressure dip in the curve in Fig. 2.1, being close to an area where cold flames appear, also exhibits delays over two different stages, with the first stage (delay Δt_1) leading to formation of such compounds as aldehydes and olefins that appear in the cold flames.

Recordings made during experiments with a rapid-compression machine show that the pressure increases at the end of delay Δt_1 (Fig. 2.5). This increase in pressure is due to a simultaneous rise in temperature and increase in number of moles. Generally, the autoignition delays for a given chemical depend on the temperature and the pressure according to the following expression:

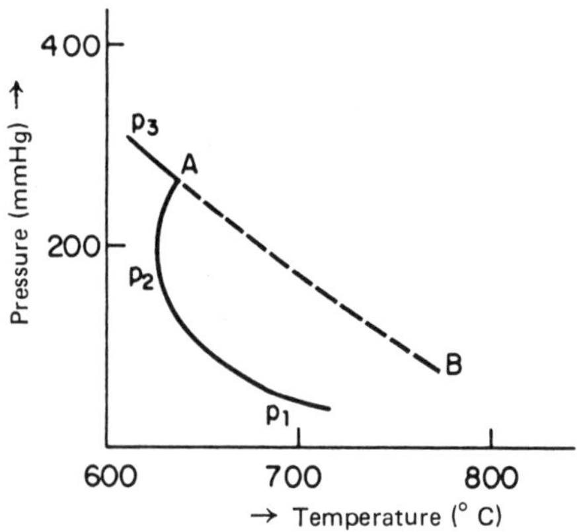

Fig. 2.3. Autoignition temperatures for mixture of *n*-hexane/air = 0.69 stoichiometric.

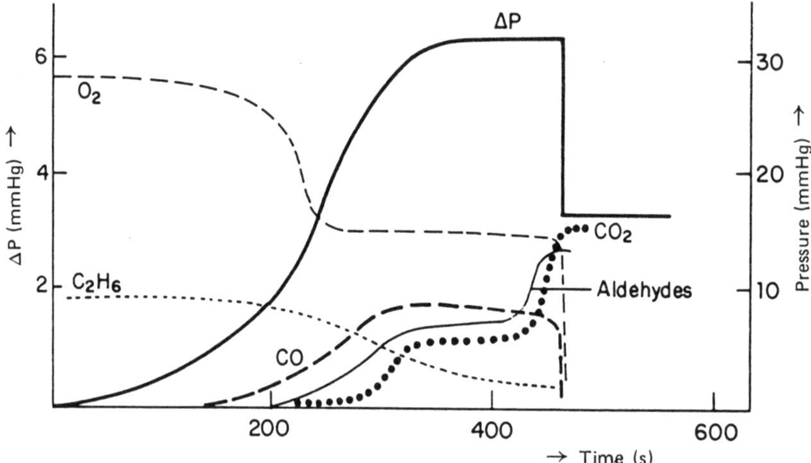

Fig. 2.4. Time-composition profile for combustion of ethane in stoichiometric admixture with air at $600°C$. (i.e. in the S-curve field of a $\dfrac{P}{T}$ autoignition diagram).

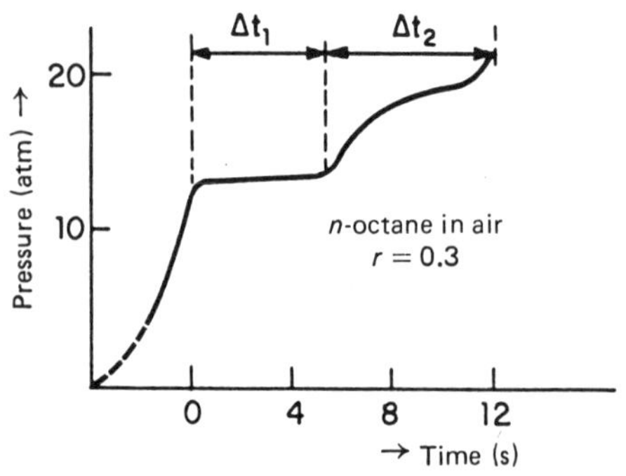

Fig. 2.5. Pressure-profile of combustion due to rapid compression for *n*-octane/air mixture, with $r = 0.3$.

$$\Delta t = K \frac{1}{p^n} \exp \left(\frac{E}{R\theta} \right) \qquad (2.11)$$

where E is an energy of activation, K is a coefficient dependent on the equivalence ratio and exponent n is dimensionless. Table 2.3 gives some values for these terms, as they have been determined by isentropic compression.

TABLE 2.3

PRE-EXPONENTIAL CONSTANT AND ACTIVATION ENERGY
FOR AUTOIGNITION DELAYS OF SOME GASOLINE-RANGE HYDROCARBONS

Stoichiometric mixture	$\lg K$	n	E (cal/mol)
Isooctane/air	− 4.77	1.49	14,900
Diisobutylene/air	− 4.83	0.95	12,700
Benzene/air	− 5.26	0.59	13,100

Certain additives in a fuel can modify the limits and particularly the delays of autoignition. Best known among such additives are the lead alkyls used to stop the knocking due to autoignition in a spark-ignition engine (Figs. 2.6 and 2.7).

Books on chemical kinetics explain the theories of the phenomena observed. Here we simply note certain tendencies implicit in the general expression for the reaction rate taken from Eq. (2.15) above:

$$V_R = Z \alpha_0 \exp \left(- \frac{E_p}{R\theta} \right) \frac{(\dot{R})}{p}$$

where $(\dot{R})$ is the number of free radicals. The reaction rate, V_R, increases either as the temperature, θ, is raised or when the number of free radicals $(\dot{R})$ increases. But since a temperature-rise also increases the number of free radicals, a temperature-rise has multiple effects on V_R. At a certain critical value of V_R the combined effect of $(\dot{R})$ and θ tends to increase rapidly so that the corresponding increase in reaction rate V_R tends to become very great, and is characterized by autoignition of the mixture, occurring at a certain threshold temperature and after a certain delay.

Furthermore, combustion reactions are generally exothermic with limited heat losses to the exterior, so that the temperature of the system increases during the induction period. Therefore, the theoretical relationships must account for different cases, according to the energy liberated and rate of heat exchanges with the environment. A mathematical relationship could be developed for each case, depending on whether the reaction is adiabatic or not, and taking into account the heat of reaction. Such specific formulas fall outside the scope of this book, and the interested reader should turn to the specialized books mentioned.

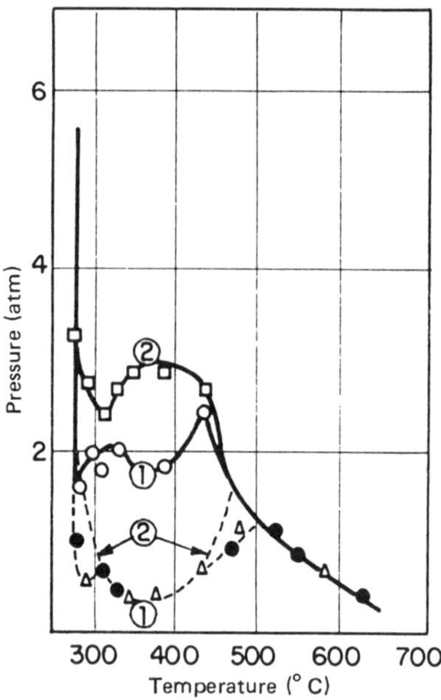

Fig. 2.6. *n*-heptane plus 1.5 parts iso-octane in air.

Fig. 2.7. iso-octane in air $r = 1.25$.

Figs. 2.6. and 2.7. Effects of tetra-ethyl lead on autoignition of gasoline-range hydrocarbons.

Curves: dotted· - - · limits of the cold flames.
solid· - - · limits of autoignition.
1. without tetra-ethyl lead.
2. with tetra-ethyl lead.

2.3. IGNITING AND PROPAGATING EXPLOSIONS

Ignition, followed by the possible propagation, of explosions is assumed to occur in homogeneous mixtures of fuel and oxidizer. The conditions will therefore be characterized by an initial temperature and pressure, as well as by an equivalence ratio (see Section 2.2).

Ignition may be caused by local heating, by a hot wall, or by a spark. From this local ignition point, the combustion spreads into the cold mixture, progressively traveling through the whole volume and transforming the initial substances into burned product gases.

Such spreading, or propagation, is observed to occur only between certain limits of equivalence ratios (called flammability limits) of the initial mixture, depending on the fuel and also on the initial temperature and pressure of the mixture. Table 2.4 gives flammability limits and autoignition temperatures for various chemicals in air.

TABLE 2.4

FLAMMABILITY LIMITS AND AUTOIGNITION TEMPERATURE
OF SOME FUELS IN AIR(1)

Compound	Formula	Flammability limits in air (vol. %)		Autoignition temperature ($^\circ$C)
		lower	upper	
Hydrogen	H_2	4.0	74.5	570
Carbon monoxide	CO	12.5	74	630
Methane	CH_4	5.0	15.0	580
Propane	C_3H_8	2.25	9.4	480
n-and isobutane	C_4H_{10}	1.85/1.8	8.4	420/480
n-and neopentane	C_5H_{12}	1.45/1.4	7.8/7.5	(290)/450
Heptanes	C_7H_{16}	(1.10)	6.7	(220)
n-octane	C_8H_{18}	(1.0)	(6)	(240)
Isooctane	C_8H_{18}	(1.0)	(6)	(670)
Ethylene	C_2H_4	(3.1)	(32)	520
Propylene	C_3H_6	(2.2)	10.5	460
Butene-1	C_4H_8	1.6	9.35	385
Butene-2	C_4H_8	1.8	9.7	435
Butadiene	C_4H_6	2.0	11.5	420
Acetylene	C_2H_2	2.5	(81)	320
Benzene	C_6H_6	1.4	(7.1)	(620)
Toluene	$C_6H_5CH_3$	(1.4)	(6.7)	(585)
Naphthalene	$C_{10}H_8$	(0.9)	(5.9)	575
Methanol	CH_3OH	(6.7)	(36)	(510)
Ethanol	C_2H_5OH	(3.3)	(19)	(490)
Acetaldehyde	CH_3CHO	4.0	56	230
Acetone	CH_3COCH_3	(2.6)	12.8	(560)

(1) Taken from A. Van Tiggelen, *Oxidations et combustions*, t.1. Editions Technip, Paris, 1968.
( ) Number uncertain.

Given the possibility for propagation of an explosion, one can study the manner in which such propagation occurs, either by a deflagration process or by a detonation process, depending on whether the rate of propagation in the cool mixture is subsonic or supersonic. Further, deflagrations can progress according to either of two types of propagation, depending on whether the conditions in the cool mixture give rise to laminar or turbulent flames.

2.3.1. Mechanism of ignition

Assuming that autoignition of a small portion of a flammable mixture has been provoked by a spark, a hot body, or an injection of hot gases, the initial burning volume must be large enough and must release more than a critical amount of energy for the explosion to spread through the whole fuel mixture. This critical volume and critical energy release will be examined separately.

2.3.1.1. Minimum volume for ignition

Ignition at some point in a combustible mixture may be assumed to form a small sphere of burned gases with a radius r. Around this sphere the flame forms an envelope of thickness, e. If the flame continues its progress out into the cool mixture after the initiating point has ceased to function, that flame will transform a small portion of cool gas into burned gases, so that the initial radius, r, of the sphere of burned gases becomes $r + dr$, (Fig. 2.8), the thickness of the flame envelope remaining equal to e_0.

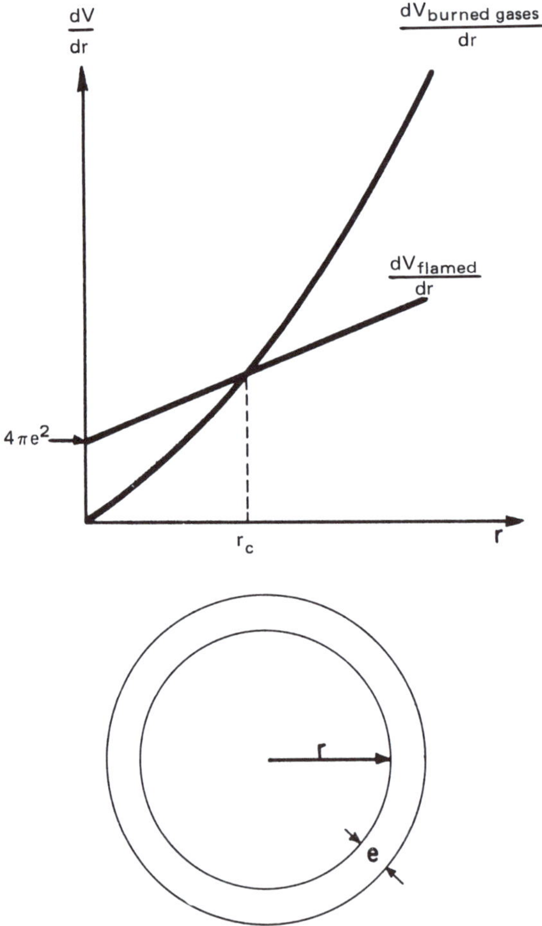

Fig. 2.8. Derivation of an expression for the minimum volume needed for ignition.

The material inside this expanding envelope of flame continues to hold energy at the same level, so that the energy available for propagating the reaction must equal the energy released by the reaction minus the incremental increase energy-content of the sphere as the cool fuel gases are transformed across the flame front into the hot combustion gases. This is to say that, for the flame to be propagated, the incremental increase in the volume of burned gases is larger than the incremental increase in volume of the flame front, or:

$$\frac{d}{dr}\left(\frac{4}{3}\pi r^3\right) \geqslant \frac{d}{dr}\left(\frac{4}{3}\pi\left[(r+e)^3 - r^3\right]\right) \qquad (2.16)$$

This reduces to:

$$r^2 - 2re - e^2 \geqslant 0 \qquad (2.17)$$

This trinominal Eq. (2.17) is positive if r is outside of the roots, and since r is necessarily positive, the condition of Eq. (2.17) is fulfilled if:

$$r_c \geqslant e(1 + \sqrt{2}) \qquad (2.18)$$

where r_c represents the critical radius of the sphere of burned gases at the moment when the energy is no longer supplied to the point of ignition.

2.3.1.2.   Minimum energy for ignition

Take an ignition system made up of an electric spark set off between electrodes with a variable gap, and with variable intensities of spark for each gap. If this ignition system is immersed in a known fuel mixture such as a stoichiometric methane-air or methane-oxygen mixture, the minimum energy needed to start an explosion can be measured for each gap between electrodes. Figs. 2.9 and 2.10 show the observed results.

The minimum energy increases extremely quickly to very high values as the gap between the electrodes is reduced to a minimum, characteristic for the pressure. This minimum corresponds to what will be called the limiting gap under certain conditions. Above this minimum gap, there is a range of gaps for which the minimum energy of ignition is constant, depending on the pressure and the equivalence ratio of the mixture (Figs. 2.9 and 2.10).

The results exhibited by this particular fuel are, in fact, typical and provide the basis for an empirical expression that is viable for the most varied fuels, as:

$$E_{min} = Kd^2$$

where d is the limiting gap.

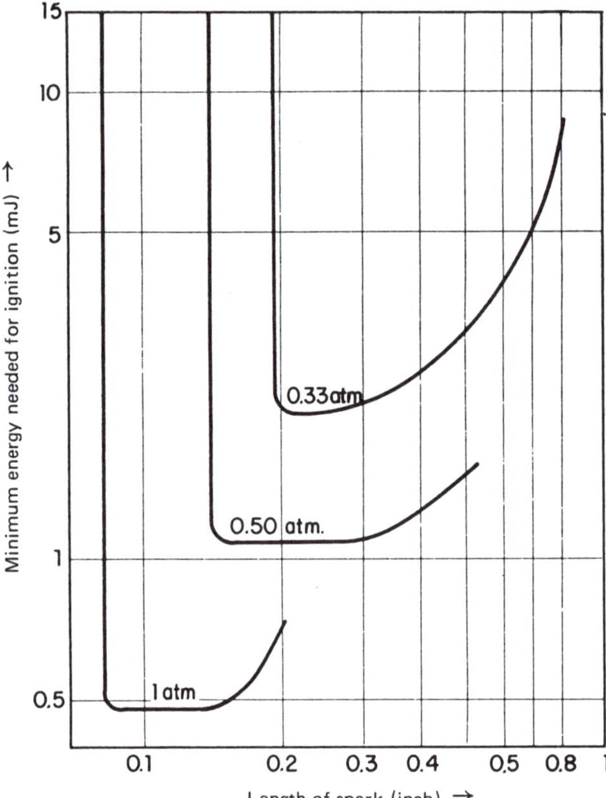

Fig. 2.9. Ignition energy versus spark necessary to ignite stoichiometric mixtures of methane in air. (From B. Lewis and G. von Elbe. *Combustion, flames and explosions of gases*, p. 328. Academic Press, New York, 1961).

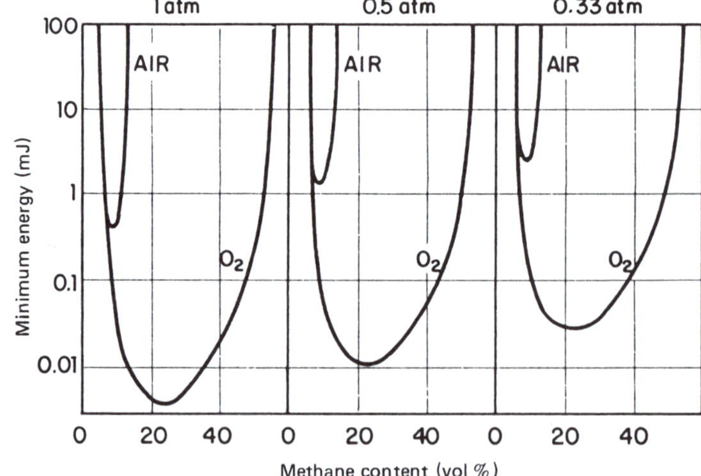

Fig. 2.10. Minimum energies for igniting various mixtures of methane in air and oxygen with an optimum gap spark. (From B. Lewis and G. von Elbe. *Combustion, flames and explosions of gases*, p. 331. Academic Press, New York, 1961).

2.3.2. Mechanisms of propagation

2.3.2.1. Deflagration in the laminar state

In a gas at rest or in laminar flow, the rate of laminar propagation, V_{nl}, of a combustion is taken as being the perpendicular component of the movement of the flame front relative to the cool gases at normal conditions of temperature and pressure.

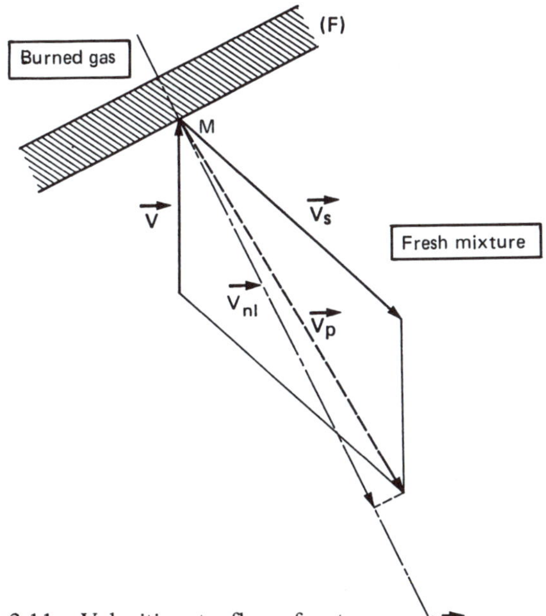

Fig. 2.11. Velocities at a flame front.

Assume a flame element M (see Fig. 2.11), characterized by a speed $\vec{V_s}$ with respect to a fixed reference system. Let $\vec{V}$ be the velocity vector of the fresh mixture at the same point M and defined with respect to the same fixed reference system. The laminar flame speed $\vec{V}_{nl}$ is then defined as the projection of the vector difference $\vec{V}_s - \vec{V}$ on a line normal to the flame front in point M:

$$\vec{V}_{nl} = \vec{V}_s\,\vec{n} - \vec{V}\,\vec{n} = \vec{V}_p\,\vec{n}$$

where $\vec{n}$ is the unit vector normal to the flame front and $\vec{V}_p$ the speed of propagation of that element of laminar flame relative to the cool gases.

In many practical cases, either the flow vector V or the vector V_s may be zero (respectively in the case of gases at rest or in the case of flames stabilized on a burner rim or on a bluff body).

$\vec{V}_{nl}$ can be measured by different methods, such as the open tube, soap bubble, spherical bomb and the Gouy burner method.

TABLE 2.5
LAMINAR FLAME SPEED OF FUEL-AIR MIXTURES
(cm/s)

Fuel	Equivalence ratio r				
	0.8	0.9	1	1.1	1.2
H_2	120	145	170	204	245
CO (wet)				28.5	32
CH_4	30	38.3	43.4	44.7	39.8
C_2H_6	36	40.6	44.5	47.3	47.3
C_3H_8		42.3	45.6	46.2	42.3
C_4H_{10}	38	42.6	44.8	44.2	41.2
C_6H_{12} (cyclohexane)		41.3	43.5	43.9	38
C_2H_4	50	60	68	73	72
C_2H_2	107	130	144	151	154
C_6H_6	39.4	45.6	47.6	44.8	40.2
CH_3OH	34.5	42	48	50.2	47.5

Table 2.5 gives values of laminar deflagration speeds, in centimeters per second, for different hydrocarbons and hydrogen mixed with air. It is apparent that the majority of the hydrocarbons exhibit a speed of laminar propagation under normal conditions of the order of 40 cm/s. Hydrogen, acetylene and ethylene, which have much higher speeds, are thus shown to be completely different fuels. Also, except for hydrogen and acetylene, the maximum speed occurs at an equivalence ratio 1 or slightly larger than 1.

In a deflagration, the flame front is the zone occurring between the cool gases and the burned gases, where the combustion reactions take place. This zone is usually very narrow and profiles of concentration and temperature across it show steep gradients between the cool gases and the burned gases. Several thermal and diffusional theories have been suggested to explain the progression of the flame front into cool gases. Certain of them lead to calculations of deflagration speeds that compare favorably with measured ones.

2.3.2.2. Deflagration in turbulent flow

When deflagration spreads in a fuel mixture already in turbulent flow, the turbulence noticeably modifies the rate of propagation; the mechanism of interaction between the flame front and the field of turbulence is still not well understood.

It will be sufficient to report here some of the experimental observations of the changes occurring in a flame front under the influence of turbulence. The thickness of the visible front increases markedly. The ionization currents that can be measured by immersing probes in the flame front are much higher than such currents measured in a corresponding laminar flame. The field of flammability seems to shrink as the intensity of the turbulence increases.

But the main difference concerns the propagation speed of the flame into the cool gases. The average normal turbulent speed in relation to the cool gases, $\vec{V}_{nt}$, is much higher than the rate $\vec{V}_{nl}$ defined above for the laminar state. Fig. 2.12, showing normal turbulent speeds measured by different authors as a function of the intensity of turbulence, indicates that the rate of deflagration increases with turbulence intensity, but that this variation also depends on the geometry of the burner.

Figure 2.13 shows the rates observed at the exit of a burner, the flame being stabilized at the sides. The apparent speed increases with the distance from the sides of the burner because turbulence is damped next to the walls.

Test results show that the normal turbulent speed stops increasing then starts to diminish beyond a certain level of turbulence. It seems that this decrease is due to extinctions resulting from a dampening effect brought on locally by the intense turbulent movement.

2.3.2.3. Propagation as deflagration or detonation

Within a fuel medium, the flame front represents a frontier between two different environments, separating on the one side burned gases from cool gases on the other. Because combustion is an exothermic phenomenon, there is a disturbance at this front due to the sudden increase in temperature. Now, it is known that compressible fluids transmit shock as waves of expansion or compression. In a perfect gas, the expansion waves are propagated at a speed less than the speed of sound, and they tend to become extended in the course of their movement.

Compression waves, by contrast, are propagated at a speed greater than the speed of sound; and during their movement, the front of the wave is progressively overtaken by the tail, which is moving through a medium already in motion. A progressive stiffening of the front of the wave results and it is transformed into a true discontinuity: that is, into a shock wave.

Considering the movement of shock waves within a fuel medium, it becomes apparent that the sudden increase in temperature occurring as the shock wave passes can cause autoignition under appropriate conditions. In such cases, the shock wave is followed by a combustion wave. From then on, the combustion supplies the energy necessary to maintain the wave phenomenon; the united shock-and-combustion is called the detonation wave; it crosses the cool gases at a speed much faster than that of sound, creating as it passes an instantaneous pressure easily equal to 20 or 25 times the initial pressure.

The supersonic propagation of a detonation wave was described in 1881 by Berthelot and Vieille. The theoretical study of the propagation of plane detonation waves was done by Hugoniot (1887) and Jouguet (1917). The compression by a shock wave of gases that do not change chemically corresponds to

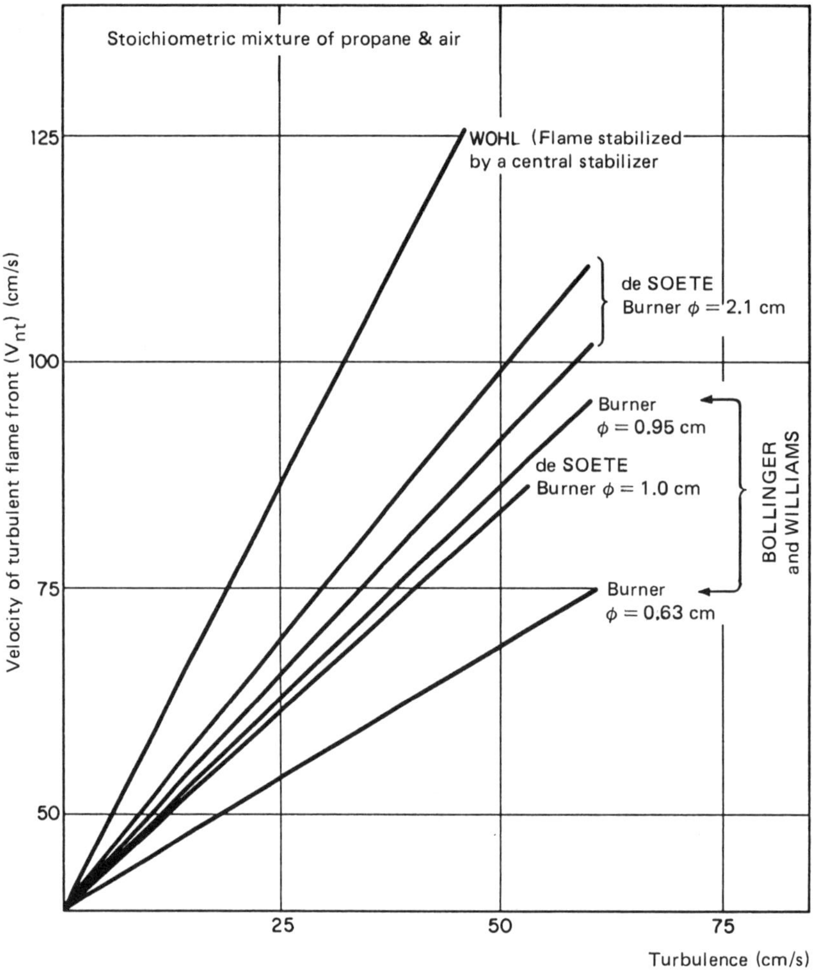

Fig. 2.12. Speeds of normal propagation of flames under turbulent conditions.

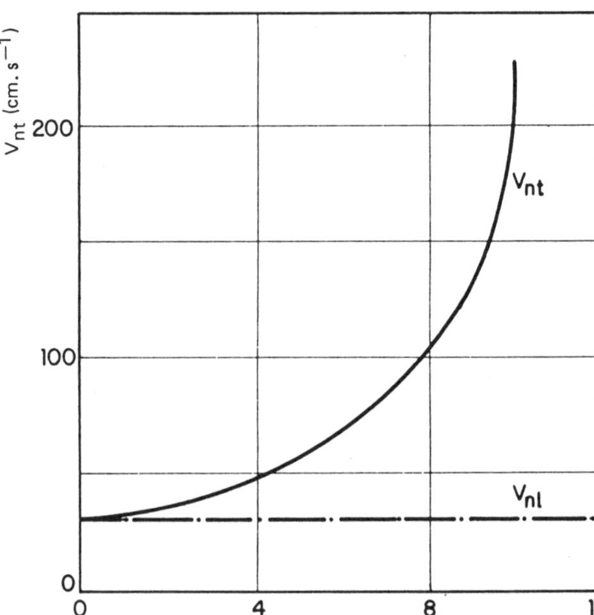

Fig. 2.13. Influence of the burner wall on the apparent speed of normal propagation of a turbulent flame. (Stoichiometric methane/air mixture).

Hugoniot's dynamic adiabatic process (Fig. 2.14). When the fluid goes from state p_1 and ρ_1 to state p_2 and ρ_2, where p = pressure, ρ = density, these states satisfy the equation:

$$\frac{p_2}{p_1} = \frac{(\gamma + 1)\dfrac{1}{\rho_1} - (\gamma - 1)\dfrac{1}{\rho_2}}{(\gamma + 1)\dfrac{1}{\rho_2} - (\gamma - 1)\dfrac{1}{\rho_1}} \tag{2.19}$$

This Equation (2.19) represents dynamic adiabatic change. In Fig. 2.14 the dynamic adiabatic transformation of the reactants is shown (Curve 1) as well as the dynamic adiabatic transformation of the burned gases (Curve 2). If point A is the initial state of the mixture, the theory establishes that by tracing the tangents to the curve for the burned gases from A, sets tangent points, B and C, which represent the state of the gases in the combustion wave, with B corresponding to detonation and C corresponding to deflagration.

In spite of the simplifying assumptions on which these relations are based, the agreement between the measured values and the calculated values is satisfactory for the pressure and the propagation speed, for example, at least for the detonation case.

The existence of a stable detonation wave in a fuel mixture requires that the mixture have a suitable composition. The range of equivalence ratio

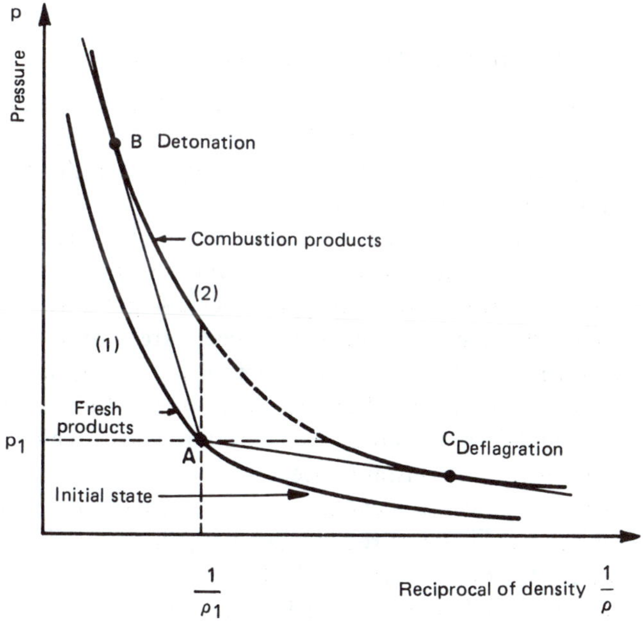

Fig. 2.14. Diagram of dynamic adiabatic changes leading to detonation.

permitting detonation is a little narrower for every fuel than is the range for ignition. Table 2.6 gives detonation limits for some fuel mixtures; these can be compared with Table 2.4.

TABLE 2.6

LIMITS OF DETONATABILITY FOR FUEL/OXYGEN MIXTURES

Fuel	Oxidizer	Fuel concentration (vol. %) for detonation	
		lower	upper
H_2	O_2	15	90
H_2	air	18.3	58.9
CO (wet)	O_2	38	90
CO (dry)	O_2	–	83
$CO + H_2$	O_2	17.2	91
$CO + H_2$	air	19	58.7
NH_3	O_2	25.4	75.4
C_2H_2	O_2	3.5-3.6	92-93
C_2H_2	air	4.2	50
C_3H_8	O_2	3.2	37
i-C_4H_{10}	O_2	2.8	31.1
$C_2H_5-O-C_2H_5$	O_2	2.6-2.7	40
$C_2H_5-O-C_2H_5$	air	2.8	4.5

2.3.2.4. The transition from deflagration to detonation

Take a sealed tube (or in general terms, a closed elongated volume containing a detonatable mixture, and ignite this mixture near one of the closed ends. The deflagration thus initiated can become a detonation after the flame front has traveled a certain distance along the tube.

The period of induction for such a detonation is characteristic of the fuel mixture, getting larger in both distance and time, as the composition of the mixture approaches its limits of detonation (Tables 2.4 and 2.6).

The movement of the visible flame front accelerates continually during the induction period, until near the end of this period a discontinuity consisting of one or more autoignitions can be observed ahead of the flame. And it is often one of these "accidents" that starts a detonation wave, which traverses the remainder of the mixture at a constant speed.

Phenomena of this type can appear in all the elongated volumes accidentally invaded by a detonatable mixture; it has happened in mine galleries, in chimneys, garbage chutes and elevators of high buildings.

The acceleration of the flame in such instances results from the movement caused by expansion of the combustion products, and the existence of compression waves ahead of the flame has been confirmed by a Schlieren optical system for compressible-flow visualization.

2.4. DIFFUSION FLAMES

Up to this point, discussion has been concerned with homogeneous mixtures of fuel and oxidizer. Since the oxidizer is usually air, homogeneity presupposes that the fuel is a gas or a vapor.

In industry, however, fuel and oxidizer are usually introduced separately to a firebox where the combustion is to take place. Because admixture must be achieved prior to combustion and because mixing is generally much slower than the physical-chemical phenomena of combustion, development of the diffusion flames is essentially dependent on fluid dynamic factors, with the chemical nature of the reactants appearing only as a secondary factor.

Fluid flow can be either laminar or turbulent, and there is a corresponding different type of diffusion flame.

2.4.1. The appearance of diffusion flames

Imagine a jet of gaseous fuel rising out of a cylindrical conduit into an atmosphere at rest (Fig. 2.15). When the diameter of the conduit is kept

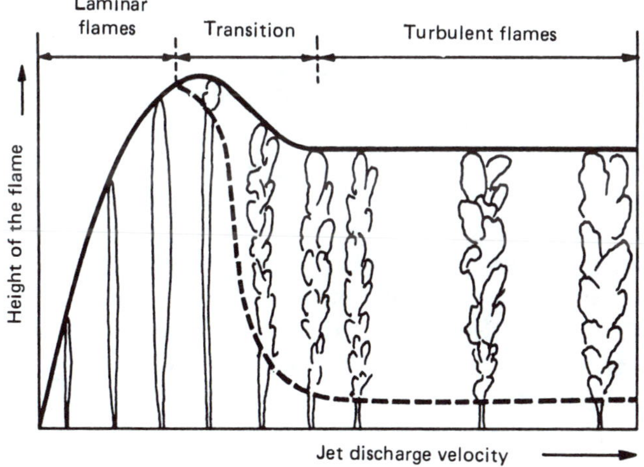

Fig. 2.15. Appearance of flames from a gas jet.
(From H.C. Hottel and W.R. Hawthorne. *Third International Symposium on Combustion.* Williams and Wilkins, Baltimore 1949, p. 254).

constant, the appearance of the flame depends on the rate of discharge and the chemical nature of the fuel. At feeble flows, the heights of the flames increase with the rate of discharge up to a critical point, after which they diminish to a level that is seen to be constant up to the rate at which the flame is blown out (Fig. 2.15).

2.4.2. The structure of the combustion zone of a laminar diffusion flame

Diffusion flames establish a reaction zone that is developed between a region rich in fuel and a region containing the oxidizer. The radial concentration profiles from the axis of the jet are observed to be rather extended for the chemicals present in the flame (Fig. 2.16). For almost all hydrocarbons, the maximum temperature corresponds to an oxidizer-fuel mixture with an equivalence ratio equal to one.

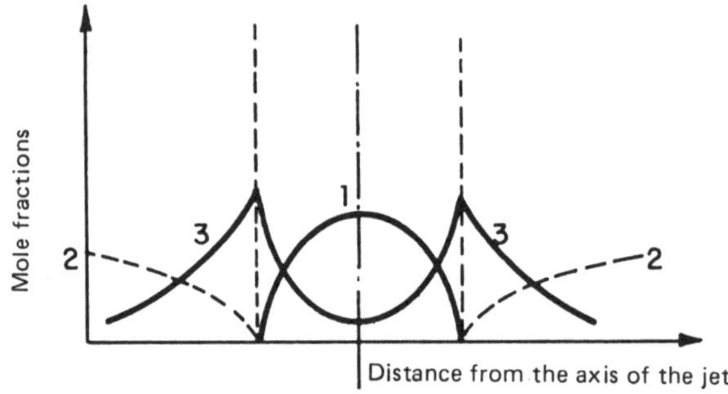

Fig. 2.16. Point conditions in a plane perpendicular to the axis of a flame.

Curves : 1. Fuel concentration.
2. Oxygen concentration.
3. Concentration of combustion products.

(From H.C. Hottel and W.R. Hawthorne. *Third International Symposium on Combustion*. Williams and Wilkins, Baltimore 1949, p. 254).

Figure 2.16 summarizes the distributions for a given flame developing in laminar flow. There are satisfactory mathematical models for such a flame. They assume that, first, the pressure is uniform for the whole region in which the flame is developed, and, second, the energy is liberated by combustion as soon as the fuel jet penetrates to a stationary oxidizing atmosphere (or that at least the streams of fuel and of oxidizer are coaxial with the parallel rates of flow). These assumptions are particularly interesting, for they bring out the relative importance of the different factors. Because the models do agree well with the experimental observations, they confirm the original hypothesis that the chemical phenomena of combustion are very fast compared to the phenomena of mixing by diffusion.

Mathematical models applicable to diffusion flames in turbulent flow have also been developed and studied with the flow of fuel taking place both in an open atmosphere and in a closed one. However, even when the description of fluid-dynamic phenomena does become satisfactory in certain well-defined cases (see Chapter 3), the prediction of the resulting flames (concentrations of chemicals, temperature, presence or absence of soot, emissions, etc.) remains uncertain because of numerous omitted factors, such as the coefficients of turbulent diffusion. The complexity of turbulent-flow phenomena requires a semi-empirical approach to problems that come up in industrial practice. This is the reason why teams have been organized to carry out experiments on models of industrial furnaces at Ymuiden in the Netherlands (*International Flame Research Foundation, IFRF*) and at Toulouse (*Groupe d'Etudes des Flammes de Gaz Naturel, GEFGN*). The experimental results achieved by those teams account for much of what follows in this book.

2.4.3. Flames of liquid fuels

Liquid fuels are the main source of industrial heat energy. Because heavy liquid fuels contain residues from refining operations, they can not be vaporized at atmospheric pressure simply through heating, without extensive thermal decomposition. Therefore, such fuels are atomized before introduction to the oxidizer in the firebox.

Atomizing is primarily done by either injecting an auxiliary fluid, usually air or steam, for spraying (Fig. 2.17), or by mechanical means of forcing the fuel through a calibrated regulator (Fig. 2.18). Atomizing by a rotary cup (Fig. 2.19) can be thought of as a combined process in which a thin layer of fuel flowing on the periphery of the cup is torn apart by axially-flowing oxidizing air.

The atomization of viscous fuels, particularly mechanical atomization, requires preheating to bring the viscosity down to 30 or 35 cSt. For numerous heavy fuels this corresponds to a temperature of 100-110° C. One way or the other, the initial liquid fuel is introduced as a mist sprayed into the firebox, and these fuel mists comprise tiny drops with varying diameters. Experiments made on atomized jets show that the diameters of the drops rather closely follow Rosin-Rammler's law. One frequently used characteristic is the jets' average Sauter diameter, which is the droplet diameter that accounts for the specific surface created by the jet, and specific surface is understood as the total surface of the drops whose combined weight is equal to unity, as cm^2/g or ft^2/lb.

In usual practice, the average Sauter diameter of jets from mechanical atomizers is on the order of 80 to 100 μ, with a statistical distribution such that a negligible fraction of the drops will have a diameter greater than 300 μ.

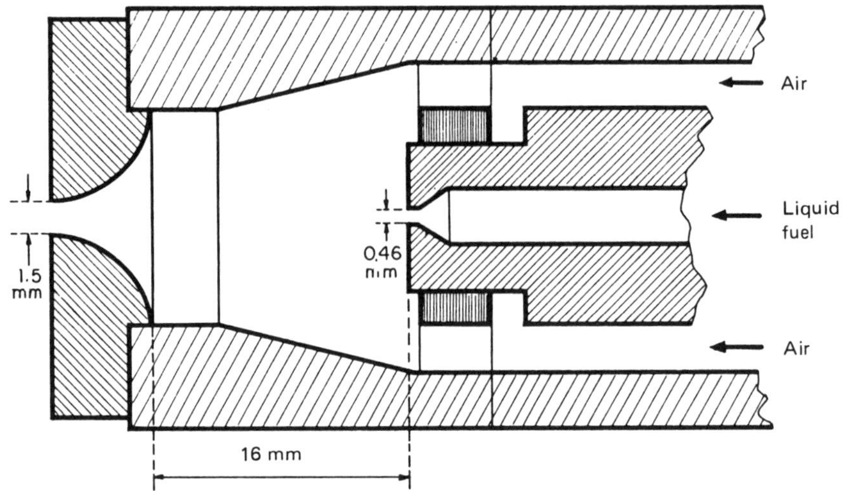

Fig. 2.17.   A typical spray atomizer for liquid fuels.

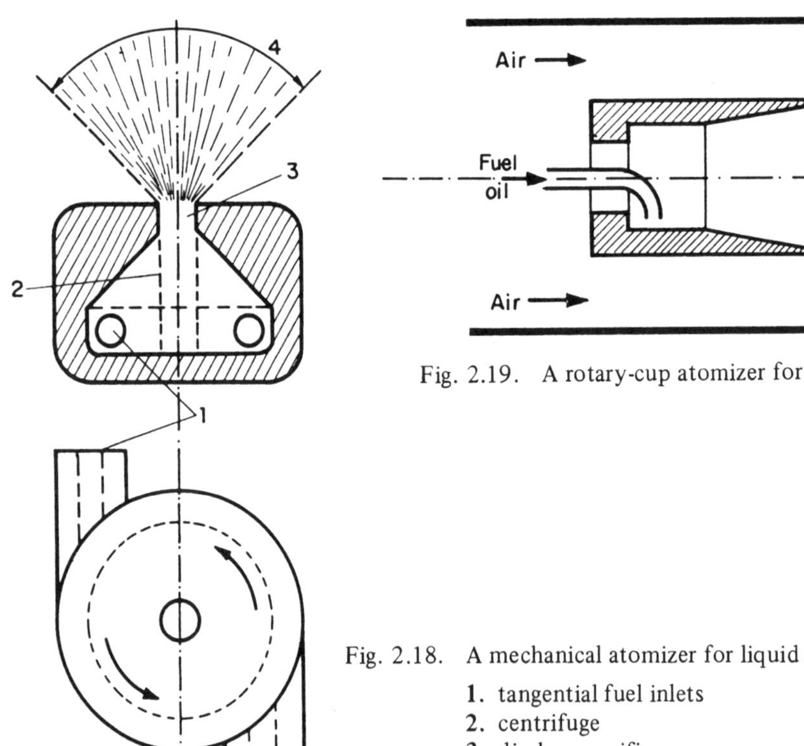

Fig. 2.19.   A rotary-cup atomizer for liquid fuels.

Fig. 2.18.   A mechanical atomizer for liquid fuels.

1. tangential fuel inlets
2. centrifuge
3. discharge orifice
4. angle of atomization.

Pneumatic atomization produces average Sauter diameters on the order of 60 to 80 μ and less variation. The least size variation occurs in atomized jets obtained with the rotating cup.

Liquid viscosity, which is a prime factor in atomizing, changes rapidly with temperature. Advantage is taken of this in heavy and viscous fuels for bringing the kinematic viscosity to within suitable limits before atomizing; i.e., to:

(a) Less than 70 cSt for burners with rotating cups or with atomization by air or steam at high pressure.

(b) Less than 45 cSt for industrial burners with air atomizers at moderate pressure.

(c) Less than 21 cSt for burners with air atomizers at low pressure and burners with mechanical atomizers at heavy flow rates.

(d) Less than 12 cSt for burners with light flow rates and for mechanical atomizers.

REFERENCES

SOETE, G., DE, FEUGIER, A. – *Aspects physiques et chimiques de la combustion.* Editions Technip, Paris 1976.

SOETE, G., DE. – *Aspects fondamentaux de la combustion en phase gazeuse.* Editions Technip, Paris 1976.

PALMIER, H.B., BEER, J.M. – *Combustion Technology: Some modern developments.* Academic Press, New York and London, 1974.

3
fluid dynamics in the firebox

The firebox, which is the essential part of all equipment to harness the energy of combustion, is the chamber in which the combustion takes place. Receiving through one or more burners the necessary fuel and oxidizer, it encloses the chemical combustion reactions whose products generally give up a part of their sensible heat to the walls (i.e., tube-walls in boilers and fired heaters) or to a charge deposited within the firebox (open-hearth smelters, reheating furnaces, etc.). After giving up this heat, the combustion products are discharged either through exchangers and heat-recovery units, or directly to the flue stack and the atmosphere. The same characteristics apply to an internal-combustion engine, with the firebox being the combustion chamber, except that combustion in furnaces and fired heaters is continuous, while combustion and energy transfer in a piston engine is by batch, according to the cycle described by Beau de Rochas as early as 1862: intake stroke of the fuel/oxidizer mixture, compression followed by combustion, exchange of energy (here in the form of expansion of the gas produced by combustion) and then discharge stroke.

A comparison with four-cycle engine combustion chambers shows the importance of fluid movement through a firebox — feeding in cool products and evacuating combustion products — fluid movements that are simultaneous and continuous in furnaces and fired heaters.

The cool products introduced through the burner to a firebox can be more or less thoroughly premixed, particularly in the case of gaseous fuels. In principle, thorough premixing may also occur with pulverized coal carried by a stream of air. In the majority of industrial applications, however, the fuel and the oxidizer are introduced side by side into the firebox where the mixture is made prior to combustion.

3.1. THE DEVELOPMENT OF A TURBULENT JET
IN A CALM ATMOSPHERE: A FREE JET

Before taking up the problem of flow in a closed chamber, it will be helpful
to review the principal characteristics of an air jet discharging into an unlimited
atmosphere that is at rest (Fig. 3.1). Assume a cylindrical injector with its
opening shaped so that all the fluid paths move at a uniform velocity parallel
to the axis. As this fluid leaves the injector and jets into the calm atmosphere,
it gives rise to turbulent exchanges of momentum at its boundaries. A characteristic nucleus forms in the shape of a cone (with its base the injector and an
angle such that its length is equal to around 5 times its diameter) in which the
fluid retains its initial velocity (Fig. 3.1). At the periphery of this cone, the
ambient atmosphere is drawn into the jet by turbulent friction. This additional
material feeds the jet, which assumes the shape of a cone inverse to that of the
nucleus. Experiments show that the angle of opening of the jet cone is constant
and equal to around 22°.

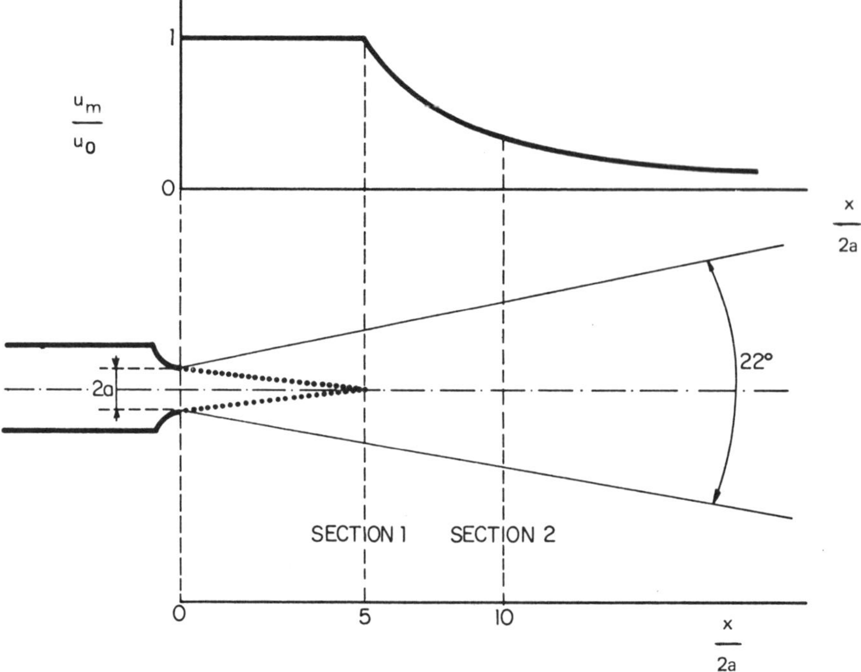

Fig. 3.1. Diagram of a free jet discharging into a calm atmosphere.

Beyond the point of the nucleus (Section 1 in Fig. 3.1) all speeds, u_m, measured along the axis are less than the initial. And the loss in speed at any point is proportional to its relative distance from the outlet of the nozzle. That is:

$$\frac{u_m}{u_0} = A\frac{2a}{x}$$

where

u = the velocity,
A = a constant,
a = the radius of the nozzle,
x = the distance along the axis,
$m, 0$ = subscripts denoting points along the axis, with m being x units downstream and 0 at the nozzle's outlet.

Across a plane perpendicular to the axis, the distribution of speeds within the jet varies as a function of the distance to the axis because of fluid friction, and beyond Section 2 (Fig. 3.1), this variation becomes uniform taking the form of a Gauss curve:

$$\frac{u_r}{u_m} = \exp\left(-k\left(\frac{r}{x}\right)\right) \tag{3.2}$$

where

u_r = the speed at the distance r from the axis,
k = constant.

Between the point of the nucleus and the appearance of Gauss-curve distributions (around 5 times the diameter of the nozzle) the relation between radial distribution and speeds is not yet established. There is, therefore, a zone extending from the end of the nozzle to section 2, a distance on the order of 10 times the diameter of the nozzle, before the flow of the jet is established.

3.2. THE DEVELOPMENT OF A TURBULENT JET IN A CHAMBER: A CONFINED JET

3.2.1. Principles and definitions

After the above brief review of free jets, consider the case of an open-ended cylinder in the entrance of which is placed a nozzle that injects the same fluid present in the cylinder. The fluid leaving the injector can be designated as motive while the fluid that enters the cylinder around the nozzle is designated as passive fluid (Fig. 3.2).

If the motive flow rate, q_0, is kept constant while varying the passive flow rate, q_a, a turbulent zone appears between the jet and the wall for every passive flow rate less than the critical value, q_c. The length of this turbulent zone increases toward the entrance of the cylinder as the passive flow rate decreases. It can be verified that for any given cylinder diameter, the ratio q_c/q_0 of critical-speed-of-passive-fluid to speed-of-motive-fluid gets much larger as the diameter of the nozzle gets smaller.

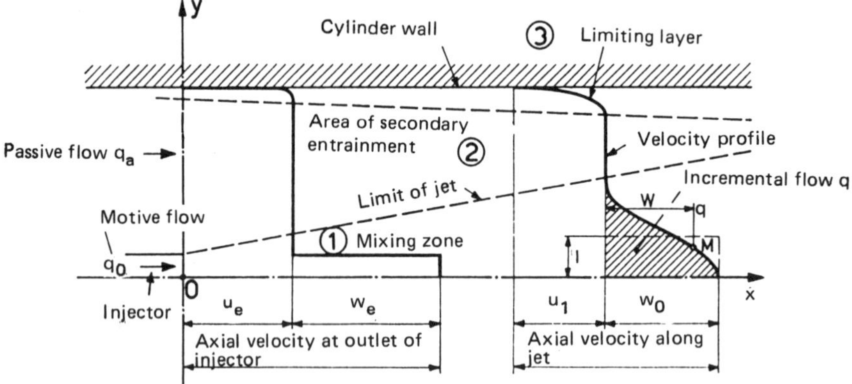

Fig. 3.2. Diagram of a confined jet with terms for correlations.

Experimental measurements of the flow rates in the body of the fluid where the mixing process occurs without swirl show that the velocities vary as in Fig. 3.2:

(1) The jet characterized by a maximum velocity along its axis, high range of velocities in an intermediate part, and then a juncture at the circumferential limit of the jet toward the region 2 (Fig. 3.2).

(2) A zone corresponding to the undisturbed region of the flow in which the velocity is constant.

(3) An enveloping zone just inside the cylinder wall, with a steep gradient of velocity towards the axis.

Let u_1 be the speed in zone 2. The static pressure in nearly constant in a cross-section of the fluid, because of the slight curvature of the current line; this pressure is a function only of the distance, x, from the nozzle. The velocity u_1 is also a function of x, decreasing as x increases. If u is the velocity at point M in the jet (see Fig. 3.2), then w is the incremental velocity by which u exceeds the constant velocity, u_1, across zone 2, and the corresponding incremental volume flow, q, (shaded area in Fig. 3.2) is given by:

$$q = \int_0^\lambda w\, 2\pi y\ \mathrm{d}y$$

where y is the distance from the axis.

Since the radius, λ, of the jet is not well defined, an effective radius, l, can be introduced, such that:

$$q = \pi w_0 l^2,$$

where w_0 is the incremental speed above u_1 along the axis of the jet.

Figure 3.3 shows the variation of reduced incremental speed:

$$\left[\frac{w}{w_0} = f\left(\frac{y}{l}\right)\right]$$

with the reduced radius of the jet. This relation of $\dfrac{w}{w_0} = f\left(\dfrac{y}{l}\right)$ is independent of distance from the nozzle, as well as from the ratio of passive to motive flows, q_a/q_0 when no swirl occurs in the considered section. The reduced curve in close to a Gauss curve, and it is seen that the reduced flow rate is negligible for a reduced diameter close to 2, that is, the effective reduced radius $L = \dfrac{l}{h}$

(h being the radius of the cylinder) cannot be greater than the value $0.5 \dfrac{y}{h},$ the maximum being $L_0 = 0.5$, which corresponds to the case where the jet reaches the wall of the cylinder, or $y = h$.

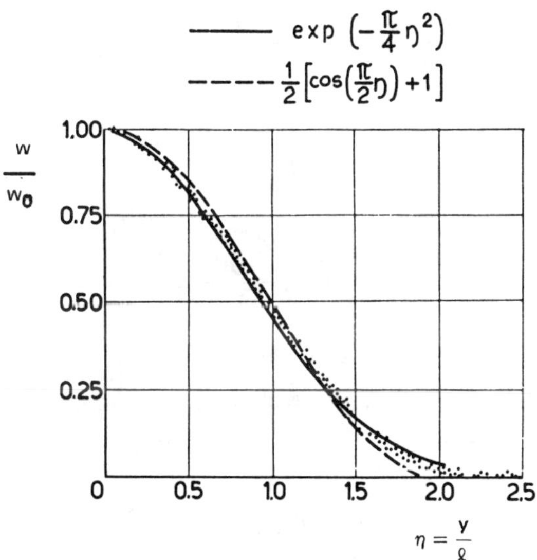

Fig. 3.3. Variation of the reduced incremental speed (Fig. 3.2) versus the reduced radius of a jet.

3.2.2. Craya and Curtet's theory

In order to develop a theory for the important industrial application of two fluids in coaxial cylindrical flow, we will adopt the work of A. Craya and R. Curtet, done in cooperation with the experimental research of the *International Flame Research Foundation (IFRF)* (Ref. 3.1).

We will keep the preceding ideas of speed and incrementally higher flow, and effective radius. The problem comes to establishing three relationships between the three variables of x (Fig. 3.2): w_0, u_1 and l.

We assume a stationary state (that is, that the derivatives in relation to time are nil). In the passive fluid, where turbulence is negligible, the Euler Equation can be written:

$$\frac{1}{\rho}\frac{\partial p}{\partial x} = -u_1\frac{\partial u_1}{\partial x} \tag{3.3}$$

The speed in the jet can be written as:

$$u = u_1 + w_0 f\left(x,\frac{y}{l}\right) \tag{3.4}$$

The Navier-Stokes Equation in the region of the turbulent jet can be written as:

$$u\frac{\partial u}{\partial x} + v\frac{\partial u}{\partial y} + \frac{1}{\rho}\frac{\partial p}{\partial x} = \frac{1}{\rho y}\frac{\partial(\tau y)}{\partial y} \tag{3.5}$$

(if τ is the shearing force per unit surface on a normal element at oy, given by the traditional relationship: $\tau = -\rho\, u'v'$).

In addition, the equation of continuity is written:

$$\frac{\partial(uy)}{\partial x} + \frac{\partial(vy)}{\partial y} = 0 \tag{3.6}$$

This system of equations can be resolved by carrying the value of transverse velocity, v, into (3.5) from (3.6), as well as the values of u and p from (3.3) and (3.4). By integrating the relationship thus found with respect to y, (3.7) is arrived at:

$$2\frac{\tau\eta}{\rho w_0^2} = \frac{(u_1^2\, w_0\, l^2)'}{u_1\, w_0^2\, l}F + \frac{(w_0^2\, l^2)'}{w_0^2\, l}kM - \frac{(w_0\, l^2)'}{w_0\, l}fF - \frac{(u_1\, l^2)'}{w_0\, l}f\eta^2 \tag{3.7}$$

with ( )$'$ derivative with respect to x:

$$F = 2\int_0^\eta f\eta\ \mathrm{d}\eta\ ;\quad kM = 2\int_0^\eta f^2\eta\ \mathrm{d}\eta\ ;\quad k = 2\int_0^\infty f^2\eta\ \mathrm{d}\eta\ ;\quad \eta = y/l$$

By taking into account the conditions at the limits, for which τ is nil on the axis and at the boundary of the jet, a new relationship is obtained between u_0, w_0 and l:

$$\frac{(u_1^2\, w_0\, l^2)'}{u_1\, w_0^2\, l} + k\, \frac{(w_0^2\, l^2)'}{w_0^2\, l} = 0 \tag{3.8}$$

With this relationship taken into account, Eq. (3.7) becomes:

$$2\,\frac{\tau\,\eta}{\rho w_0^2} = k\,(M - F)\,\frac{(w_0^2\, l^2)'}{w_0^2\, l} - \frac{(u_1\, l^2)'}{w_0\, l}\, f\eta^2 - \frac{(w_0\, l^2)'}{w_0\, l}\, fF \tag{3.9}$$

Previously, it was seen that the curve of the reduced speed is independent of the distance from the nozzle, x. Expressed differently, the function f, in $\dfrac{w}{w_0} = f\!\left(\dfrac{y}{l}\right)$, does not depend on x. The functions k, M and F are thus only functions of η and the second member of the equation includes three terms in the forms $\varphi\,(\eta)\,\psi\,(x)$. The functions $k\,(M - F)$, Ff and $f\eta^2$ are generally close enough so that the affinity of the speed profile takes with it the affinity of the profile of unit frictions.

Integrating Eq. (3.9) in relation to η gives:

$$\sigma = -\,a\,\frac{(w_0^2\, l^2)'}{w_0^2\, l} + \frac{1}{4}\,\frac{(w_0\, l^2)'}{w_0\, l} + b\,\frac{(u_1\, l^2)'}{w_0\, l} \tag{3.10}$$

σ, a coefficient of shearing intensity, is from here on a function of x alone.

Finally, another accurate relationship between the variables u_1, w_0 and l is furnished by considering that the sum of motive and passive flow, Q, is constant, as:

$$Q = \pi w_0\, l^2 + \pi u_1\, (h - \delta)^2 \tag{3.11}$$

where δ is the thickness of the boundary layer of displacement.

The group of Eqs. (3.8), (3.10) and (3.11) makes up a system of differential equations that permits one to obtain the functions sought by integration when the elements q, h, δ, f and σ are fixed.

Equation (3.8) can be written with $u_1 = \dfrac{Q - q}{\pi h^2}$ neglecting δ and with $w_0 = \dfrac{q}{\pi l^2}$.

$$u_1 q' + 2u_1' q + k\left(\frac{q^2}{\pi l^2}\right)' = 0$$

Integration of this equation gives:

$$-\frac{3}{2}R^2 + R + k\,\frac{R^2}{L^2} = m \tag{3.12}$$

where

R = ratio of incremental volume flow to total flow,

$= q/Q$

L = effective reduced radius,

$= l/h.$

This integration constant, m, is thus a function only of the entrance conditions; and being dimensionless, it is a parameter of dynamic similitude for turbulent jets.

By carrying over into Eq. (3.10) the values of L taken from Eq. (3.12) and of u_1 taken from Eq. (3.11), it is possible to integrate Eq. (3.10) and calculate the family of curves in Figs. 3.4 and 3.5, where values of L and R are plotted as functions of $\xi = \int_{x_0}^{x} \dfrac{4\sigma}{h}\, dx$ for values of m.

Speed, u_1, in the undisturbed zone 2 does change as a function of the distance, x, from the nozzle, so that its variation can be expressed in terms of R by the formula:

$$\frac{u_1}{U} = 1 - R$$

$\left(U \text{ is the average speed in the cylinder and defined as } U = \dfrac{Q}{\pi h^2} \right)$. Thus, the speed, u_1, decreases as R increases and is zero for $R = 1$.

If the lines of current are traced along the mixing cylinder, there will be a certain point N, beyond which all of the passive flow is carried along in the primary jet (Fig. 3.6). In this region, that motive energy still held by the jet tends to entrain more fluid, but can do so only at its own expense, thus giving rise to a turbulent current of recirculation — a current whose existence is due to u_1 becoming nil.

Also, since the jet can not be extended beyond the wall of the cylinder, the reduced radius, L, can not surpass a certain value less than one; but since the theory is viable only if there is a zone of constant speed, u_1, L is further limited to a value L_0, which is seen to be about 0.5 from the Gauss curve in Fig. 3.3. Consequently, when the curve for $L_0 = 0.5$ is drawn on the diagram $R(\xi)$ in Fig. 3.5, it defines an area beyond which the theoretical values have no physical meaning. This diagram is drawn for a Gaussian distribution of the incremental speeds, i.e., $k = 0.5$.

Inside this area, certain curves reach the ordinate $R = 1$, and the curve for $m = 1.5$ crosses a point $L_0 = 0.5$ and $R = 1$. For values of $m > 1.5$ there will be recirculating turbulence in the mixing chamber, but not for $m < 1.5$.

The recirculating turbulence is reflected in a network of current lines like that in Fig. 3.6. The zone of constant speed, u_1, exists in the region MN and u_1 is positive; however, u_1 is equal to zero in section N. A particular current line $NN'PCN$ surrounds a region where the current lines close in on themselves and

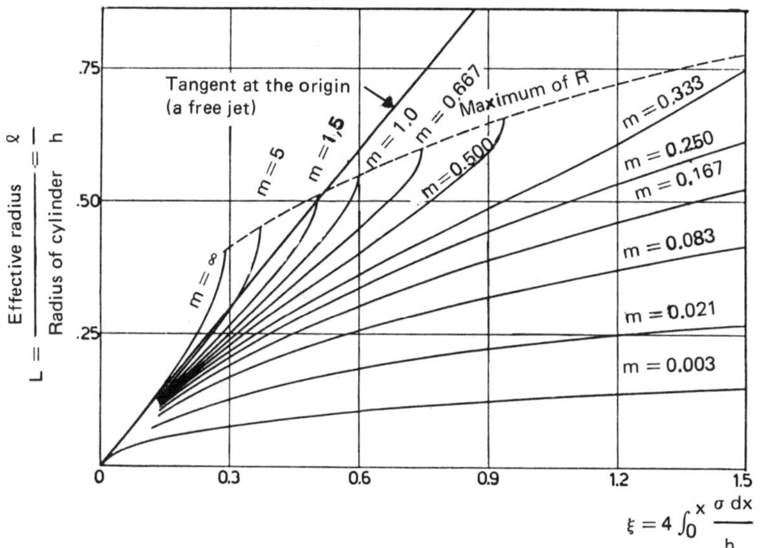

$$\xi = 4 \int_0^x \frac{\sigma\, dx}{h}$$

Fig. 3.4. Integration of Eq. (3.10) for effective reduced radius, L.

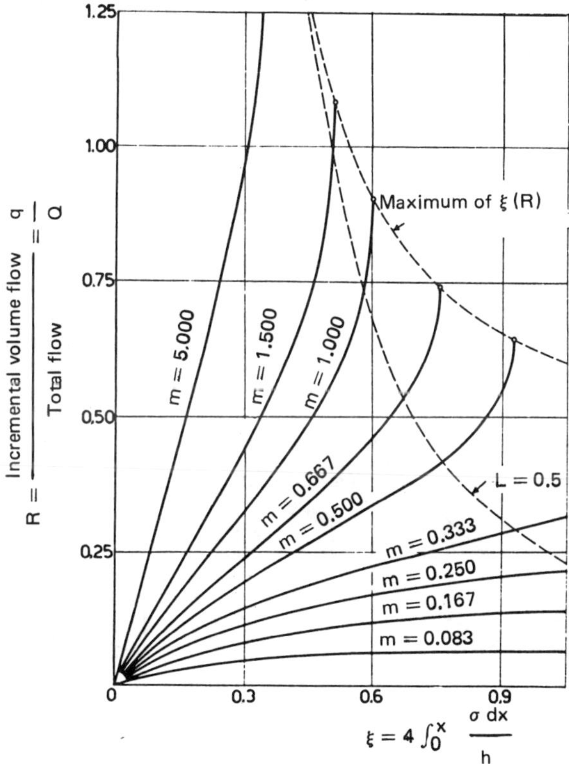

$$\xi = 4 \int_0^x \frac{\sigma\, dx}{h}$$

Fig. 3.5. Integration of Eq. (3.10) for values of R.

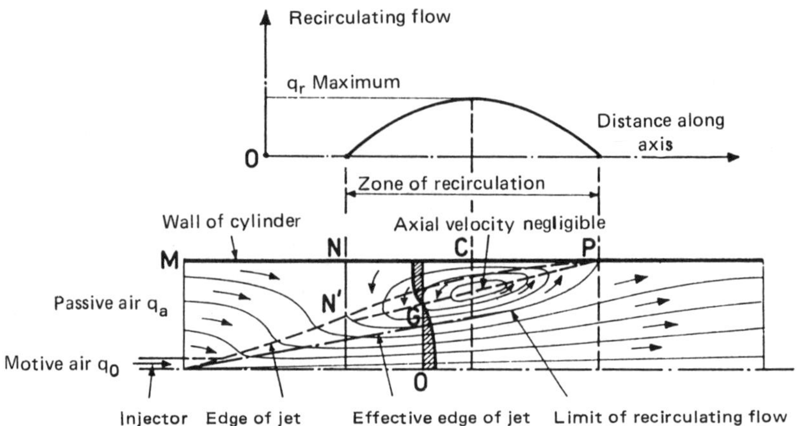

Fig. 3.6. Flow patterns of a confined jet.

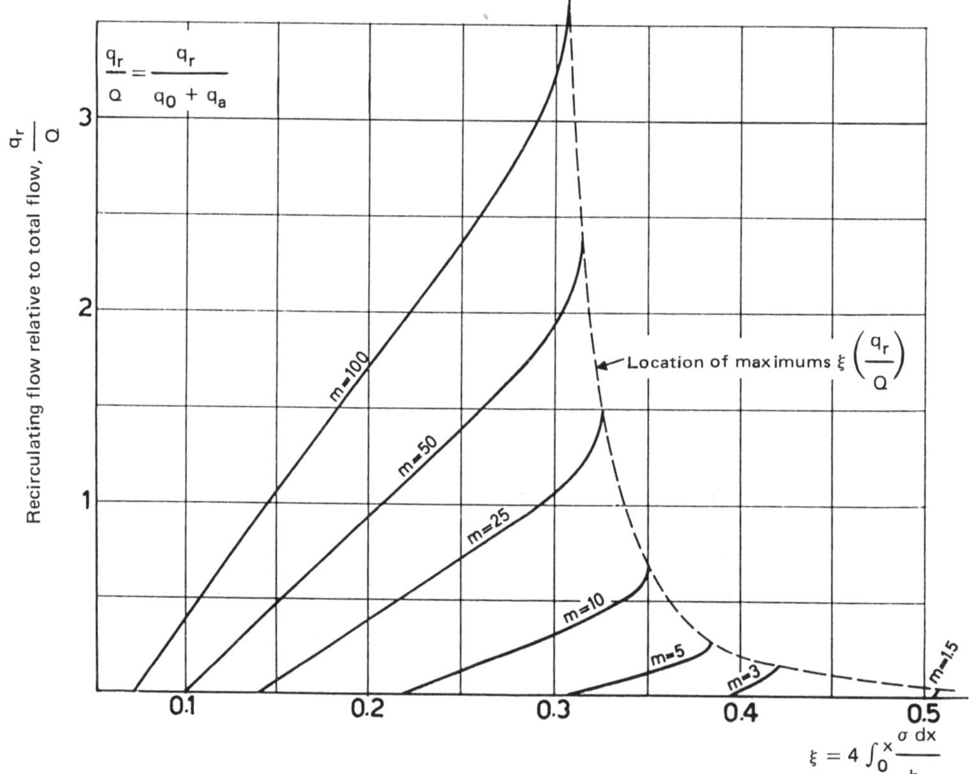

Fig. 3.7. Calculated relative recirculated flow.

constitutes the limit of the recirculation zone. It is useful to distinguish this line from the line where the component of longitudinal speed is zero.

The profile of longitudinal speeds in the recirculation-region looks like that in Fig. 3.6; and since longitudinal speed is zero at G, the volume flow, q_r, across circular area of radius OG, corresponds to an incremental speed that is this time greater than the actual speed of the jet.

Experience confirms that the profile of reduced incremental speed follows the data closely enough to permit applying the preceding theory. R. Curtet showed that the recirculated rate of flow, q_r, is related to the other variables as:

$$\frac{q_r}{Q} = (R - 1) \left\{ L^2 \left(1 - \ln \left[\frac{L^2(R-1)}{R} \right] \right) - 1 \right\} \tag{3.13}$$

The relative recirculated rate of flow can accordingly be calculated as a function of m and the reduced term ξ. Results of this calculation are plotted in Fig. 3.7, where the recirculation flow increases as a function of ξ up to a maximum corresponding to a critical point beyond which the calculation can not be pursued.

It can be seen from Fig. 3.7 that the recirculation flow tends toward a maximum as a function of the reduced term ξ for each value of m. This theoretical maximum flow corresponds to a maximum flow obtained through experiments. The value of q_r/Q at point C (Fig. 3.6) corresponds to the reduced ξ for this maximum.

3.2.3. Functions exhibiting similitude

The highly useful Craya-Curtet theory permits predicting the existence of a recirculation zone in the flow of a jet inside a confined circular enclosure; it permits determining the recirculating flow, and it permits determining the dimensions of the region where this recirculation begins — all through knowledge of a single parameter of similitude, m.

Parameter m is defined by Eq. (3.12). It is a constant for a given configuration, depending only on the conditions at the entrance of the cylinder. If in addition the profile of speeds is uniform across the nozzle outlet, then $k_e = 1$, at the entrance of the chamber (Eq. 3.2).

From then on, the value of R_e for Eq. (3.12) at the entrance of the cylinder can be calculated from the flow rates of the nozzle q_0 and the passive fluid, q_a. If a is the radius of the nozzle (Fig. 3.1) we have:

$$R_e = \frac{\text{incremental rate} \cdot \pi a^2}{q_0 + q_a}$$

that is:

$$R_e = \frac{q_0(h^2 - a^2) - q_a a^2}{(q_0 + q_a)(h^2 - a^2)}$$

from which m is determined by the equation:

$$m = -\frac{3}{2}R_e^2 + R_e + \frac{R^2}{a^2}h^2 \tag{3.14}$$

Parameter of similitude, m, can thus be determined from Eq. (3.14). Nevertheless, it has a physical aspect that is important to bring out.

The integral of Eq. (3.8) can also be rendered as:

$$u_1 w_0 l^2 + k w_0^2 l^2 - \frac{w_0^2 l^4}{2h^2} = C^{te} = m\frac{Q^2}{\pi^2 h^2} \tag{3.15}$$

In addition, the theorem for turbulent movement can be applied upstream of section N (Fig. 3.6) as:

$$\int_s (p + \rho u^2)\ dS = P$$

This integral can be calculated with Eq. (3.4); and static pressure, p, being constant in a straight section, Bernoulli's theorem applied to the region of the flow where the constant speed, u_1, exists, gives:

$$\frac{p}{\rho} + \frac{u_1^2}{2} = H_a = \text{Const.}$$

The integral P is calculated easily and taking into account Eq. (3.15) the following equation is arrived at:

$$P = \pi\rho\left[H_a h^2 + \frac{Q^2}{2\pi^2 h^2} - m\frac{Q^2}{\pi^2 h^2}\right]$$

From this comes:

$$m + \frac{1}{2} = \frac{\pi^2 h^2}{Q^2}\left[\frac{P}{\pi\rho} - H_a h^2\right]$$

which can be written:

$$m + \frac{1}{2} = \frac{\pi h^2}{Q^2}\int_0^b\left(u^2 - \frac{u_1^2}{2}\right)2\pi y\ dy \tag{3.16}$$

This equation can finally be made explicit as a function of the quantities of movement inside the entrance section of the chamber, in the following form:

$$m + \frac{1}{2} = \frac{G_0}{G} + \frac{G_a}{G} - \frac{1}{2}\frac{G_a}{G}\frac{h^2}{h^2 - a^2} \tag{3.17}$$

In this equation:

$$G_0 = \rho u_0^2 \, \pi a^2 = \rho \frac{q_0^2}{\pi a^2} \qquad = \quad \text{quantity of movement of motive fluid.}$$

$$G_a = \rho u_a^2 \, \pi(h^2 - a^2) = \rho \frac{q_a^2}{\pi(h^2 - a^2)} \qquad = \quad \text{quantity of movement of passive fluid.}$$

$$G = \rho U^2 \, \pi h^2 = \frac{\rho(q_0 + q_a)^2}{\pi h^2} \qquad = \quad \text{quantity of movement of mixed fluids}$$

since q_0 and q_a are volumetric flow rates.

In order to fully exploit these results, the coefficient σ, for tangential shearing intensity should be known. Curtet found that for values of m over 0.25 the average value of σ is close to 0.0285 in flows that were axially symmetrical.

3.2.4. Thring and Newby's theory

Another parameter of similitude, developed by Thring and Newby, is often used in the study of these recirculation phenomena. For this, it is assumed the dimensions of the furnace are such that the jet follows the laws of a free jet until it meets the walls. They then used the experimental results obtained by Hinze and van der Hegge Zijnen (Ref. 3.3) on free jets, and especially the equation allowing the flow rate of ambient gas carried along by a jet, to be expressed as a function of the variable, q, i.e.:

$$q = q_0 \left(0.20 \, \frac{x}{a} - 1\right)$$

The flow of the passive fluid inside a cylinder is then carried along by the motive fluid to a distance x_0 such that:

$$\frac{x_0}{a} = 5 \frac{q_0 + q_a}{q_0}$$

A second variable x_{00} is defined as that where the jet meets the wall; it is given by the approximate relationship:

$$x_{00} \, \# \, 4.5 \, h$$

(i.e., a half-angle at the peak of the cone of the jet equal approximately to 12°).

They propose defining the recirculated flow rate as the difference between the flow rate carried to x_{00} and at x_0. This recirculated flow rate is then equal to:

$$q_r = q_0\left(0.20 \, \frac{x_{00}}{a} - 1\right) - q_a$$

from which:

$$\frac{q_r}{q_0 + q_a} = 0.20 \; \frac{q_0}{q_0 + q_a} \; \frac{4.5\,h}{a} - 1 = 0.9 \frac{q_0}{q_0 + q_a} \; \frac{h}{a} - 1$$

The parameter of similitude then proposed by Thring and Newby is the group:

$$\theta = \frac{q_0 + q_a}{q_0} \; \frac{a}{h} \tag{3.18}$$

This parameter is none other than:

$$\theta = \sqrt{\frac{G}{G_0}} \tag{3.19}$$

If a is small relative to h, it is seen that the parameters θ and m hold a simple relationship, and Eq. (3.17) is reduced to:

$$m + \frac{1}{2} \simeq \frac{G_0}{G} + \frac{1}{2} \frac{G_a}{G_0}$$

In addition, in a furnace with diffusion flames, the flow, G_a, of the passive air, is often small relative to G_0, the motive fluid, so that:

$$m + \frac{1}{2} = \frac{1}{\theta^2} \tag{3.20}$$

is obtained.

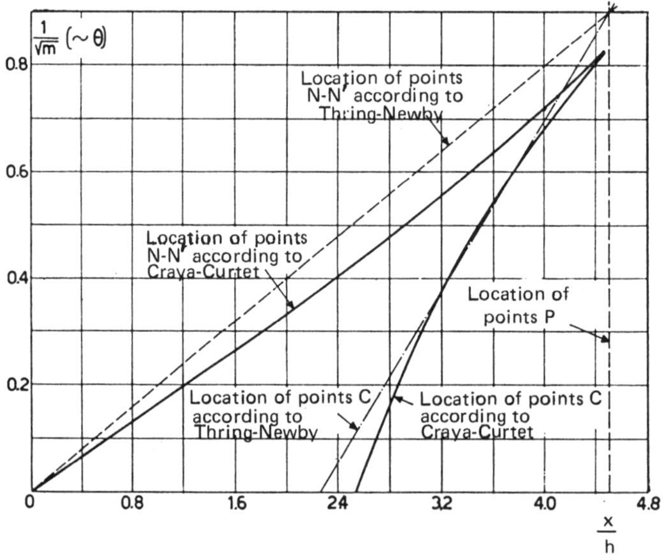

Fig. 3.8. Comparison of results from Thring-Newby and Craya-Curtet (See Fig. 3.6).

In order to further compare these two parameters, consider the values for the points N and C derived from these two theories and plotted in Fig. 3.8. A good agreement is observed.

3.3.   THE FREE JET IN AN INFINITE ATMOSPHERE

Curtet (Ref. 3.1) has shown theoretically that when a jet is in a confined environment and associated with recirculation (e.g., for the values of the m parameter over 1.5) the jet can be slightly more open or slightly more closed than a free jet, depending on the m value. Nevertheless, for jet travel up to about three times the radius of the cylinder, the laws of free jets can be applied without discernible error over the range of m-values in industrial furnaces. Thus it seems of interest to return to the general laws of a free jet.

The theoretical studies of Curtet and Craya can be applied to a free jet evolving about its axis in an infinite atmosphere at rest. In Eqs. (3.3) to (3.17) and Fig. 3.2, this implies extending h toward infinity, while u_1 is reduced toward zero. Under such conditions, Eq. (3.8) is written:

$$(kw_0^2 \, l^2)' = 0 \tag{3.21}$$

This equation is viable both in infinite atmosphere, where the distribution of speeds follows Gauss' law ($k = \text{Const.} = 0.5$), and in the zone immediately after the nozzle outlet, where the distribution of speeds depends on the variation of Gaussian constant, R, with x. In the outlet nozzle, there is a constant radial-speed profile, that is, $k = 1 = k_e$.

The integration of Eq. 3.21 for this condition gives:

$$kw_0^2 \, l^2 = k_e \, w_e^2 \, a^2 \tag{3.22}$$

where subscript e signifies the dimensions at the nozzle outlet.

In the same way, Eq. (3.10) can be reduced to:

$$\sigma = \frac{1}{4} \frac{(w_0 \, l^2)'}{w_0 \, l}$$

and taking into account Eq. (3.21)

$$\sigma = \frac{1}{4} l'$$

Experience has shown that the term σ is effectively constant beyond a certain length of travel, so that according to the established relationships:

$$l = 4\sigma x = 0.114x \tag{3.23}$$

$$\frac{w_0}{w_e} = \frac{a}{4\sigma x}\sqrt{\frac{k_e}{k}} \tag{3.24}$$

This last equation can also be written to take into account that the distribution of the speed is uniform in the nozzle, i.e., $k_e = 1$; then remembering that $k = 0.5$ and $\sigma = 0.0285$:

$$\frac{w_0}{w_e} = \frac{6.2}{\dfrac{x}{2a}} \tag{3.25}$$

The axial speed in the jet varies as the reciprocal of its length of travel.

Also, Eq. (3.23) indicates that the effective radius, l, of the jet is proportional to its length of travel. But with conditions established according to the Gaussian distribution of speeds, it is also known that this effective radius is close to half of the radius to the point where the speed is nil, i.e., $L_0 = 0.5 = l/x$. In other words, turbulent free-jets have a constant cone structure over the established range of conditions.

These relationships are made to resemble more traditional concepts by considering the radii at a section of the jet where the speed is no more than half of the speed measured along the axis; this radius will have the symbol:

$$r_{0.5v}$$

At such a point, Eq. (3.23) takes the following conventional form:

$$r_{0.5v} \sim x \tag{3.26}$$

and the proportionality constant is close to 0.10.

It is also possible to calculate the increase in speed of the fluid being entrained by the jet. The total rate of flow of the jet at any one point along its path is related to the nozzle flow rate by:

$$\frac{Q}{q_0} = \frac{w_0 l^2}{w_e a^2} = \sqrt{\frac{k_e}{k}}\, 4\sigma\frac{x}{a} = 0.32\,\frac{x}{2a} \tag{3.27}$$

These last three equations obviously are viable only under the prescribed conditions, since they assume that k corresponds to a Gaussian distribution of speed across the jet, and that σ is constant. While the profile of speeds at the nozzle outlet is generally uniform ($k_e = 1$), there is a transition zone before the profile across the jet is established. Since the preceding equations apply only to an established speed, the length of travel should be counted only from the point of the jet's nucleus (Fig. 3.1). In order to take the length of this nucleus into account, a length-of-travel, x_0, is chosen to represent the distance from the nozzle outlet to the nucleus' point. This distance x_0 is generally between 0 and 1.6 the radius of the nozzle, a. In order that the symbol x may keep its original definition (length of travel from the nozzle outlet) the last equations are re-written:

$$\frac{w_0}{w_e} \sim \frac{2a}{x - x_0} \tag{3.28}$$

$$r_{0.5v} \sim x - x_0 \tag{3.29}$$

$$\frac{Q}{q_0} \sim \frac{x - x_0}{2a} \tag{3.30}$$

These equations are substantiated through experience with free jets in an infinite atmosphere (Refs. 3.3, 3.4 and 3.5).

3.4. EXPERIMENTAL STUDY OF THE TYPICAL FLUID DYNAMICS IN A FIREBOX

The foregoing theory permits stating precisely that, if a jet with a volumetric flow, q_0, enters a firebox, this jet will develop at its beginning in a manner similar to a free jet in an infinite atmosphere; subsequently, it will disintegrate by giving birth to inverse currents, or currents of recirculation. The theory has shown the importance of a Curtet-Craya parameter of similitude, m, which uses the idea of quantity of movement defined in Section 3.2.3, and which permits calculating the principal characteristics of flow. It is thus possible to predetermine the variables described in Fig. 3.6 as a function of this parameter m, as well as of a parameter, θ, called the Thring-Newby parameter. Referring to Fig. 3.6, one should note:

(a) The abscissa of point N, that is, the length of jet travel where recirculation begins. This abscissa is nil if the jet comes out of an orifice centered in a plate that stops up the rest of the entrance to the cylinder, since this corresponds to the situation when $q_a = 0$. In this case, the parameter of similitude is reduced (according to Eq. 3.17) to:

$$m = \frac{h^2}{a^2} - \frac{1}{2} \quad \text{or} \quad \theta = \frac{a}{h}$$

If q_a is different from zero, point N is determined by the values $\xi(m)$ corresponding to the zero ordinate in Fig. 3.7.

(b) The abscissa of point C, which is the length of travel to the heart of the recirculation turbulence. This abscissa is given by the network of curves in

Fig. 3.7 and corresponds to the maximum of ξ for the value of m under consideration. In order to calculate the actual abscissa, take $\sigma = 0.0285$ which is assumed constant over the range of flow and speed. The abscissa in relation to the radius of the cylinder is then given by:

$$\frac{x}{h} = 8.77 \ \xi$$

(c) The value of the relative flow rate recirculated, q_r. Fig. 3.7 permits determining the evolution of this recirculated flow rate as well as the value of the maximum for different parameters of m as a function of the abscissa.

It must be pointed out, nevertheless, that this theory applies in practice only when the recirculation currents can be established freely, that is, when the cylinder is long enough.

Numerous experimental verifications of this theory have been made as part of the work done by the *International Flame Research Foundation (IFRF)* (Refs. 3.8 to 3.16). We review here the principal results, obtained as well for the experimental furnace at Ymuiden as on the cold scale model with a cylinder 225 mm in diameter.

In the cold tests, the burner is simulated by a tube (of variable diameter from 6 to 20 mm), placed in the axis of the model (motive fluid, q_0). The passive fluid is distributed uniformly around this tube while taking all precautions such that its rate is parallel to the axis.

3.4.1. Evolution of speeds along the axis

We will use the symbol V_0 to designate the average speed at the outlet of the burner and $V_{\max}$, the speed along the axis inside the chamber; this to avoid confusion that might arise with symbols u and w which were used in the theoretical section.

Figure 3.9 shows the evolution of $V_0/V_{\max}$ in a cold model as a function of the distance along the axis for different values of the Thring-Newby parameter, θ. For comparison Eq. (3.25) for an unconfined free jet has been used to calculate that curve. Except for very low values of θ, these observed curves are very close to that of the free jet up to a length of travel that can reach five times the radius of the chamber.

This comparison, predicted by Curtet's theory, is also viable for the furnace. Nevertheless, the fluid dynamics of a burner are not as well defined generally as for an air jet flowing out of a pipe or tube. Also, the ideas of equivalent speed, V_e, and equivalent radius, a_e, are introduced. These dimensions are defined as the speed and the outlet radius of a simple jet that would have the same quantity of axial flow and the same mass rate, m_0, as the actual jet leaving the burner, i.e.:

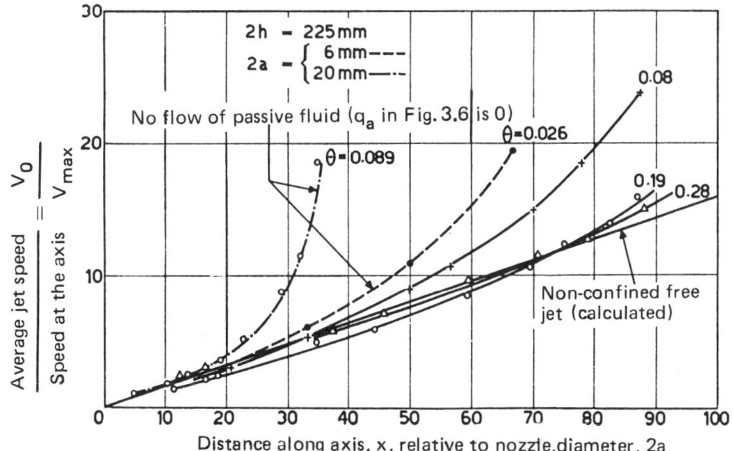

Fig. 3.9.   A comparison of relative jet speed versus relative distance along its axis by theory and by experiment.

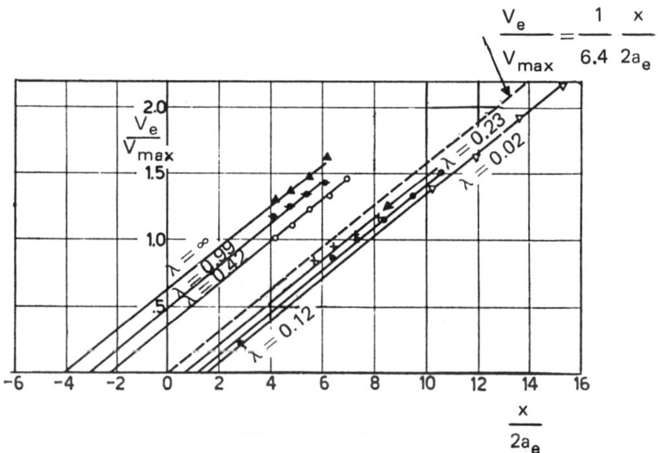

Fig. 3.10.   Dissipation of the velocity of a jet from a wall into a furnace. The ratio of :

$$\frac{\text{Equivalent velocity at jet outlet}}{\text{Velocity along the jet axis}} = \frac{V_e}{V_{max}}$$

is plotted against the ratio of:

$$\frac{\text{Axial distance from the jet outlet}}{\text{The equivalent diameter of the jet at its outlet}} = \frac{x}{2a_e}$$

$$\rho V_e \pi a_e^2 = m_0$$

$$\rho V_e^2 \pi a_e^2 = G_0$$

from which are produced the equations for defining these equivalent dimensions:

$$V_e = \frac{G_0}{m_0} \qquad a_e = \frac{m_0}{\sqrt{\pi \rho G_0}} \tag{3.31}$$

Figure 3.10 shows the variation of the ratio V_e/V_{max} for a burner having two coaxial flows at different speeds discharging from the inlet wall of a furnace. The ratio of the speed of the outside-jet to the speed of the central-jet is symbolized by λ. It is seen that after a certain length of travel, the resulting jet acts like one single jet coming from a source whose length of travel can be measured from in front of or behind the entrance plane of the furnace. This example illustrates the significance of the abscissa, x_0, introduced in Eq. (3.28). It will be noticed that the curves parallel a straight line corresponding to a value of 6.4 for the constant coefficient of Eq. (3.28) (instead of 6.2 for the theoretical value).

3.4.2. Evolution of radial speeds

Figure 3.11 shows the radial distribution of flow rates. The ratio of the local velocity, V, at coordinates x, r to the velocity V_{max} along the axis is plotted against the ratio of radii as defined for Eq. (3.26). The experimental results correspond to the flame from a burner consisting of a fuel lance with a mechanical atomizer, around which air is introduced from concentric rings through a nozzle. This burner is located in the inlet wall of a cylindrical combustion chamber.

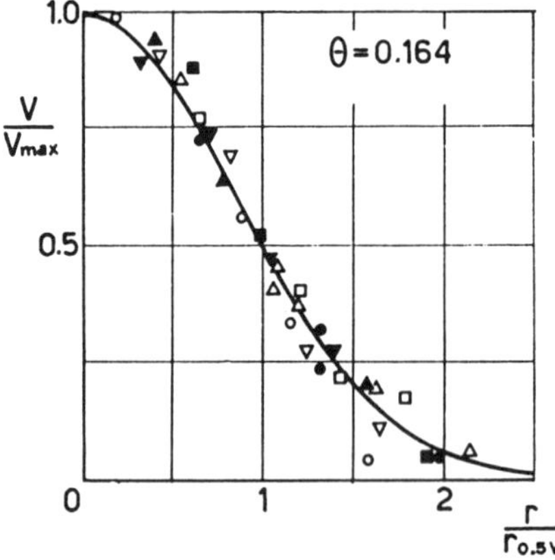

Fig. 3.11.   A jet into a furnace: Radial variation of speeds parallel to the axis (see Eq. 3.2 and Fig. 3.3).

These tests show that the radial velocity profiles satisfy a Gauss curve. The distance covered along the jet corresponds to about four times the radius of the cylindrical combustion chamber.

3.4.3. Form of the jet

Figure 3.12 shows the variation in diameter of a jet for the same experimental conditions as Fig. 3.11, that is, for conditions of a free jet, Eq. (3.29). After a transition period, the jet widens proportionally with the length of travel up to about four times the radius of the cylindrical combustion chamber. Beyond that, the jet practically does not widen, due to the influence of the chamber's walls.

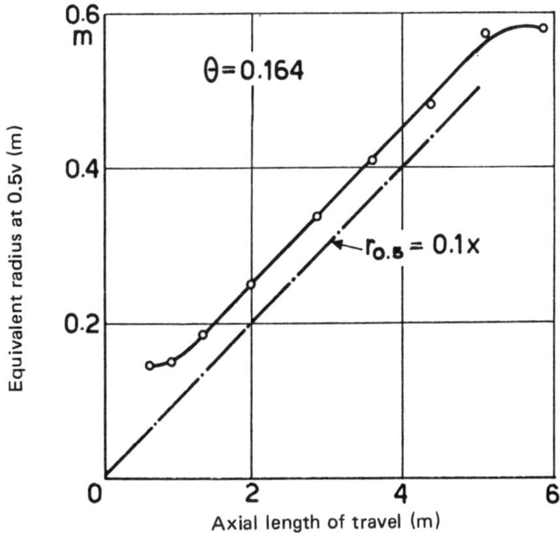

Fig. 3.12. Expansion of a jet stream.

These few examples thus confirm that when a jet discharges into a chamber, its development beyond a certain distance is very close to that of a free jet in an infinite atmosphere. However the constants of Eqs. (3.29) and (3.30) are slightly modified, and the relation is viable up to a distance of 12-15 times the diameter of the jet. Beyond, the evolution of the jet is affected by the walls of the chamber, that is by the zone of recirculation of burned gases.

3.4.4. The study of recirculation

Numerous studies have confirmed both qualitatively and quantitatively the flow of gases recirculated by a jet within an enclosure.

The results obtained by *SOGREAH* (*Société Grenobloise de Recherches Aérauliques et Hydrauliques*) from a cold test model with uniform feed from the inlet end of a cylinder have been compared to the theory by Curtet (Ref. 3.1). As shown in Fig. 3.13, the study concerns the development of recirculated flow as a function of the distance from the nozzle. It is seen (Fig. 3.13) that the experimental curve rises close to the theoretical curve and that the calculated maximum flow conforms well with the experimental one.

Figure 3.14 compares experimental recirculating flows with the Craya-Curtet theory for motive and passive air feed, in both the cold model and the furnace. It is seen that, for m greater than 4, results are grouped around the following linear relationship:

$$\frac{q_r}{q_0 + q_a} = 0.430 \left(\sqrt{m} - 1.65 \right) \tag{3.32}$$

If the value of m is further increased (for example, by reducing the diameter of the nozzle) it is seen (Fig. 3.14) that the length of the furnace can affect the value of q_r. This means that in a short chamber the nucleus of recirculation is not able to develop freely and finds itself to be truncated toward the bottom; it follows that the recirculation flow rates are diminished.

The experiment shows that, in practice, the path of travel at which maximum recirculations take place is approached toward the upper part where θ values decrease, but it remains effective at 3.5-4 times the radius of the chamber. This path is longer than can be derived from the Craya-Curtet theory, as has already been shown in Fig. 3.13. Figure 3.7, however, shows the theory predicts that the path varies little for practical values of m (that is, x/h taken between 2.7 and 3 for m varying between 100 and 10).

3.4.5. Visual tests

For many studies in fluid mechanics it is useful to visualize the currents. Indeed, visual study has often been able to highlight a particular aspect of the phenomenon.

Visualization tests have been done by Barchillon and Curtet (Ref. 3.16) on a hydraulic apparatus. A mixing cylinder with a diameter of 160 mm was connected to two reservoirs; a valve system permitted control of the flow of water into the cylinder (flow rate, q_a); a nozzle with a diameter of 12 mm permitted injecting a flow, q_0, of water at one end of the cylinder along its axis. Visualization was achieved by injecting air bubbles with a diameter of 0.1-0.2 mm at the nozzle outlet. The parameter θ varies between 0.075 and 0.941 (that is for m between 176.7 and 1.05 respectively).

Figure 3.15 shows the lines of current flow for four values of the parameter, θ, as well as the distribution curves of the average speeds (obtained by measuring

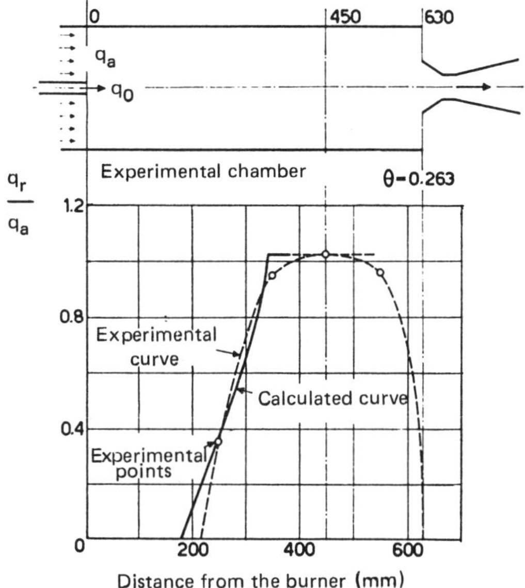

Fig. 3.13. Recirculation around a jet into a chamber: Experimental and calculated recirculation ratios as a function of distance from the jet nozzle.
(Source: *SOGREAH*)

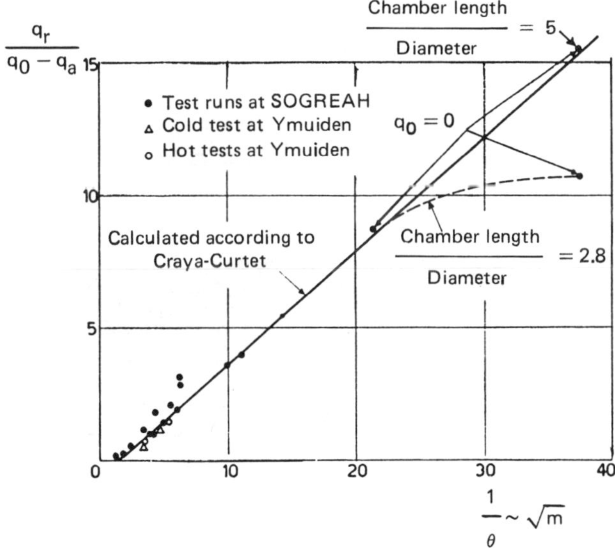

Fig. 3.14. Development of maximum recirculation around a jet.

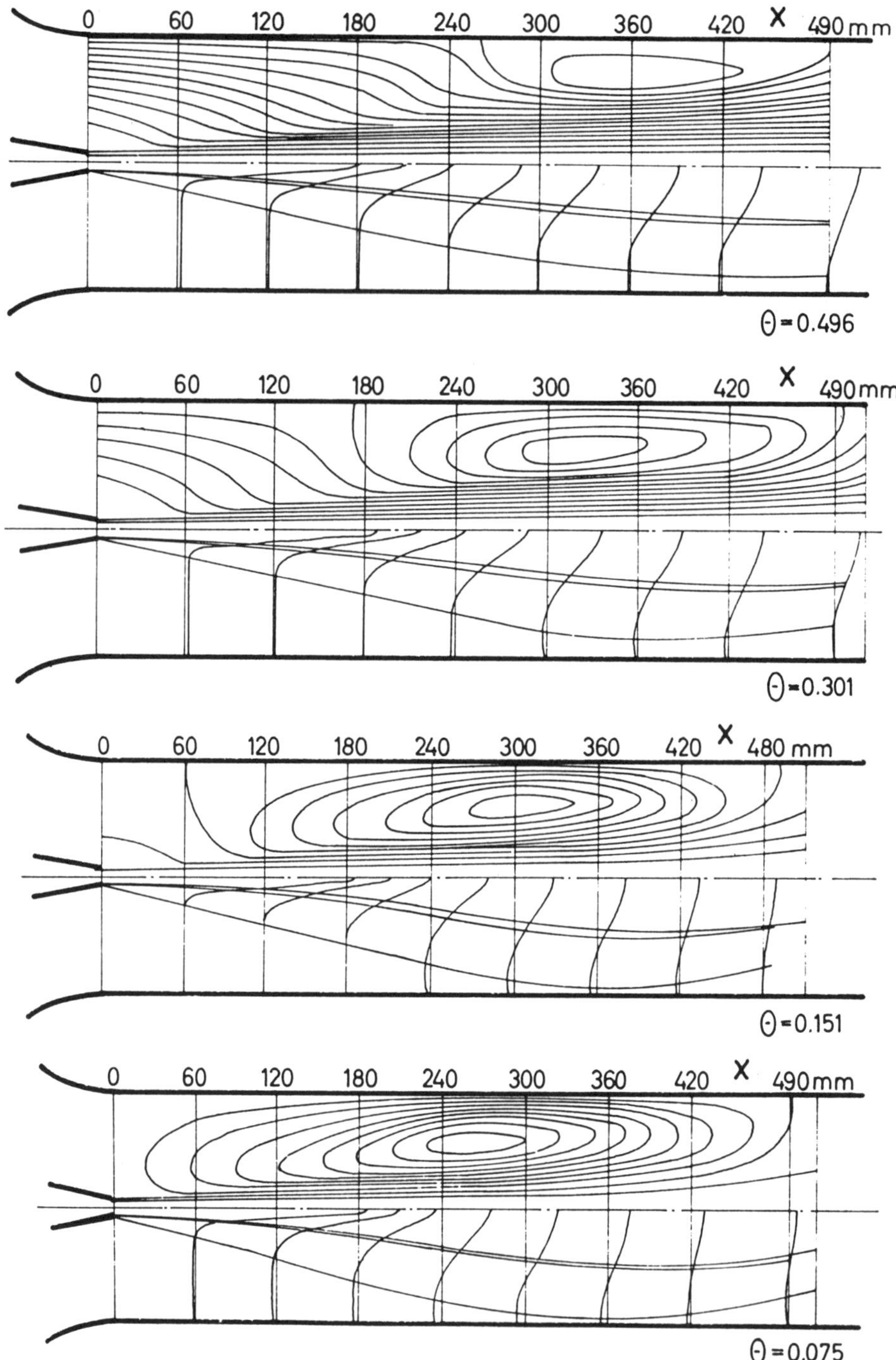

Fig. 3.15.   Visualized flow patterns around a turbulent jet.

total pressure and static pressure at the wall) and the diameter of half-rate, $r_{0.5v}$, effective diameter, and the normal limit of the jet, as given by points where the rate is equal to the ambient rate, u_1. All these dimensions have been described previously.

The configuration corresponding to $\theta = 0.075$ was that for which the passive flow rate q_a, was nil, as in the case of a heater with a burner situated on its inlet, which is 13.33 times as wide as the diameter of the burner. It will be noticed also that the heart of the recirculation current (point C in Fig. 3.6) goes slightly toward the inlet as θ becomes smaller.

For typical operating values, this hydraulic test shows that the length of the jet remains between 3.5 and 4 times the radius of the cylinder which confirms results obtained previously for gaseous fluids. In addition, the length to the end of the recirculation zone (point P in Fig. 3.6) is remarkably steady, no matter what the value of θ and is equal to about six times the radius of the chamber. These relations, which depend on the size of the chamber, require that the chamber be big enough for the flow to develop freely.

A complement to these observations is the variation of static pressure at the wall (Fig. 3.16). Two regions are thus revealed, one with pressures lower than the inlet pressure of the ambient fluid, and another with higher pressures. The low-pressure region, which is upstream of the heart of the turbulence, exhibits little variation in pressure; but the downstream region shows a static pressure that increases as the inverse of parameter θ.

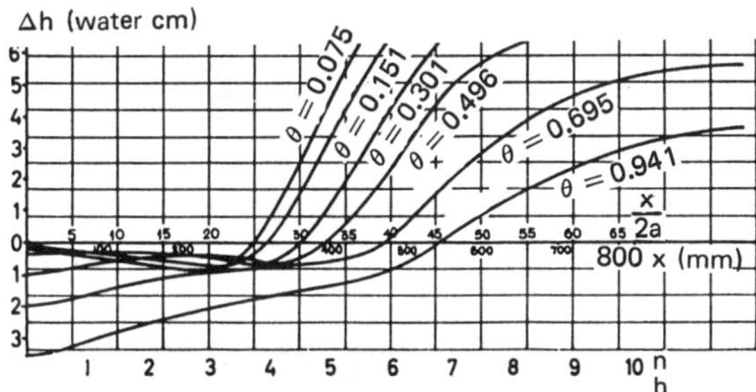

Fig. 3.16.   Variations in static pressure along the cylinder wall during experiments for Fig. 3.15.

Figure 3.16 also permits calculating the ambient flow with the Bernoulli equation; and it is seen that this flow tends to increase beyond the heart of the turbulence.

3.4.6.   Instantaneous fluctuations of flow

The above study has been completed by its author with an instantaneous observation of the flow. The current lines in Fig. 3.15 represent lines of mean

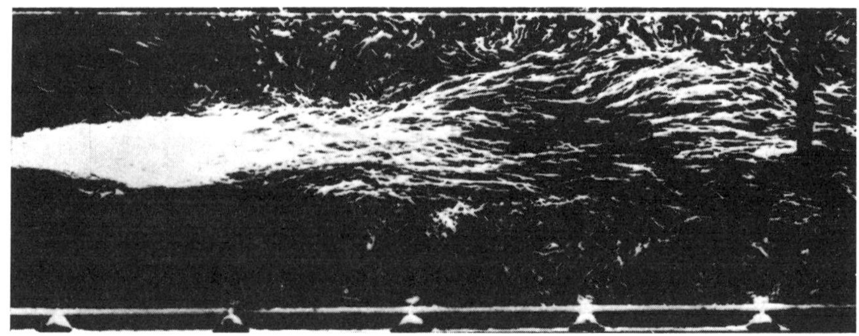

$\theta = 0.266$

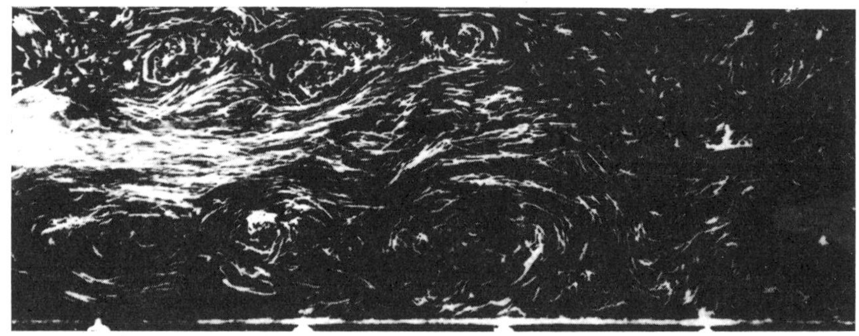

$\theta = 0.151$ (detail)

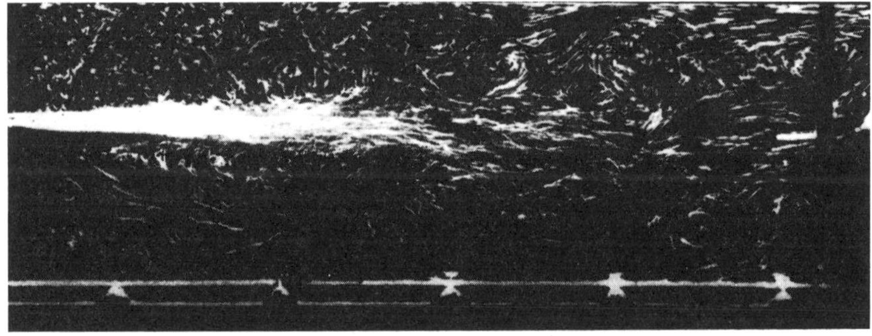

$\theta = 0.075$

Fig. 3.17.   Short-lived eddy currents formed during turbulent flow of a liquid jet.

flow during a time of the order of one minute. For a shorter time (about 1/10 to 1/20 s) a family of swirls appears.

Photos in Fig. 3.17 correspond to 1/20 s. It can be seen that the recirculation occurs through elementary swirls alternatively above and under the jet. On the contrary, along the wall, the back flow is steadily established. When $\theta = 0.075$, it comes back up to the entry section.

Measurements of turbulence intensity made on an air model with a heated wire anemometer have given indications on the intensity of this swirling movement.

In the jet axis, the turbulence is similar to that observed in a free jet. The turbulence is 10 to 15% higher in the confined jet upstream the heart of recirculation. Downstream this point, turbulence increases quickly up to 80% that of the free jet.

It has been observed also that near the point where the flow speed is nil, the back flow occurs by puffs of recirculated fluid which mixes with ambient quasi non turbulent fluid, with a periodicity of some cycles per second.

3.4.7.　Influence of the radius of the burner

As a first approximation for chambers of sufficient length, only one parameter, θ, or m, is needed to predict the configuration of the flow. However, tests carried qut on a cold model (Refs. 3.8 and 3.11) have shown that the recirculated rates of flow are not identical for the same parameter when the radii of the burner are different. Experiments show that the parameter a/h actually plays an important role as well.

An important consequence of this observation involves a detailed analysis of those points in Fig. 3.14 that correspond to large values of θ. These are shown in Fig. 3.18 where it is seen that the recirculated flow can be much higher than the theory predicts. For example, for $a/h = 0.158$ and $q_a = 0$ (that is, $1/\theta = 6.32$), the rate of flow predicted by the theoretical curve is only 70% of the measured flow rate. Considering, as previously, a linear variation as a function of $\sqrt{m}$, the experimental points satisfy the equation with the form:

$$\frac{q_r}{q_0 + q_a} = A\left(\sqrt{m} - B\right)$$

A and B being functions of a/h, given in Fig. 3.19.

In the same way, it has been seen that, if the burner radius increases, the heart of the recirculation zone approaches the entrance of the chamber. Thus, for a burner with a diameter that is 60% of that of the chamber, this distance is about the dimension of the chamber's radius. Figure 3.20 shows variations of the relative jet length as a function of θ and of the parameter a/h. We have also drawn the curve that corresponds to a sudden disconnection ($q_a = 0$).

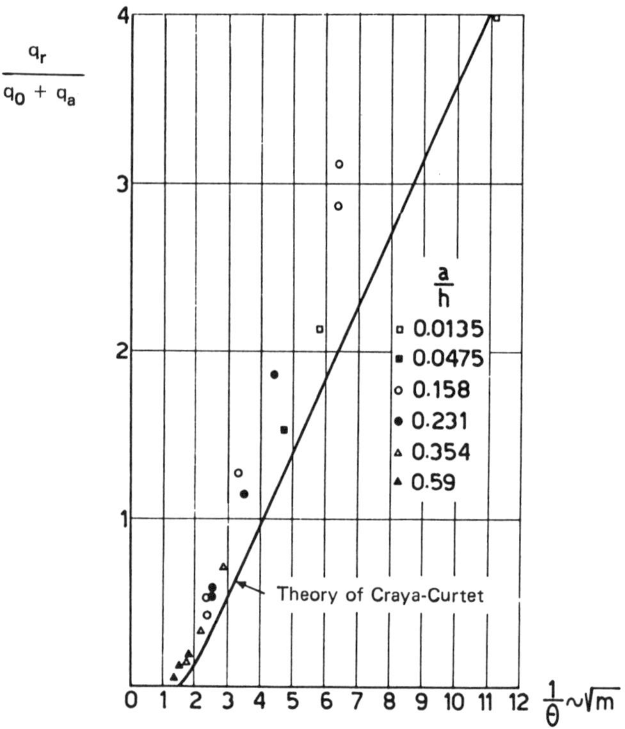

Fig. 3.18.   Divergence of experimental from calculated relative recirculation flow rates, $\dfrac{q_r}{q_0 + q_a}$.

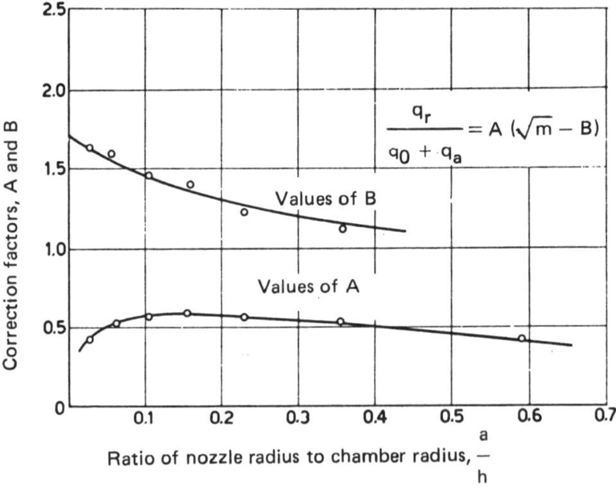

Fig. 3.19.   Correction factors for applying the Craya-Curtet number
to recirculation calculations.

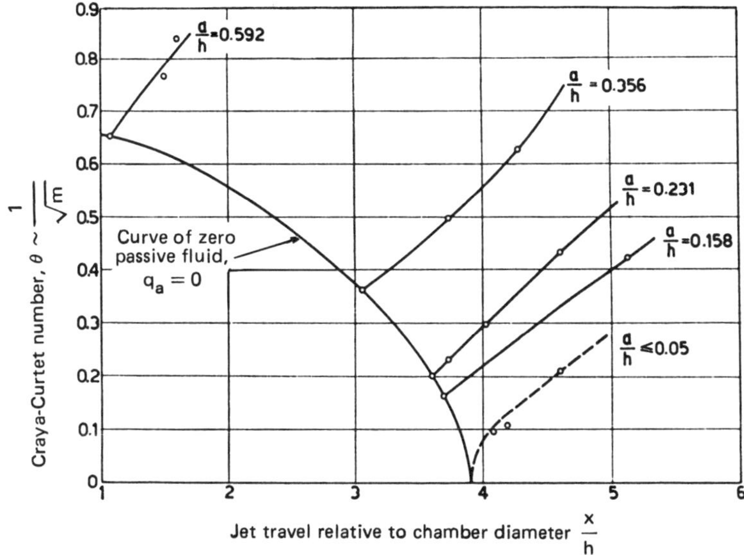

Fig. 3.20. The effects of the Craya-Curtet number, m, and the ratio of nozzle radius to chamber radius, a/h, on the location of the zone of maximum recirculation along the path of jet travel relative to the chamber diameter.

These results can be explained in part by the importance of the transition zone of the jet at the outlet of the nozzle. The recirculation zone is developed in that portion of the chamber where the transition takes place.

3.4.8. Conclusions

In concluding this study in fluid dynamics of furnaces and heaters which are fed by a burner that gives an axial jet, we recall with Chedaille, Leuckel and Chesters (Ref. 3.14) the schematic configuration of possible flow (Fig. 3.21). Two extreme cases corresponding to practical configurations are shown. The upper part of Fig. 3.21 shows a closed chamber equipped at its inlet with an axial burner that feeds both air and fuel in two coaxial jets. The rates of flow are practically the same in the center and in the ring and the speeds are similar. Along the line of travel occurs a first zone corresponding to a first mixture of the two gas currents leading to a jet with the characteristics of a free jet. At the frontier of the ring jet occur exchanges with the recirculated burned gases coming from below as a result of the turbulence.

These exchanges are completed in a seond zone corresponding to development of a jet similar to a free jet. This second zone extends (Fig. 3.21) to a section

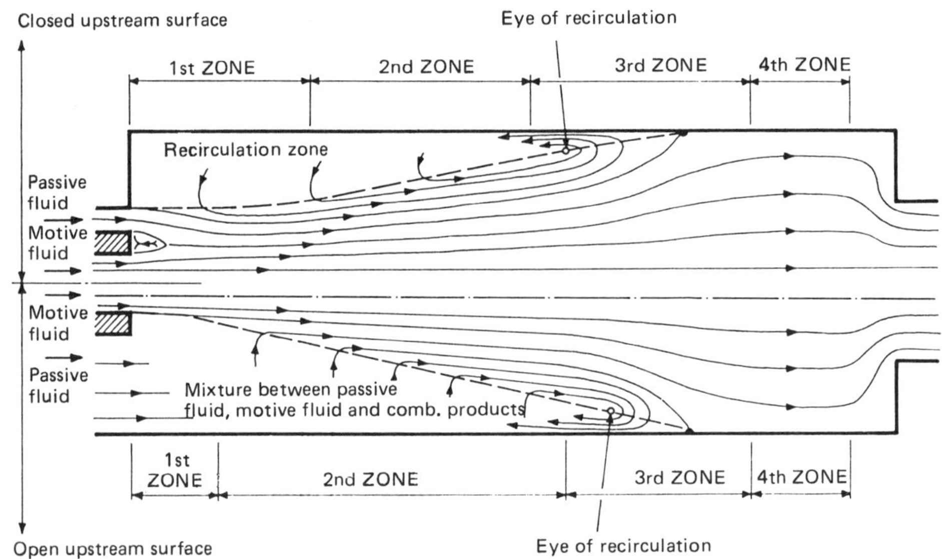

Fig. 3.21. Model for gas flows within a cylindrical combustion chamber.

corresponding to the heart of the recirculation. Then, in a third zone, the static pressure increases up to the end of the recirculation, while a current with a low rate is established below.

The lower part of Fig. 3.21 corresponds to the case of a chamber fed with secondary air across its entire inlet face. The burner furnishes a single axial jet with a high speed. The speed of the secondary air being low, its movement is negligible. A first flow zone corresponds to establishing a Gaussian speed profile across the jet and mixture with cool air from the outside. The second zone can be considered as corresponding to the development of the jet and admixture of secondary air with recycled combustion products, this mixture serving to feed the jet. Finally, there is again found a third zone in which the maximum recirculation is obtained, followed by an increase in static pressure and possibly a fourth zone with a low axial flow speed.

These models are obviously schematic and correspond to a situation where the chamber is fed by one single burner. They also apply when several burners are placed in a burner wall. But, before applying the general rules described above, certain approximations should be made. If, for example, the burners are arranged side by side close to the center of a burner wall, they can be considered as a single jet and the recirculated flow determined on that basis. On the other hand, the problem is more difficult if the burners are spaced so that each jet keeps its own individuality for a long time. Similarly, if an axial obstruction is placed downstream in a jet, the acceleration of the gases passing between this body and the side walls must be taken into account; it is conceivable that such an arrangement disturbs the flow since this acceleration can inhibit the establishment of the recirculation because of the lowering of the static pressure thus created in the downstream part.

Referring to the preceding models, however, it can be said that mixtures between air and fuel, on the one hand, and recycled combustion gases, on the other, occur very close to the burner outlet. The study of the mixing conditions in these regions is thus important, since a more or less rapid combustion results from this mixture.

This is the subject of the following chapter, which considers the case of one single jet. The case of several adjacent burners will simply be surmised.

REFERENCES

3.1   CURTET, R. – "Sur l'écoulement d'un jet entre parois". Imprimerie du Ministère de l'air, P.S.T. No. 359, Paris 1960 (See also: *Combustion and Flame*, Dec. 1958, vol. 2, 4, 383.)

3.2   THRING, M.W., NEWBY, M.P. – "Combustion length of enclosed turbulent jet flames". IVth *Internat. Symp. on Combustion*, p. 789. Williams and Wilkins, Baltimore, 1953.

3.3   HINZE, J.O., VAN DER HEGGE ZIJNEN, B.G. – "Transfer of heat and matter in the turbulent mixing zone of an axially symmetrical jet". *Appl. Sci. Res.* 1949, A-1, p. 435-461, Martinus Nighoff, the Hague.

3.4   CORRSIN, S. – "Investigation of flow in an axially symmetrical heated jet of air". *N.A. C.A.A.C.R.*, 3123 WR, 94.

3.5   REICHARDT, H. – "Gesetzmässigkeiten der freien Turbulenz". *VDI Forschungsheft*, 1942, 414, B, vol. 13.

3.6   COHEN DE LARA, G., PERRIN, R. – "Etude aérodynamique des phénomènes de recirculation dans un modèle schématique de fours à flammes de diffusion de section circulaire". *IFRF*, No. F 17/b/3.

3.7   COHEN DE LARA, G., PERRIN, R. – "Etude de la recirculation dans les modèles schématiques de fours à flammes de diffusion. Influence de la longueur du four et des modes d'alimentation en air". *IFRF*, No. F 61/b/2.

3.8   COHEN DE LARA, G., POUSE, J., PERRIN, R. – "Etude de la recirculation dans les modèles schématiques de fours à flammes de diffusion. Influence du diamètre du brûleur et de la conicité de la chambre de mélange". *IFRF*, No. F 61/b/3.

3.9   CHEDAILLE, J., LEUCKEL, W. – "Etude des flammes d'huile obtenues avec des brûleurs à pulvérisation mécanique et à rotation d'air, terminées par un divergent". *IFRF*, No. F 31/a/40.

3.10 COHEN DE LARA, G., CURTET, R. – "Etude de la recirculation dans les modèles schématiques de fours à flammes de diffusion". IIIe *Journée d'Etudes sur les flammes*, Feb. 1958. Paris.

3.11 COHEN DE LARA, G., RIVIERE, M. – "Etude expérimentale des paramètres géométriques et dynamiques intervenant sur le phénomène de recirculation dans un modèle de four de section circulaire". IVe *Journée d'Etudes sur les flammes*, April 1961, Paris.

3.12 BEER, J.M., CHIGIER, N.A., LEE, K.B. – "Modelling of double concentric burning jets". *IFRF*, No. C, 50/a/2, May 1962 et IXth *Internat. Symp. on Combustion*, p. 892, Academic Press, New York, 1963.

3.13 RIVIERE, M. − "Presentation des résultats de la 2e série d'essais sur le mécanisme de la combustion". *IIe Journée d'Etudes sur les flammes*, Dec. 1955, Paris.

3.14 CHEDAILLE, J., LEUCKEL, W., CHESTERS, A.K. − "Aerodynamic studies carried out on turbulent jets by the *IFRF*". *IIIrd Symposium on flames and industry*, Oct. 18-19 1966, London.

3.15 HUBBARD, E.H. − "Recirculation in cold models of furnaces: a review of work carried out at *SOGREAH*". *J. Inst. Fuel*, April 1962, XXXV, p. 160-173.

3.16 BARCHILLON, M., CURTET, R. − "Structure détaillée d'un jet en présence de recirculation", *Ve Journée d'Etudes sur les flammes*, Nov. 1963, Paris.

4

the stabilization of flames

4.1. GENERAL

All the flows described in the preceding chapter are organized within the firebox; in industry they are most often controlled in the vicinity of the burner through designs to favor stabilization of the flames.

4.1.1. Conditions of stability

A statement that a combustion reaction is stable implies that the reaction remains fixed in a space through which flow the inflammable fluids that take part in the combustion.

To simplify, assume a homogeneous premix of air and fuel flowing at rate $\vec{V}$ (Section 2.3.2.1 and Fig. 4.1); assume there is a flame front separating the cool mixture from the hot burned combustion gases, and isolate a small element ab of the flame front, which is assumed linear because of its smallness. Now, consider the conditions of existence and permanence of this element.

The speed of flow of the mixture, $\vec{V}$, in Fig. 4.1 may be decomposed into $\vec{V}_n$, the speed normal to surface ab, and speed $\vec{V}_t$ parallel to that surface. The position of the flame front will be fixed in space if the rates $\vec{V}_n$ and $\vec{V}_t$ are equal and opposite to the corresponding rates of propagation of the combustion.

A primary condition for stability therefore, is that the speed of normal propagation of the combustion front be equal and opposite to the speed of flow, which is turbulent in the majority of cases. The result is:

$$V_o = V \cos \alpha$$

where α is the angle between $\vec{V}_n$ and the resultant, $\vec{V}$, of $\vec{V}_n$ and $\vec{V}_t$ (Fig. 4.1).

In addition, surface particles are transported from *a* to *b* under the influence of the component $\vec{V_t}$. Therefore, the particles going away from *a* must be part of a flow of similar particles in order for the flame front to be constant. Finally, when the initial region of the flame front is considered, there must be a constant source of ignition to keep the reaction moving opposite to the flow.

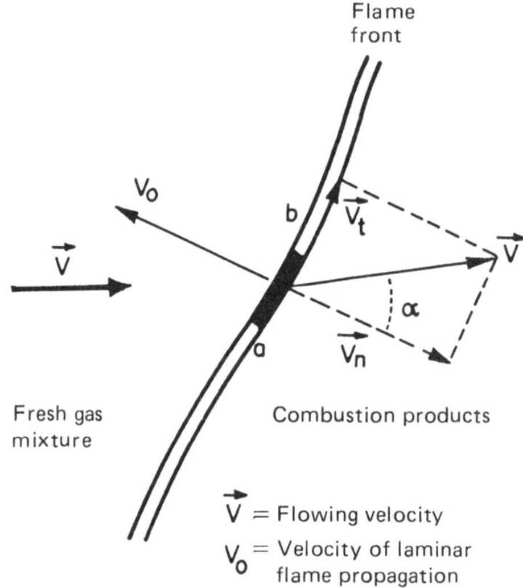

Fig. 4.1. The dynamics of a flame front.

Consequently, a stable combustion front is maintained in a flowing combustible mixture, when:

(1)   A sufficiently intense source of heat exists in a precise place.

(2)   The fundamental rate of turbulent combustion is equal to the normal rate of flow of the cool mixture:

$$V_o = |\, V \cos \alpha \,|$$

4.1.2.   Methods for stabilizing a flame

In order to stabilize a flame, it is necessary, first, to arrange a source of heat that is at once sufficiently intense, sufficiently constant, and in a sufficiently precise region for initiating the combustion reactions. This heat can be obtained from the combustion gases themselves as long as there is a method of fluid mechanics that permits recycling the hot combustion products as necessary to a favorable place in the vicinity of the burner.

Several stratagems answer this need, as follows:

(a)   An obstacle, placed in the flow perpendicular to its average rate, creates a wake in which the recycle currents are established toward the upstream area (Fig. 4.2). Such an obstacle could possibly be fluid, and be made by a jet directed perpendicularly or countercurrently to the principal flow.

(b)   A rotation of the fluids within the burner's nozzle creates at the burner outlet a diverging flow with a decrease in the static pressure along the axis downstream of the burner (Fig. 4.2). This reduced pressure in turn creates reverse currents with recirculation toward the burner.

These two methods of stabilization will be examined before taking up stabilization with outside heat by a pilot flame or with pilot oxygen.

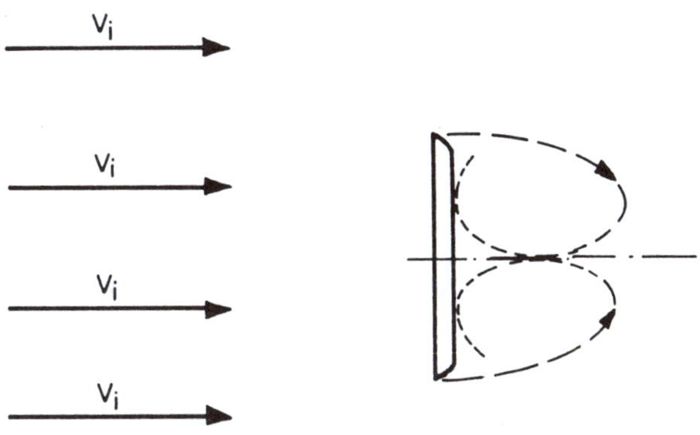

Downstream of an obstruction

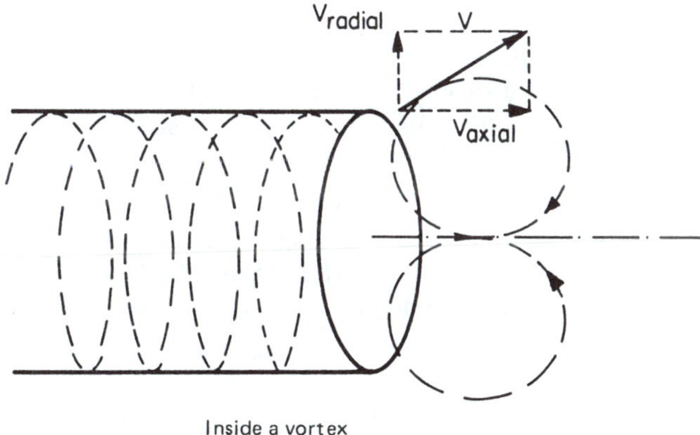

Inside a vortex

Fig. 4.2.   Methods for recirculating hot combustion gases.

4.2. STABILIZATION WITH AN OBSTACLE

4.2.1. The experimental study at Ymuiden (Ref. 4.1)

The burner is made up of an injection tube of liquid fuel, centered on the axis of a combustion-air delivery pipe in the study at Ymuiden. A metallic disc with radius r, welded around the end of the fuel injection rod, permits reducing the passage for the combustion air (Fig. 4.3).

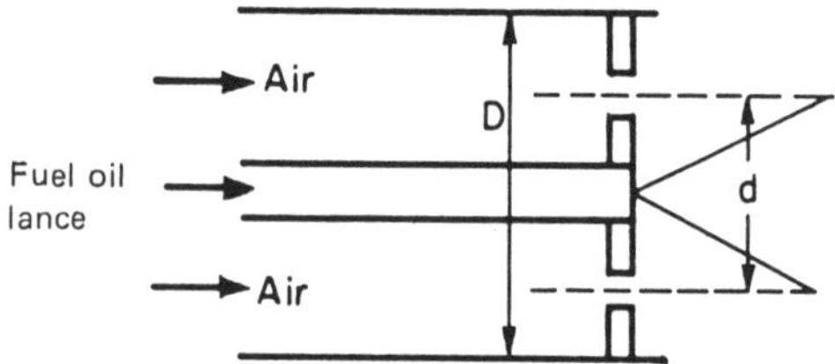

Fig. 4.3.   Diagram of a burner with stabilizing disc such as was used
at Ymuiden.

A study of the air flow, made cold downstream of the burner, permitted measuring the static pressure all across the stabilizer disc and along the axis of the burner. Data from these measurements are shown in Figs. 4.4a and 4.4b; these data correspond to longitudinal speeds of air of 40 m/s and 70 m/s. The pressure is expressed in terms of $\rho\, U_0^2$, so that it is proportional to the mass rate of flow of air. (U_0 is the longitudinal rate of flow; ρ is the density of air.)

In Fig. 4.4.b, the distance downstream from the burner is expressed in terms of the radius, r_0, of the stabilizer disc. The static pressure along the axis is lower than the ambient pressure up to a distance about equal to the diameter, and it shows a marked minimum at around 0.7 diameters.

Figure 4.5 shows the current lines forming a recirculation torus downstream of the burner. Negative flows of fluid along the axis reach about 5% of the longitudinal flow of air at the outlet of the burner. In operation, the nozzle projects the fuel oil as a fog into the torus of recirculation. The very finest drops are drawn into the recirculation movement. They evaporate there and feed a flame at that edge of the torus that hangs onto the edges of the disc.

Figure 4.6 is a photograph of this burner in operation. The liquid jet forms a dark cone centered on the disc. Fuel that has been drawn out into the torus after expanding over the disc is reconverged by the ring-jet of air. Fed with fuel in this way, the recirculation zone is surrounded by a flame that appears as a

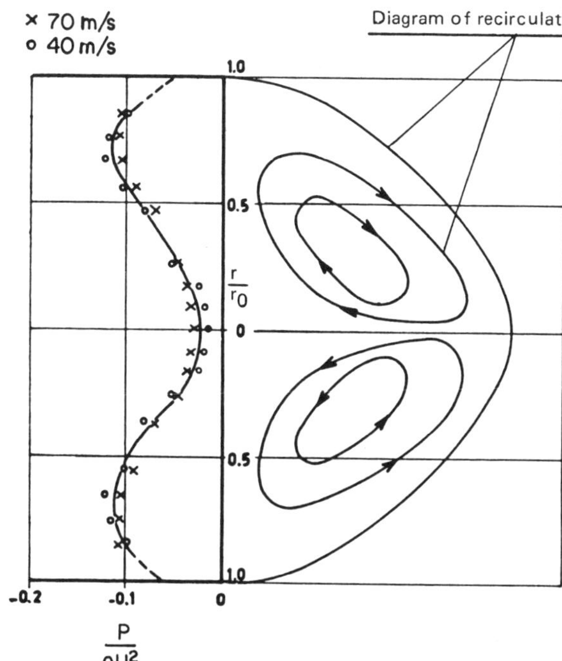

Fig. 4.4a. Pressure profile accross the downstream face of a stabilizing disc.

P = gage pressure.
ρ = density of air.
U_0 = mass rate of flow.
r_0 = radius of disc.

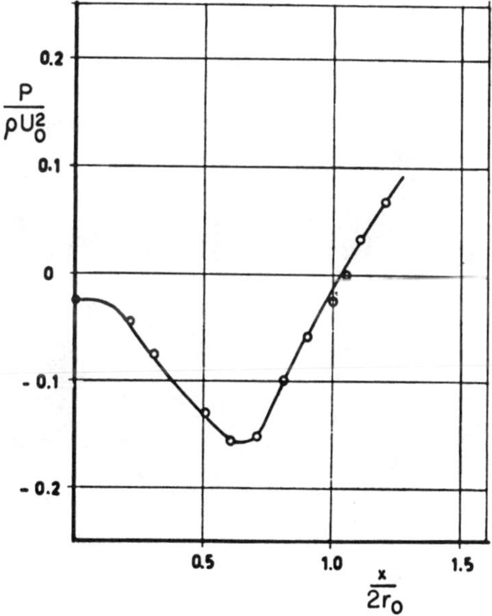

Fig. 4.4b. Pressure profile along the axis of flow downstream of a stabilizing disc.

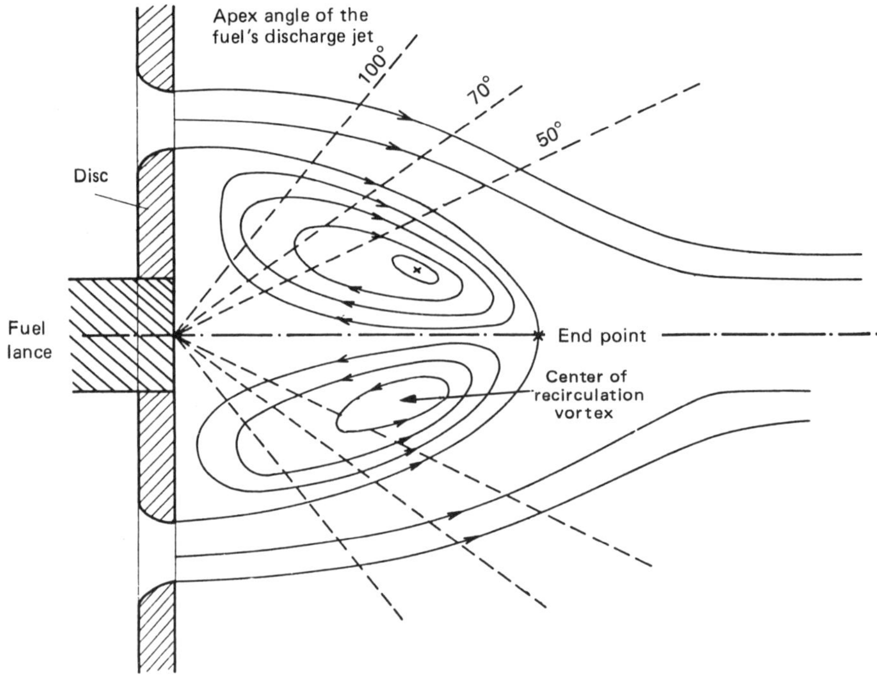

Fig. 4.5. Diagram of fluid currents and the recirculation vortex downstream of a stabilizing disc.

Fig. 4.6. Enlargement photo of experimental flame. Note jet of fuel oil and scattered flame due to recirculation. Refer to bottom right photo of Fig. 4.7 for comparisons. (Photograph by *IFRF*).

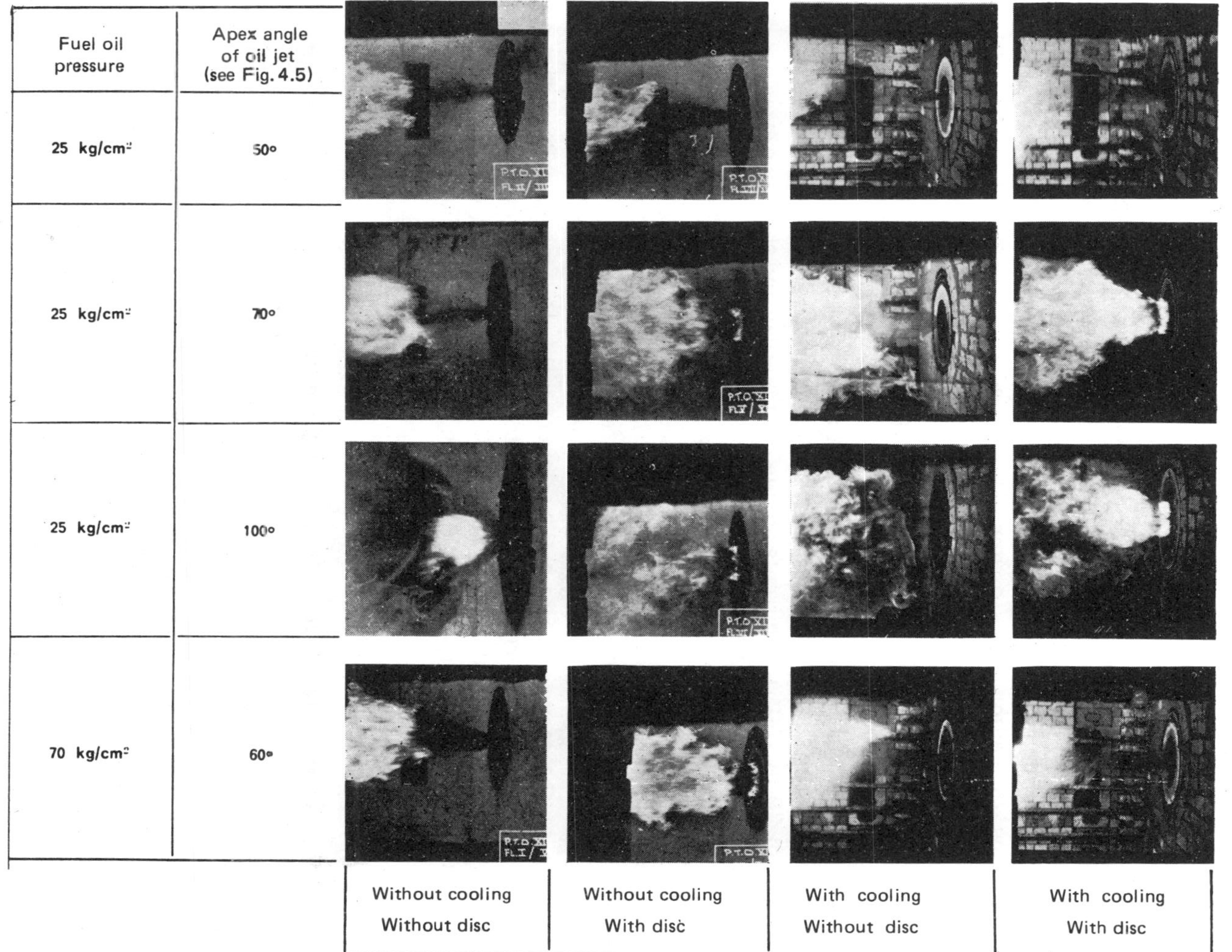

Fig. 4.7. Experimental flames both with and without cooling in the firebox and with and without stabilizing discs. (Photograph by *IFRF*).

white ring in Fig. 4.6. In addition, the photograph shows that neither the flame, nor the fuel jet perceptively modifies the flow of air downstream of the disc, compared to the conditions studied with the cold model.

Of course, the appearance of a stable flame supported by a torus of recirculation is based on an adequate atomization of the fuel that permits feeding to the currents a certain percentage of the fuel as fine drops easily drawn along by the gas, the angle of the conic jet of liquid being chosen accordingly. Figure 4.5 shows that the liquid jet crosses the torus of recirculation on a path that is more or less long, depending on its opening. Therefore, the angle of the opening of the liquid jet appears as one of the factors taking part in the stabilization of the flame.

The photographs in Fig. 4.7 were taken of the experimental *IFRF* furnace at Ymuiden. They compare flames with Fig. 4.6 (lower right corner of group), with and without the disc, and for different angles of jet opening. With these flames, the longitudinal flow rate of air to the burner is 10 m/s. For jets with an angle of 70 to 100° the currents are well fed by fuel, and the flame attaches itself up close to the stabilizing disc. For an angle of 50°, the torus of recirculation is not fed enough by fuel and the effect of the disc is completely hidden. These photos also show the effect of cooling the experimental furnace with water tubes distributed along the side walls; they are very apparent in the two columns of photographs on the right. Such cooling lowers the temperature of the smoke that mixes with the combustion air at the burner level, resulting in a cooling of the recycled gases and a consequently slower combustion. The effect is most apparent in comparing the photographs at the bottom of Fig. 4.7, which are labeled "without cooling, with disc" and "with cooling, with disc". In these photos, fuel is injected under a pressure of 70 bars, and the angle of the spray is 60°.

4.2.2. The experimental study at Toulouse (Ref. 4.2)

The experimental group at Toulouse *Groupe d'Etude des Flammes de Gaz Naturel (GEFGN)* conducted a study of a stabilizing disc on flames from natural gas. The burner (Fig. 4.8) was about the same as the burner at Ymuiden except that the air passages at Ymuiden were not shaped, whereas the obstacles used at Toulouse had the shape of a disc or of a cylinder or of a cone, welded to the end of the gas inlet tube.

This gas intake tube (Fig. 4.8) divides the current of natural gas into two jets, an axial jet discharging at high speed from a 7 mm orifice (a speed on the order of a few hundreds of meters per second or even sonic speed depending on the pressure) and a concentric ring jet discharging at a slow speed on the order of 10 to 20 m/s because of its large cross-section.

In Figure 4.9, related to cold-performed experiments the central jet traverses

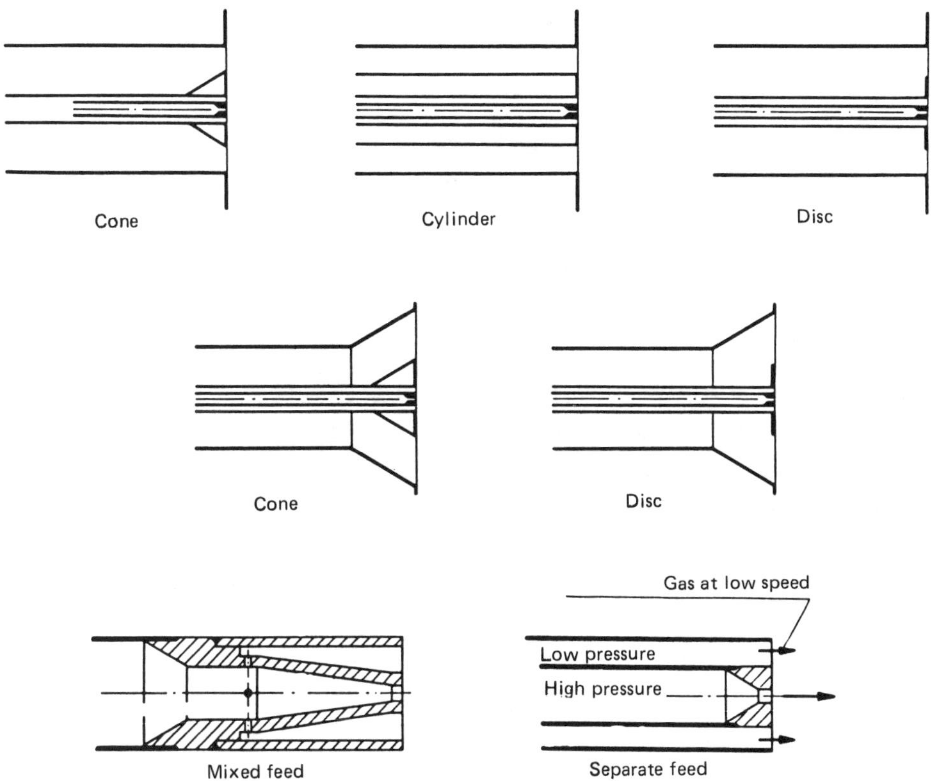

Fig. 4.8. Experimental burner outlets used at Toulouse to study downstream fluid-flow currents.

the recirculation zone at a high speed while the gas coming from the ring jet is carried along in the wake behind the disc, mixing with the combustion air. This explains why a stable flame is installed below the disc. It results from the formation of a combustible mixture of air and ring-jet gas in the torus of recirculation. The flame is maintained by this torus, while gas from the central jet is ignited downstream, as soon as it is mixed with enough air. Therefore, once the impulse is strong enough two flames often appear from this device, one formed in the vicinity of the disc and another slightly downstream where the rest of the gas burns.

Comparable experiments have shown that the shape of the upstream obstacle plays practically no role in the stabilization of the flame. In certain experiments with the disc replaced by the cylinder or even by the cone (Fig. 4.8) it was observed that the wake is essentially the same as that caused by a thin disc. And this similarity parallels similarities in the mixture of gas and air as well as in the stabilization mechanism of the flame.

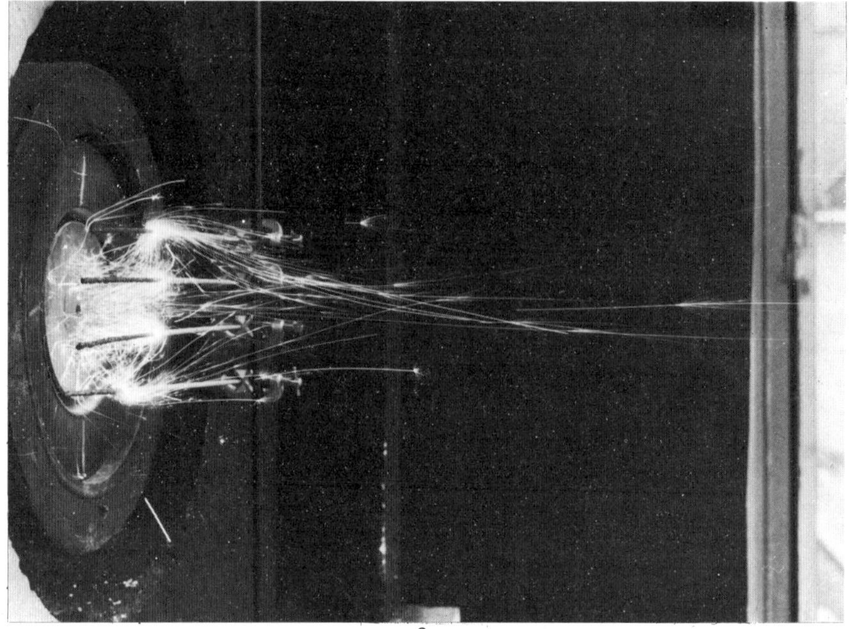

a

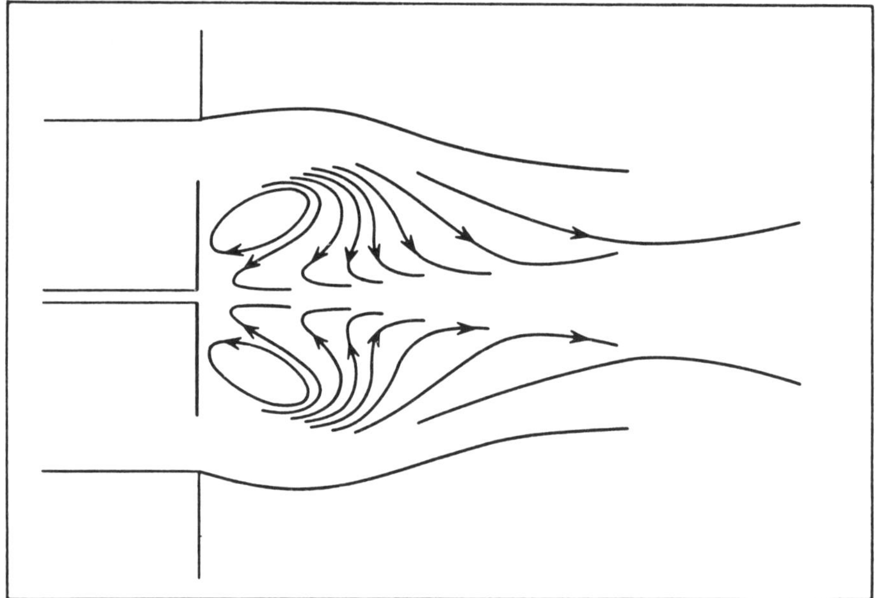

b

Fig. 4.9. Study of cold flows downstream of a burner made with fireworks sparklers. The photograph shows currents as identified by the sparks; the diagram summarizes extended observations of the spark flows.
(Photograph by *GEFGN*).

We learn from these experiments that the important feature for stabilization with an obstacle is the dimension of the plane surface on the downstream side of the obstacle. When the obstacle is in a cylindrical air feeder, the characteristic dimension is the ratio of the obstacle's diameter to the diameter of the cylindrical feeder.

4.2.2.1. Effectiveness of stabilization with an obstacle

The researchers at Toulouse have determined the effectiveness of disc-stabilization at different flows for a given proportion of gas to air. Results of numerous experiments were reduced to a single curve (Figs. 4.10 and 4.11) by plotting the specific impulse over the diameter of the burner, $G/Q \times 1/D$, against the downstream air speed divided by the square of the disc diameter, U/D^2. The terms had the following units:

G = impulse of the gas jet (N).
Q = mass flow of natural gas (kg/s).
U = speed of the air (m/s).
D = the diameter of the obstacle disc (m).

This curve (Fig. 4.11) corresponds to the following empirical equation:

$$\frac{G}{QD} = -\ 1.36\ 10^{-3}\ \left(\frac{U}{D^2}\right)^2 + 2.57\ \left(\frac{U}{D^2}\right)$$

It must be remembered that, in these experiments, the air speed is low relative to the movement caused by the central jet of natural gas. Figure 4.8 shows details of the tube feeding natural gas. The flow of gas varied from 70 to 200 m³/hr. The obstacles were discs with diameters of 108, 120, 210, 266 and 280 mm, or cylinders with diameters of 108, 266 or 280 mm, or a cone with a base diameter of 280 mm. Different air speeds varying from 7 to 50 m/s were created around the obstacle.

Figure 4.10 shows that, for the same rate of flow of natural gas, stabilization with the disc increases with an increase in diameter of the disc. This diameter determines the dimension of the torus of recirculation extending downstream to a distance practically equal to the diameter. Of course, the results reported here depend closely on feeding the natural gas so as to cause formation of a combustible mixture at the periphery of the torus of recirculation. In this case, the gas tube with a double jet fulfills this condition. In the case of liquid fuels, the size-range of the droplets formed by atomization allows the finest to be drawn into the torus of recirculation to burn there.

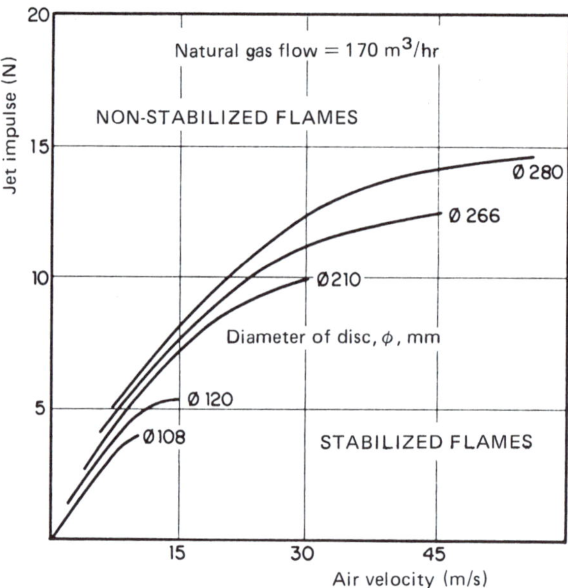

Fig. 4.10. The effectiveness of a stabilizing disc as a function of its diameter.

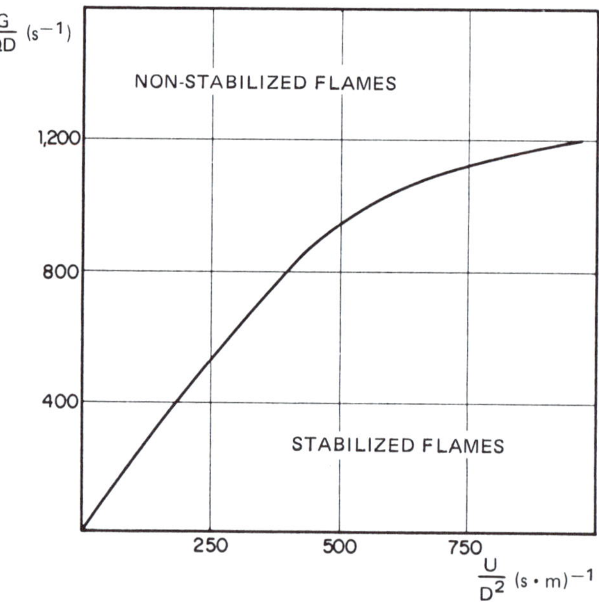

Fig. 4.11. Empirical correlation for effectiveness of a stabilizing disc as a function of its diameter.

G = impulse of the gas jet (N).
Q = mass rate of flow (kg/s).
U = air speed (m/s).
D = diameter of the stabilizing disc (m).

4.3. STABILIZING BY ROTATING THE FLUIDS
AT THE BURNER OUTLET
(Refs. 4.3 and 4.4)

The principles for stabilizing by fluid rotating at the burner outlet are similar to the principles for stabilizing with an obstacle. The main idea is to create a low-pressure zone downstream from the burner along its axis. Flow paths are thus bent and when the feed is properly adjusted, a recirculation torus is formed, where fuel and air mix in a proportion suitable for peripheral burning. This acts as the ignitor of the principal flame.

This low-pressure zone is obtained by impelling the combustion air and possibly the fuel to rotate around the axis in combination with the axial flow rather than by forcing the air around an obstacle. This rotating creates centrifugal forces that cause the fluids to accumulate at the periphery as static low pressure is established along the axis.

4.3.1. The cause of fluid rotation

Starting fluids to rotate usually involves the combustion air, either entirely or partly. In fact, when liquid or atomized fuels are used, only the air carries an appreciable amount of kinetic energy such as can be transformed into energy of rotation. When the fuel is gaseous, it is usually furnished at considerable pressure in the pipeline network, and it is economical to use this pressure energy to organize the chosen flow pattern. Further on, we will give some precise notes on such use of pressure energy.

Figure 4.12 illustrates a common type of burner. This burner has a cylinder for conveying the combustion air. The fuel (fuel oil or gas) is supplied through a tube in the axis of the cylinder. The rotating movement is first imposed on the air by passing it over fins that cover at least part of the cylinder, the end of which can be variously shaped into a nozzle. The means of attaching the burner outlet to the wall of the furnace has a very direct influence on the flow pattern, as we will see later. The position of the end of the fuel nozzle which can be more or less deeply inserted into the furnace also plays a role in the organization of the flows. On one hand, this nozzle can act as an obstacle to bend current lines; on the other, its position determines the way that the fuel is introduced into the flowing air.

As indicated for the burner in Fig. 4.12, the fins can involve all of the air flow or merely direct passage of a part. If the fins are well laid out, they transform

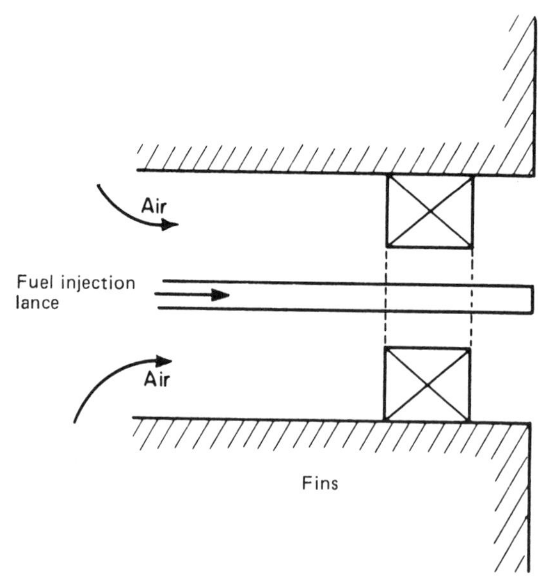

Fig. 4.12. Diagram of a burner with the air rotated by axial fins.

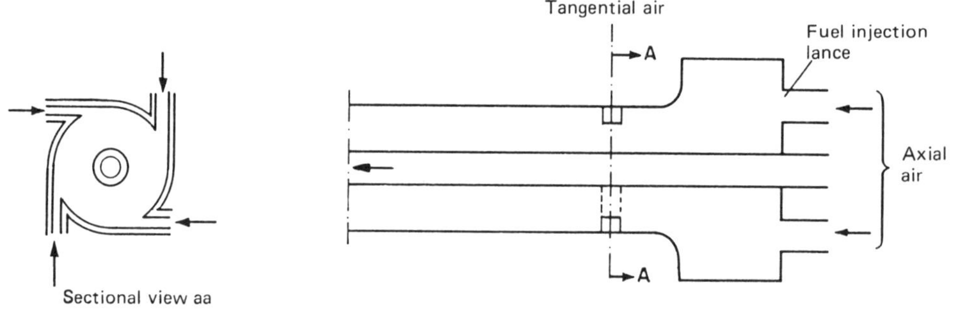

Fig. 4.13. Diagram of a burner with the air rotated by tangential
inlets.

the air's axial flow into rotating flow at a slight cost of friction and consequently
a limited pressure drop.

For their studies, the engineers at Ymuiden and Toulouse adopted experimental burners in which a part of the combustion air is introduced tangentially
(Fig. 4.13). The advantage of this design lies in the simplification it brings to
calculations for rotating flow. They depend only on the rate of flow of air in the
tangential entrances and the radius of gyration, which is fixed by the design.
This way of introducing the air effectively dissipates kinetic energy through
friction inside the burner.

4.3.2.   Distribution of speeds and pressures in rotating flow

Rotating flow can be characterized by the ratio of the rotating flow rate, G_Φ, to the axial flow rate, G_x. These terms are expressed as:

$$G_\Phi = \int_0^{r_0} 2\pi \rho\, UW r^2\, dr$$

where

G_Φ = the rotating flow rate (kg $\cdot$ m^2/s^2)
G_x = the axial flow rate (kg $\cdot$ m/s^2)
ρ = density of air,
r = the distance of the point r from the axis of the burner and r_0 the radius of the burner barrel,
U and W are the axial and tangential velocities, respectively, at point r.

As long as the variations in pressure are slight in the burner, ρ is considered as practically constant, which allows:

$$G_\Phi = 2\pi \rho \int_0^{r_0} UW r^2\, dr$$

The expression for axial flow rate is written:

$$G_x = \int_0^{r_0} 2\pi \rho U^2 r\, dr + 2\pi \int_0^{r_0} \frac{\partial p}{\partial x} r\, dr$$

that is:

$$G_x = 2\pi \rho \int_0^{r_0} U^2 r\, dr + 2\pi \int_0^{r_0} \frac{\partial p}{\partial x} r\, dr$$

The second term in this expression for G_x is a pressure term, and is of little importance if the decline in static pressure along the axis of the burner is minor. The experiment shows that $\partial p/\partial x$ is not only small but practically uniform in a plane perpendicular to the axis.

To convert the ratio G_Φ/G_x to a dimensionless number, G_x is multiplied by a pertinent length, as for example, $r_0 G_x$; and the dimensionless number S is called the swirl number:

$$S = \frac{G_\Phi}{r G_x}$$

often used to characterize the intensity of the movement of rotation.

Using a hemispheric probe with five holes in an experimental burner with tangential air inlets, the researchers at Ymuiden measured the vectoral components of the speed of the air in planes perpendicular to the axis of the burner.

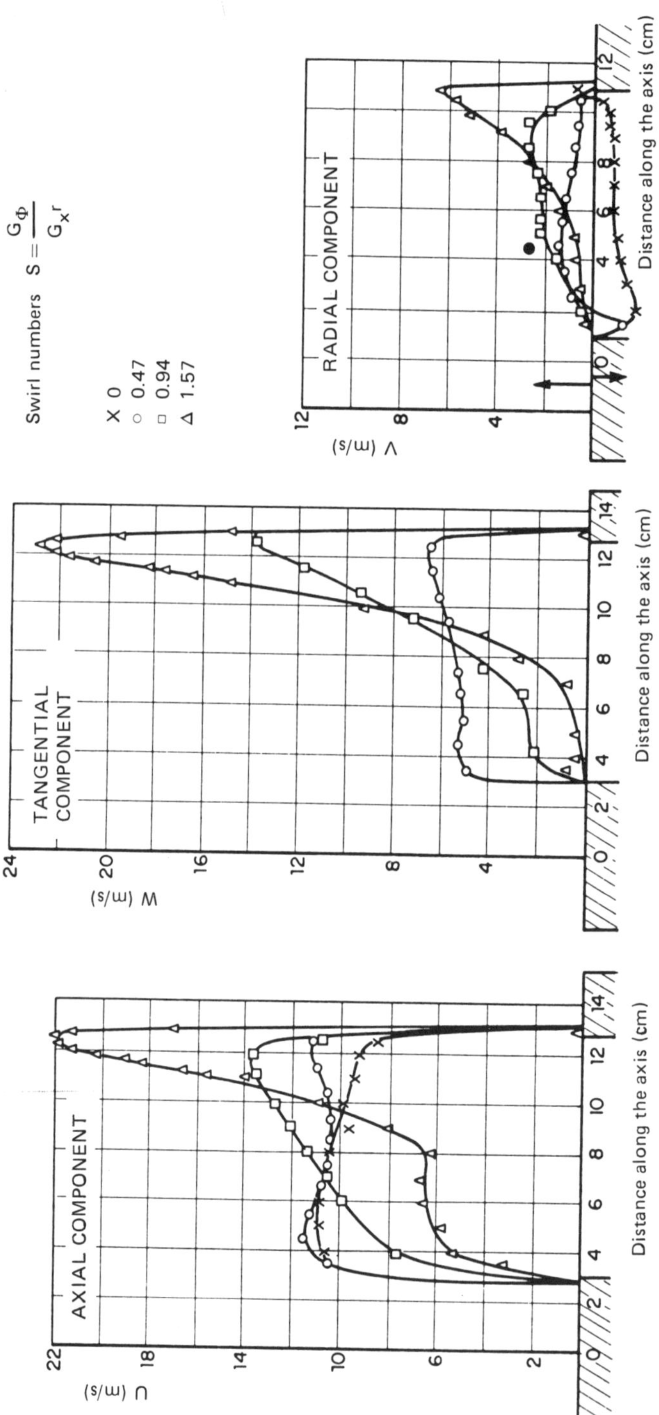

Fig. 4.14. Experimental data from a burner with the air rotated by tangential inlets.

U = axial velocity
W = tangential velocity.
V = radial velocity
G_Φ = rotating flow rate $(\mathrm{kg \cdot m^2/s^2})$.
r = distance to axis of burner (m).
G_x = axial flow rate $(\mathrm{kg \cdot m/s^2})$.

Figure 4.14 shows the values observed in a plane perpendicular to the axis when located 41 cm downstream from the tangential air inlets. In this figure, U is the axial component, W the tangential component, and V the radial component of the velocity. For each one of these components, four different curves drawn for four different values of S illustrate the influence of the amount to which the air is rotated.

As far as the axial rate is concerned, it is about uniform across the burner barrel for a value of S equal to zero. As might be expected, this speed becomes zero in the immediate vicinity of the walls because of friction. However, measurements near the walls always require a great deal of care, because the measuring instrument tends to disturb the flow.

For S-values up to 1.57, on the other hand, Fig. 4.14 clearly shows the appearance of a maximum axial speed next to the outside wall at the outlets of the tangential air injector. The centrifugal force of rotation has concentrated the air near this outside wall where the air flows as a fluid cylinder lining the exterior wall.

The curves of the tangential velocity component, W, also reach a maximum next to the outside wall. The maximum becomes more marked as the intensity of the rotation (indicated by the swirl number, S) becomes greater. As shown in Fig. 4.14, concentrating the air into a peripheral layer by centrifugation tends to increase both the maximum axial and tangential velocity components.

The distribution of radial velocities completes this description. Without any rotation ($S = 0$), the radial flow is directed toward the axis (negative values in the figure) which induces an axial velocity, U, that is slightly larger along the wall of the fuel lance than next to the outside wall. For $S = 1.57$, by contrast, the radial component has a maximum next to the wall of the burner barrel which corresponds to a displacement of the air toward this peripheral zone.

Figure 4.15 shows the range of static pressures along the axis when taken next to the barrel wall and along the wall of the lance. Readings were taken with a burner barrel that ended either with a convergence or a convergence-divergence. Fig. 4.15 shows that inside the barrel the radial variations in pressures depend very little on the shape of the outlet, except for the immediate vicinity. For a high amount of rotation, the effects of internal friction are of little importance compared to centrifugal force; and the authors have verified that in a plane perpendicular to the axis, the theoretical equation:

$$\frac{\partial p}{\partial r} = \rho \, \frac{W^2}{r}$$

relating the pressure to distance from the axis, r, the calculated value of tangential velocity, W, does conform well with experience, (p = density of the fluid).

In summary, the rotation of air in a burner barrel is started by the potential energy of the upstream pressure. This pressure energy is transformed into

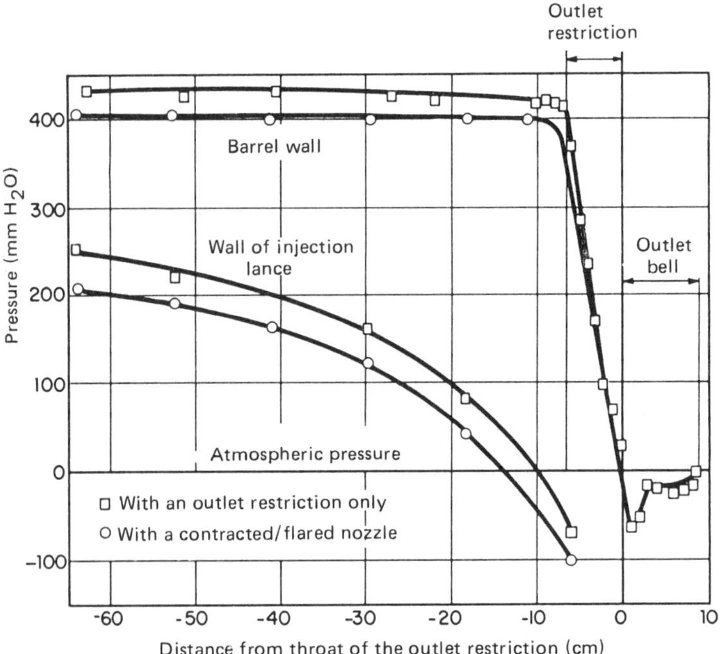

Fig. 4.15. Static pressures along the inside of a burner gun, in which the air is rotated to a high relative speed.

kinetic energy with velocities having important tangential components, either by means of a baffle, or by means of air entering tangentially. The baffles permit carrying out this transformation with slight friction losses, as long as the baffles are well placed. As soon as rotating flow develops, the air accumulates by centrifugal force into a peripheral layer that flows next to the interior wall of the burner barrel. In the experiments described, the presence of a fuel injection lance along the axis prevented the formation of recirculating currents at the center of the burner. The distribution of the fluid flowing in the barrel is not much affected by the shape of the outlet of the burner.

4.3.3.   Flow in a chamber fed by a burner with swirl

At the outlet of the burner, the constraints of the barrel wall ends, and through inertia the air currents flow along a vector whose radial and tangential components are equivalent to the centrifugal force and axial momentum established inside the barrel. Thus, the currents are divergent and this creates a reduction in pressure along the axis downstream of the burner. Here again, a central depressurized zone encircled by a peripheral flow is found just beyond the burner, as in the wake of an obstacle.

Events farther downstream depend not only on the initial orientation of the fluid currents, but also on the shape of the chamber, as that determines the organization of recirculation currents throughout the firebox. Three different types of flow have been observed (Fig. 4.16) from a burner with a contracted and flared outlet. With this configuration, the fuel-injection lance discharges at the throat of the restriction.

Three types of flow are observed, depending on the intensity of the rotation inside the barrel. The shape in Fig. 4.16A corresponds to a mild rotation. The jet comes away from the wall of the outlet bell immediately downstream of the throat and expands in the chamber along an axial pattern with an angle of opening slightly larger than that which would be observed without any rotation. In this configuration, recirculation currents are established in the combustion chamber all along the walls around the full central jet.

The curve in Fig. 4.16C corresponds to an extreme case in which the intensity of the rotation carries the flow to the walls of the flared bell. On entering the chamber, the flow is modified by a small axial velocity component and passes along the wall as a general recirculation is organized at the center of the chamber. This flow pattern is favored by the presence of the flared opening at the outlet of the burner.

The flow pattern in Fig. 4.16B represents an intermediate shape of the ring jet (Fig. 4.5) except that this has a central nucleus of recirculation and a peripheral zone along the walls of the cylindrical firebox. This flow is similar to that obtained by the action of a disc.

Establishment of one or another of these three types of flow depends both on the intensity of the rotation (S'-value) and on the position of the fuel lance. Figure 4.17 identifies the parameters studied at Ymuiden and Fig. 4.18 summarizes the observations relative to the stability of patterns A and B of Fig. 4.16. In the experiments, the ratio of the diameter of the central lance to the diameter of the throat of the nozzle was 0.48. Very similar results were also obtained with a diameter ratio of 0.39. If this ratio is 0.65 or larger, however, the obstruction of the fuel injector becomes very important and flow in shape of Fig. 4.16B is always obtained, even without any rotation, as the A-flow disappears. The S' factor to which Figs. 4.18 and 4.19 are related is calculated as follows:

$$S' = \frac{\text{radial flow from kinetic moment}}{\text{axial flow rate} \times d'}$$

where d' is the diameter of the mouth of the flared bell (Fig. 4.17).

Figure 4.19 shows the S'-values at which flows of Figs. 4.16A and 4.16B appear and disappear as a function of the half-angle of the flared opening (α in Fig. 4.17). For α of 35° and higher, the critical value of S' is nearly constant at about 0.5; for α of 30° and smaller, the critical value of S' tends to increase rapidly. These results correspond to a fuel injector with diameter ϕ such that $\phi/d = 0.48$ (Fig. 4.17).

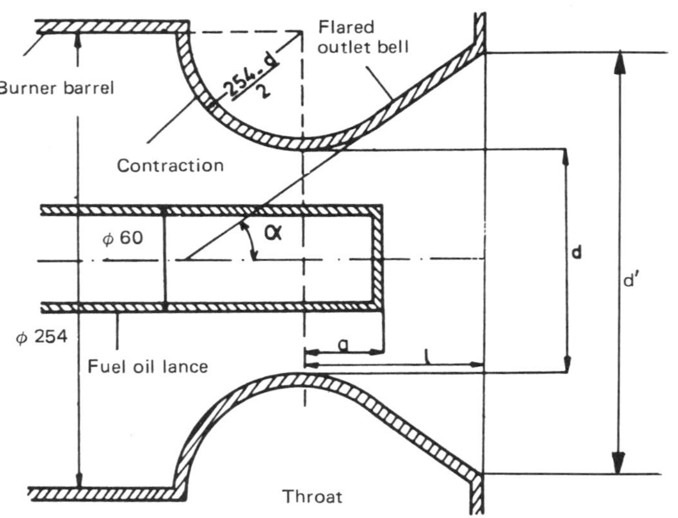

Fig. 4.17.   Diagram and critical dimensions for the outlet of a burner with rotated combustion air.

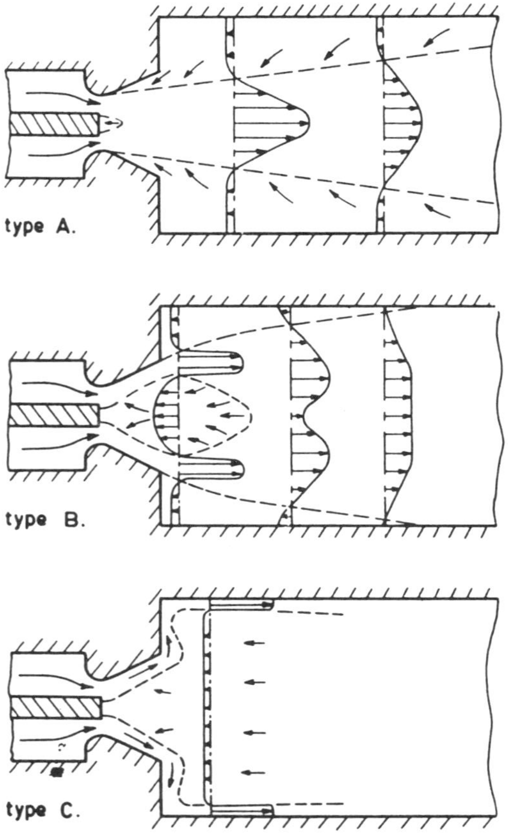

Fig. 4.16.   Characteristic flow patterns at the outlet of a burner in which the air is rotated at varying intensity.

Type A: Mild rotation inside burner.
Type B: Intermediate rotation inside burner.
Type C: Intense rotation inside burner.

(See Fig. 4.17 for a description of the burner outlet.)

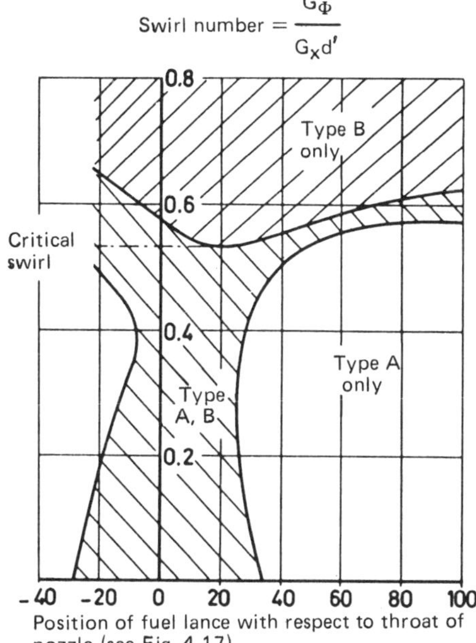

Fig. 4.18. Effects of air rotation and fuel-lance location on flow patterns at the outlet of a burner (see Fig. 4.16 for descriptions of flow patterns).

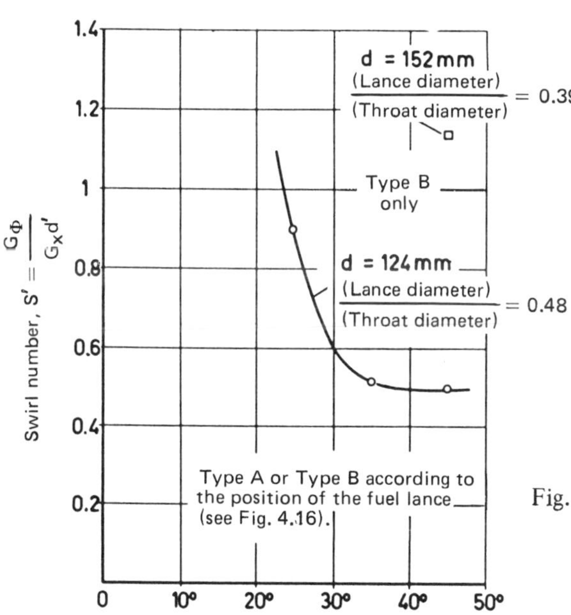

Fig. 4.19. Effects of air rotation and angle of nozzle's outlet bell on the flow-pattern at the outlet of a burner.

For a thinner rod ($\phi/d = 0.39$), the experimental data-point obtained for $\alpha = 45°$ is much higher ($S' \cong 1.1$), demonstrating the importance of the fuel-rod's role as an obstacle (Fig. 4.19). The transition from B-type flow to C-type occurs with wide bells on the order of $45°$, at which the two forms were able to coexist in the preceding experiments. Also, in order to obtain wall-flow, much depends on the junction between the bell of the nozzle and the wall of the chamber. The preceding results correspond to a sharp-edged junction. The rounding of this junction tends to favor type C flow even without any rotation ($S' = 0$).

4.3.4. Influence of swirl on the nucleus of recirculation

The simplest basis for comparisons consists of a burner with a cylindrical nozzle and no modifications at the outlet (Fig. 4.20). Even without induced rotation, the presence of the fuel-injection rod in this nozzle presents an obstacle for which a small zone of recirculation can be observed. As S increases, the effect of this obstacle is pronounced, due to the outward movement of the fluid currents and a nucleus of recirculation is developed downstream of the burner. Flows are of the type shown in Fig. 4.16B.

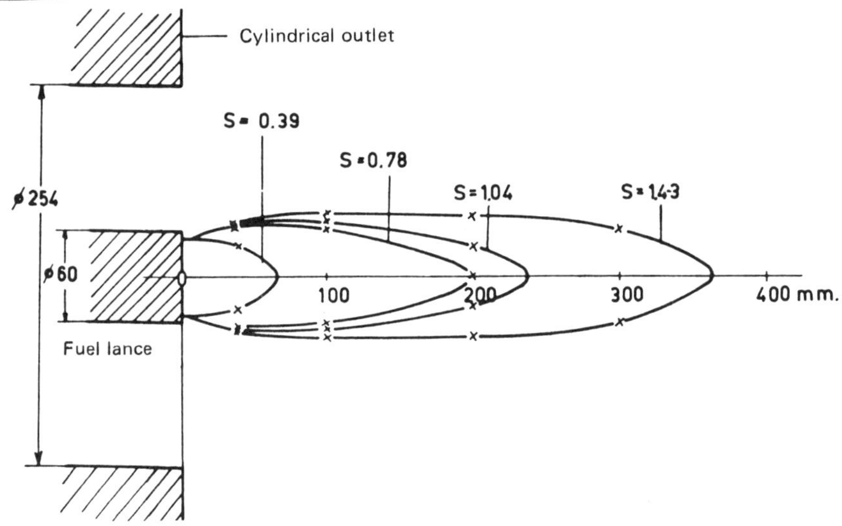

Fig. 4.20. Dimensions of the fields of recirculating gases at the outlet of a straight-sided burner for various intensities (swirl numbers) of air rotation.

By flaring the burner outlet with a contracting-expanding bell (Fig. 4.21) the expansion of the fluid currents in the combustion chamber is increased. Figure 4.21 shows the development of the recirculation zones as a function of the swirl number, S, for a bell with a $35°$ angle. It should be noted that for $S = 0$, a metastable zone of recirculation can be observed when the fuel-injection rod is advanced slightly beyond the throat of the bell as illustrated in Fig. 4.21.

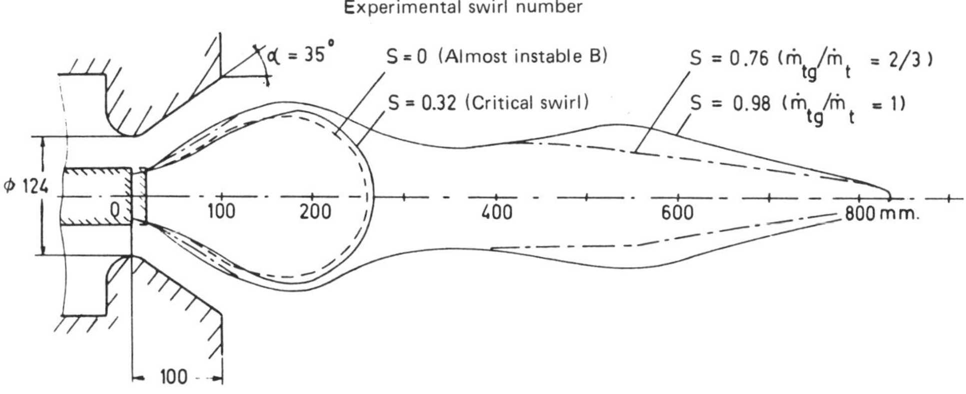

Fig. 4.21. Dimensions of the fields of recirculating gases at the outlet of a nozzle-ended burner for various intensities (swirl numbers) of air rotation inside the burner.

The angle of the nozzle's bell plays an important role in development of downstream flow. Figure 4.22 summarizes the observations made at Ymuiden for an experimental gas burner with a contracting-expanding bell, the angle of the bell's mouth capable of being varied. In this burner the height of the bell cone was equal to the diameter of the throat:

$$\left(\frac{L}{D} = 1\right)$$

The fuel-injection rod ended at the throat and the S number for intensity of the upstream rotation was maintained constant at 0.85.

For these experiments the variable (ordinant in Figs. 4.22, 4.23, and 4.24) was taken as the ratio of the axial velocity of the gas to the velocity of the air at the throat of the burner, G. As the gas flow varied, the velocity of the air at the throat was maintained constant, equal to 50 m/s. For high gas velocities the flow pattern (and the flame that it created) was that of Fig. 4.16A with recirculation around the axial jet. For low rates of gas, the flow resembled Fig. 4.16B, with a central recirculation zone. For each angle of the nozzle's bell, there was a critical value of G called, G_c, at which both type A and type B flows appeared. This transition zone is shaded in Figs. 4.22 and 4.23. It can be seen from these figures that G_c does not vary for opening-angles larger than about 20°, but that it increases rapidly for low angles approaching a cylindrical outlet. The rate limit for the gas under the conditions of the experiment is the speed of sound, being 383 m/s, and does not allow for observing the point that corresponds to $\alpha = 0$.

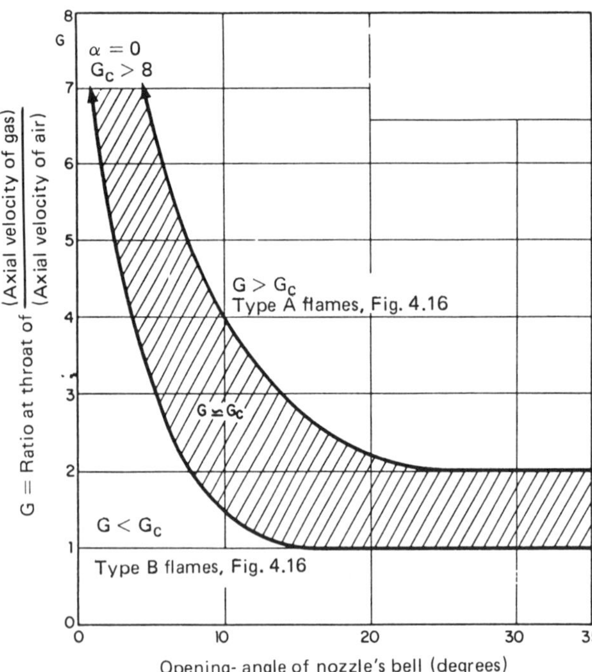

Fig. 4.22. Effect of opening-angle of nozzle's bell on flames from a burner. Type A flames resulted from a straight jet, whereas Type B flames resulted from rotation of the air inside the burner.

Axial velocity of air $= 50$ m/s.

$$\frac{\text{Length of bell}}{\text{Diameter of throat}} = 1.$$

$S = 0.85$.

Fuel lance ends at the throat.

Fig. 4.23. Effect of opening-angle of nozzle's bell on flames from a burner with its fuel lance ending in the outlet plane of the nozzle's bell.

Axial velocity of air $= 50$ m/s.

$$\frac{\text{Length of bell}}{\text{Diameter of throat}} = 1.$$

$S = 0.85$.

Fuel lance ends in the outlet plane of the nozzle's bell.

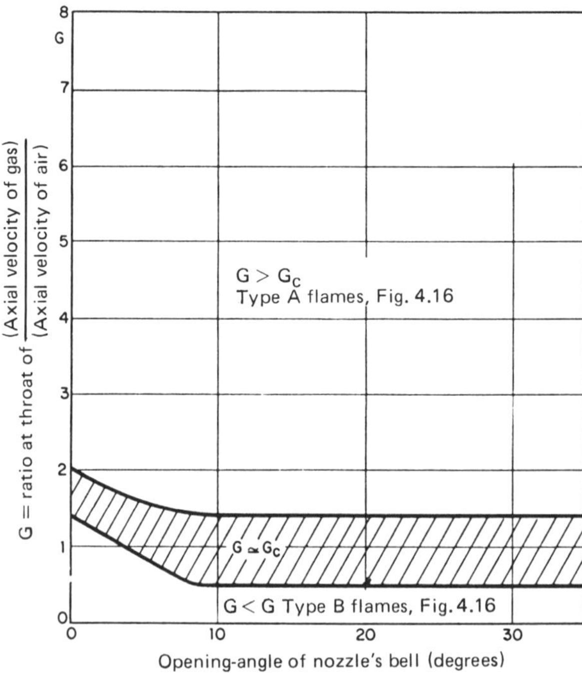

These results are profoundly modified if the gas injection rod ends at the mouth of the bell. The results of tests carried out under the conditions at Ymuiden on this configuration are summarized in Fig. 4.23. Under these conditions, the critical ratio G_c is only doubled when the bell is replaced with a cylindrical exit.

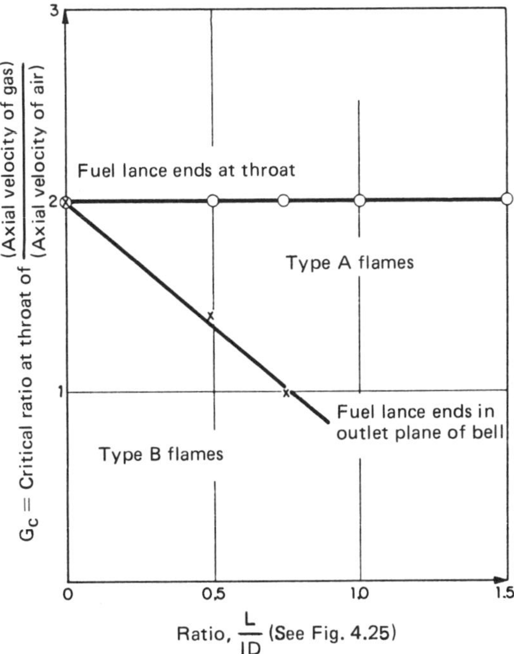

Fig. 4.24. Effect of the length of an outlet bell on the flames from a burner.

Opening angle of bell $= 25°$.
Internal Diameter (ID) of burner barrel $= 131.3$ mm.

Another important geometric parameter for burners with a bell opening is the length of the bell L (Fig. 4.24). Lengthening the bell favors expansion of the flow currents. If the fuel is injected at the throat, it participates with the air in this expanding flow and mixing conditions are independent of variations in L. If the end of the fuel-injection rod is advanced to the mouth of the bell, by contrast, the velocity of the air at the end of the rod depends on L, and the mixing conditions thus vary as a function of L. This is reflected in a variation of critical ratio, G_c, as a function of L (Fig. 4.24). When the fuel-injection rod is at the mouth of the bell, lengthening the bell favors formation of flame type A along the axis, while flames of type B are observed only for low axial velocities of gas. The data shown in Fig. 4.24 correspond to a bell-opening angle of $25°$.

4.3.5. Comparing the stability offered by flow patterns
(Type Fig. 4.16B vs. type Fig. 4.16C)

Systematic tests were run at Toulouse by *GEFGN* researchers to determine the conditions under which type-B and type-C flows occurred with different angles of nozzle opening.

These tests were run on a burner with a sharp-edged conical outlet from a cylinder with a sharp-edged juncture to the cone. In cold tests with this model, there was never any fuel-injection lance placed in the center of the cylinder, since the obstruction afforded by this lance is minor as long as its diameter is

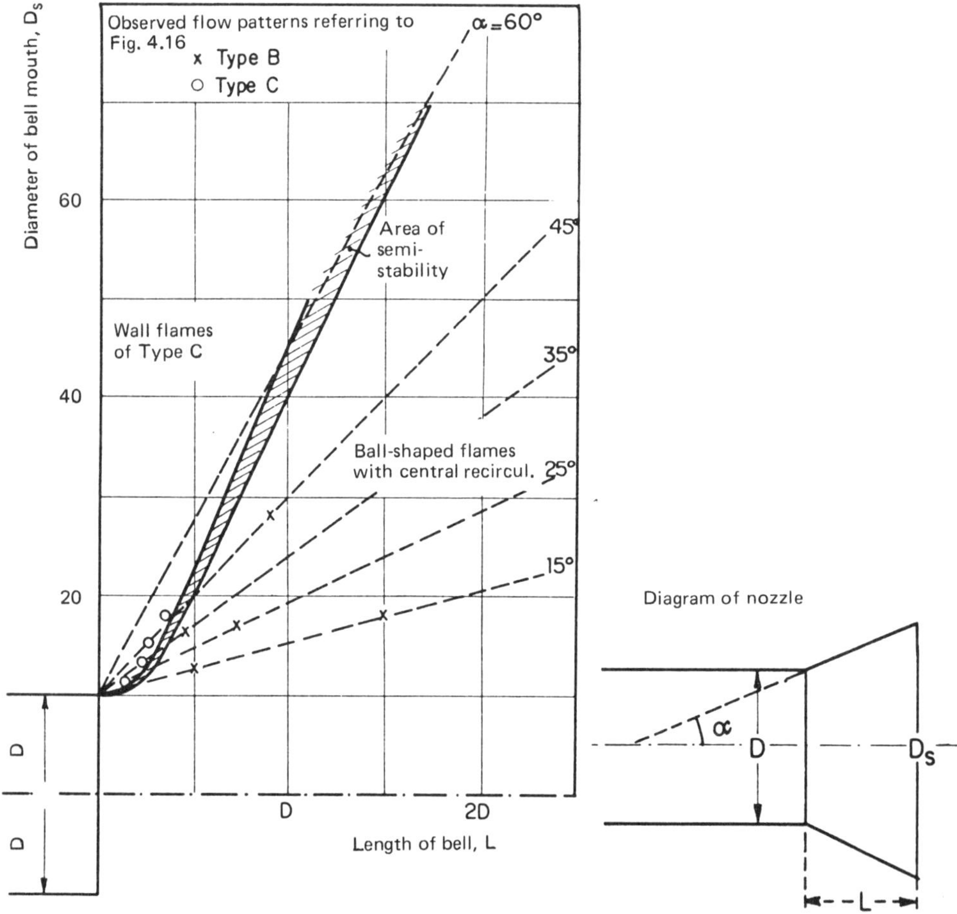

Fig. 4.25. Observed relationships between dimensions of a burner's nozzle and the shapes of the flames produced. (Experimental observations correspond to conditions at the outlet of the nozzle's bell.)

small relative to that of the cylinder. Under these conditions, the experimenters observed that the flow pattern depended on the parameter (P):

$$P = \frac{D_s}{L}$$

where

 D_s = diameter at the mouth of the bell (Fig. 4.25),
 L = length of the bell.

The observed limits are the following:

$$P < 4 \quad \text{type B flow}$$

$$4 < P < 4.5 \text{ equally stable flow B or C}$$

$$P > 4.5 \text{ type C flow}$$

Figure 4.25 summarizes results from cold tests on the model. They were confirmed by tests run in the furnace where the flames corresponded to flow patterns as follows:

Type B: flame in a ball caught at the outlet of the burner.
Type C: flat flame spread out on the chamber wall.

4.3.6. Gas burners: fluids rotated by the gas

Many installations get natural gas from pipelines at high pressures of 3-6 bars. It seems interesting to study the possibilities for using the pressure-energy of the gas to rotate the fluids in the burner, despite the relatively small mass of the gas as compared to that of the air (1 to 16 for methane). *GEFGN* devoted a whole series of experiments to this technology at Toulouse. Two types of flames were systematically studied, flames of form B (in a ball) and of form C (along the wall), according to the definition given above in Section 4.3.3 and reviewed in Fig. 4.16.

4.3.6.1. Flames in a ball

Figure 4.26 shows a drawing of the *GEFGN* type of fuel nozzle used by the experimenters at Toulouse to obtain flames in a ball. Figure 4.27 shows this fuel nozzle in a burner with air, as well as the way the gas is fed.

By means of this equipment a central jet of gas is rotated around the axis of the burner and discharged into a combustion chamber where it is surrounded by a cylindrical flow of combustion air, the axial rate of the air flow being on the order of 4 to 5 m/s.

The intensity of the rotating motion varied in three ways:

(a) For a constant diameter of the tangential injectors, as a function of the rate of flow of the gas.

(b) For a given rate of flow, as a function of the diameter of the tangential injectors.

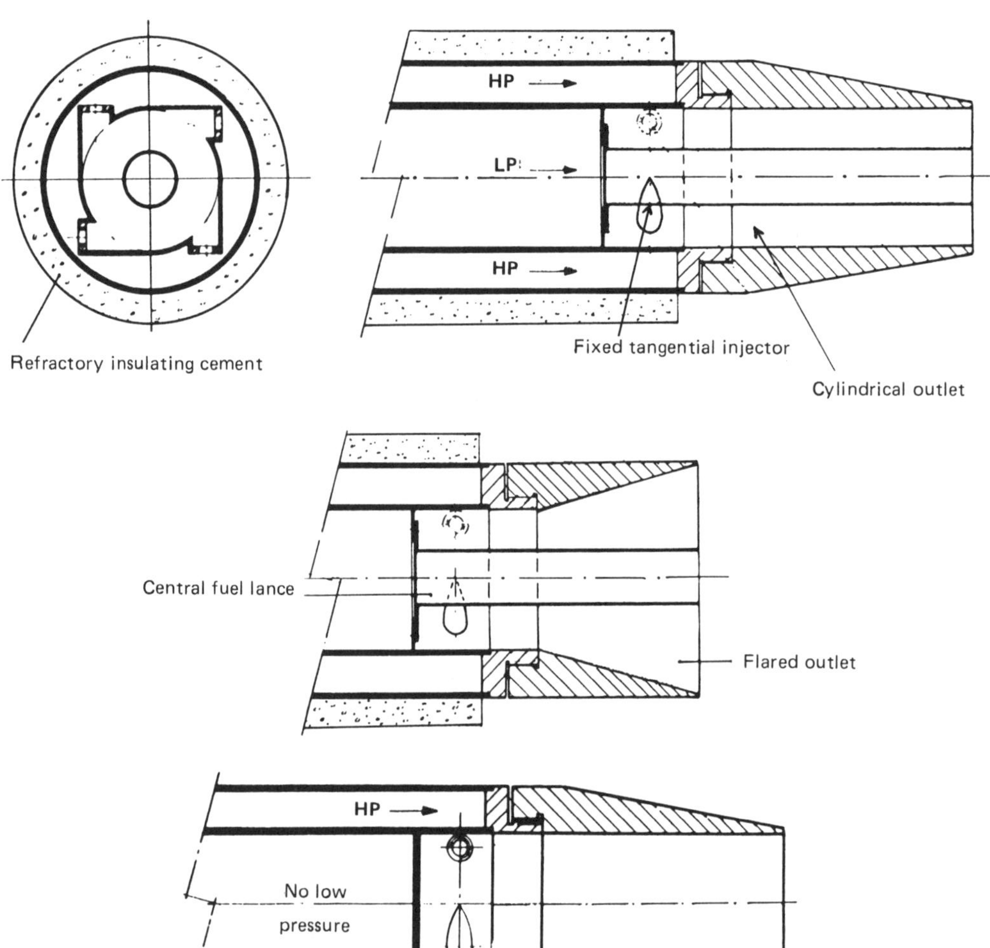

Fig. 4.26. Diagrams of burner-nozzle designs to impel rotation with the kinetic energy of high pressure natural gas. (See Fig. 4.27 for burners.)

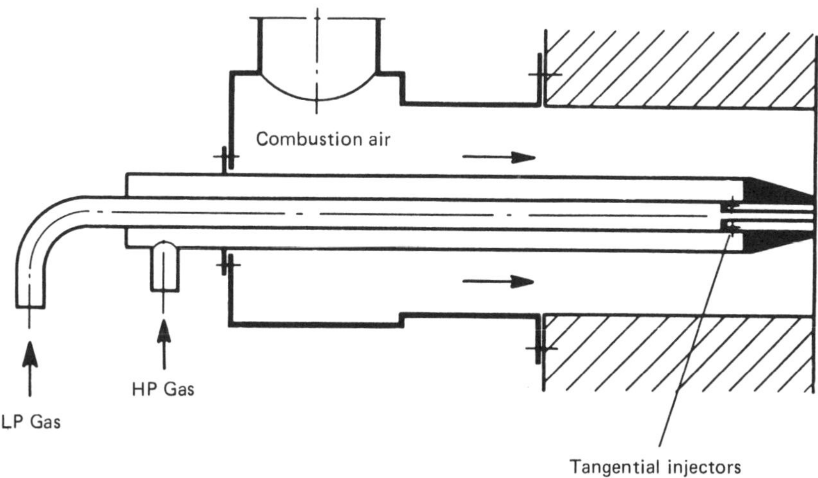

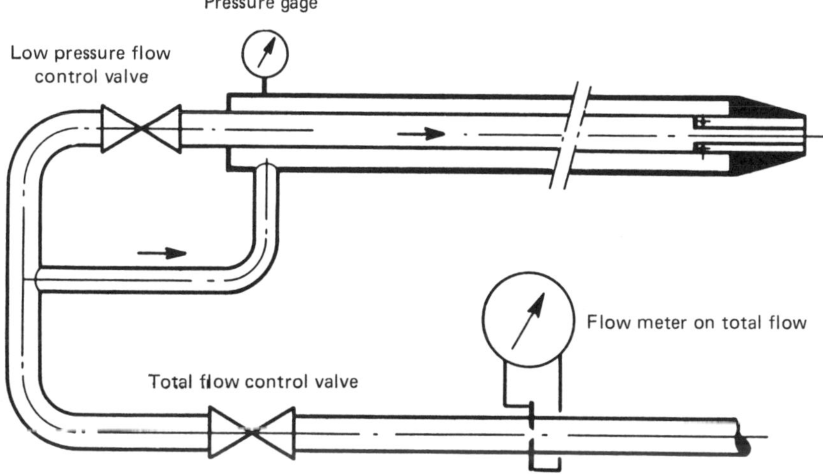

Fig. 4.27. Arrangement and flow diagrams for experimental burner imparting rotation with the impulse from high pressure natural gas.

(c) For a given diameter of injector and a given rate of gas flow, as a function of the radius of gyration at different inside diameters of the nozzle.

The following results were observed on a cold model fed with air (Ref. 4.3):

(a) For given injectors, the angle of the jet downstream from the burner remains practically constant as long as the flow of gas to the injector is subsonic.

(b) With subsonic flow of gas, the angle of opening of the jet increases with the diameter of the injectors, for a constant rate of flow.

(c) This same angle of opening of the jet increases, on the other hand, with the rate of flow of the gas when it leaves the injectors at the speed of sound.

(d) Finally, even within sonic conditions, the angle of opening of the jet was always less than 110°.

Figure 4.28 summarizes the observations of *GEFGN* at Toulouse.

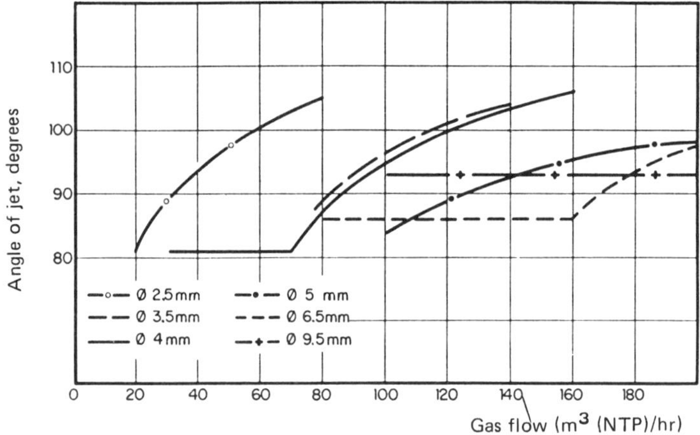

Fig. 4.28.   Test data taken with the experimental burner shown in Figs. 4.26 and 4.27.

Taking these observations to Q_v/G, the ratio of the volumetric rate of flow of gas to the impulse due to the gas (i.e., to a quantity inversely proportional to the specific impulse) these same results are shown again in Figs. 4.29a and 4.29b. In subsonic conditions (Fig. 4.29a) the angle of opening α of the flow tends to close slightly as the specific impulse increases. In supersonic conditions with tangential injectors, when the specific impulse increases (Fig. 4.29b) the angle of opening of the jet, α, increases and varies, from about 80° (i.e., $\alpha/\alpha_0 = 1.0$) in subsonic conditions, to 110° in supersonic conditions, which corresponds to the maximum of specific impulse of the gas from Lacq at ambient temperature (i.e. $Q_v/G = 0.00175$).

In the nucleus of recirculation established at the center of the rotating jet, the velocity of gases flowing along the axis toward the burner is considerable. The experiments at Toulouse have shown that this central recirculating flow can

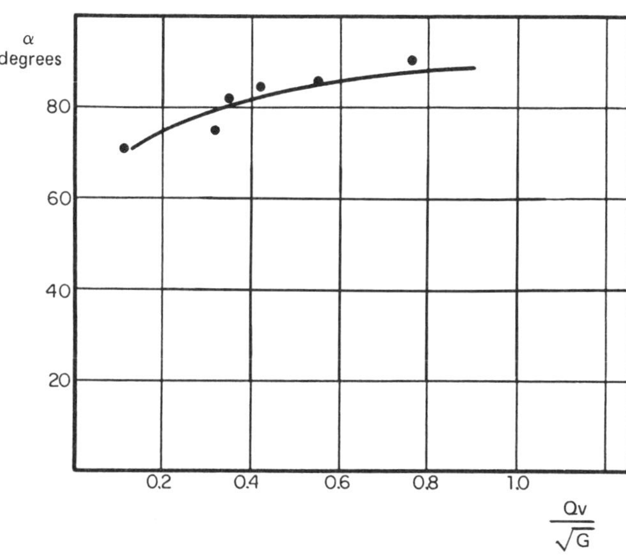

Fig. 4.29a. Jet angles for subsonic speeds at the tangential
gas inlets.

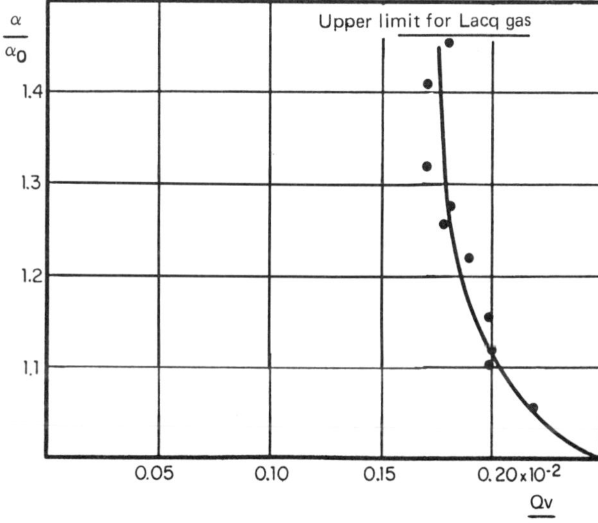

Fig. 4.29b. Ratios of (jet angle
for supersonic tangential inlet
speed) / (jet angle for subsonic
tangential inlet speed) for the
same values of gas flow and
specific speed.

Fig. 4.29. Jet angles for the burner of Fig. 4.27 as a function of
the total gas flow, Q_v (m^3/s) and the specific impulse
imparted by the gas, G (N). Combustion air was at 15° C.

equal or even more than double the rate of flow of fuel gas leaving the burner. This nucleus of recirculation is established equally well in an unlimited atmosphere as in a confined chamber. It carries both unburned gas and combustion products, a mixture that is at rather high temperature, thus assuring the constant ignition of the air-gas mixtures downstream of the burner. Moreover, this operating mechanism is not very sensitive to any external recirculation of the jet, thus it is not sensitive to the temperature of the walls as that affects the temperature of the external recirculation. The result is that this method for stabilizing flames in a ball is effective not only in fireboxes operating at high temperature but also in fireboxes with tube walls that are relatively cold.

4.3.6.2. Wall flames

The pressure-energy of natural gas can also be used to form flames Fig. 4.16C in the immediate vicinity of the burner wall. Here again we cite the studies made at Toulouse. Fig. 4.16C shows flow according to the Coanda effect, by which a fluid flowing along a wall pulls away from the wall only when there is a marked break in the slope of the wall abetted by an impulse sufficient to create a very open angle from the wall.

Figure 4.30 illustrates the burners used to make such flames at Toulouse. In these burners the burner barrel is connected to the chamber wall by a refractory-lined opening similar to the so-called burner-blocks of industrial furnaces. This opening has a rounded outlet to streamline the transition of fluid currents to the chamber wall. In Fig. 4.30a, air to the burner is rotated by tangential inlets similar to those described in Fig. 4.13. The fuel-gas is also introduced tangentially, but at a low velocity, into the rotating air. Fig. 4.30b employs axial delivery of combustion air with tangential injections of gas at high speed through the walls of the burner-block. In this burner a cylinder of refractory cement along the axis of the burner-block restricts the thickness of the air-current layer downstream of the gas injectors, and thus promotes rapid mixing between the layer of air and the gas from the tangential injectors. Finally, in Fig. 4.30c, the gas is discharged at high speed in a plane perpendicular to the flow of air through a series of injectors arranged around the extremity of a central rod. The movements of the air and the gas are therefore at right angles, and under proper conditions, the impulse of the gas fixes the whole airstream to the rounded outlets of the burner-block. In all three cases, a mixture of air and gas flows along the wall to create a wall flame.

4.3.7. Rotating combustion air with fins (Ref. 4.5)

Figure 4.31 shows a combustion-air pipe that carries along its axis a fuel-injection tube equipped with fins at the end. These fins act like axial-flow impellers to impart a rotating movement to a part of the combustion air without

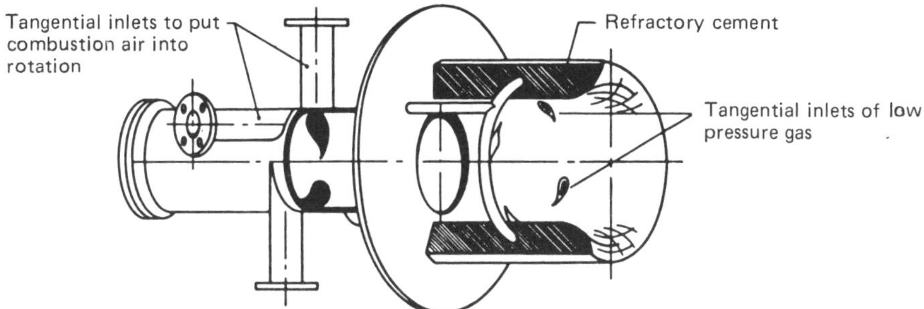

Fig. 4.30a.   Rotation by combustion air.

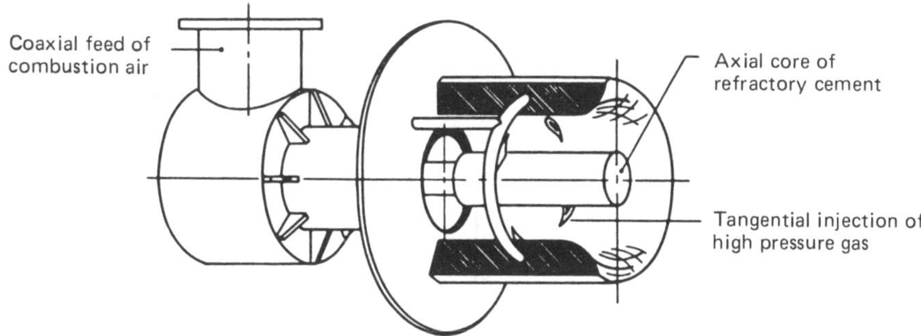

Fig. 4.30b.   Rotation by fuel gas.

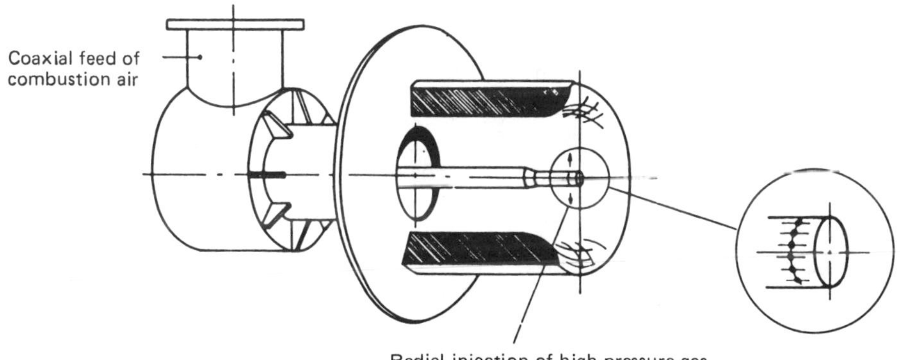

Fig. 4.30c.   Radial flow by fuel gas.

Fig. 4.30.   Experimental burner designs for creating wall flow as described in Fig. 4.16C.

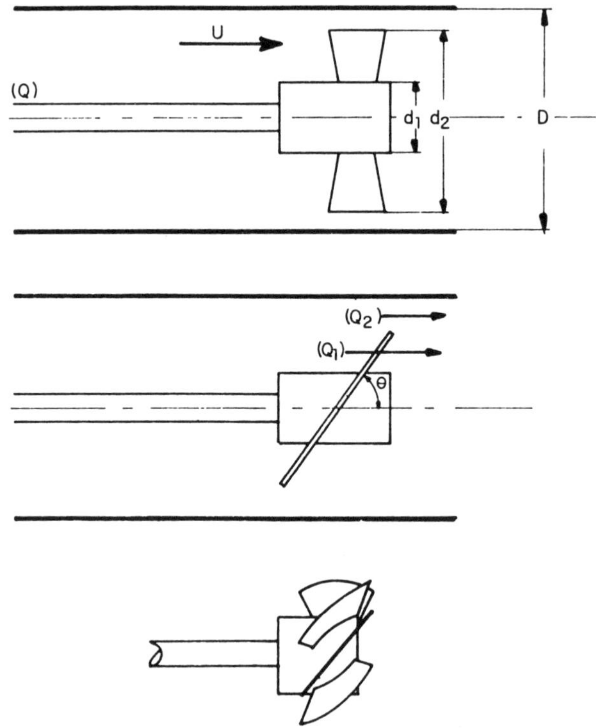

Fig. 4.31.   Designs for rotating combustion air in a burner barrel by
means of axial-flow impellers.

incurring a great loss of kinetic energy. If Q is the volumetric flow of air passing through the deflector and U is the velocity of approach of that air parallel to the axis of the pipe, as it passes through the fins, the air is accelerated at the rate:

$$V = \frac{U}{\cos \theta}$$

The components of V are:

U, the velocity along the axis.

W, the tangential velocity equal to $U \tan \theta$.

Under these conditions one can create the term G_x, reflecting the axial movement, as:

$$G_x = QU$$

and the term G_Φ, reflecting the rotational moment as:

$$G_\Phi = Qr_g U \tan \theta = G_x r_g \tan \theta$$

where r_g is the radius of gyration associated with the deflector.

If d_1 and d_2 are the internal and external diameters, respectively, of the fin deflector and r_1 and r_2 are the corresponding radii; and if external diameter, d_2, is taken as the characteristic dimension associated with G_x, the expression for swirl number, S, becomes:

$$S = \frac{G_\Phi}{G_x d_2} = \frac{r_g}{d_2} \tan \theta$$

It remains to determine r_g. By definition, we have:

$$G_\Phi = Q r_g W = \pi \rho U W r_g (r_2^2 - r_1^2)$$

For a differential ring with thickness dr, at distance r from the axis, we have:

$$dG_\Phi = \rho 2\pi r \, dr \, U r W$$

with ρ taken as the effective air density for the whole deflector section:

$$G_\Phi = 2\pi \rho U W \int_{r_1}^{r_2} r^2 \, dr = \frac{2}{3} \pi \rho U W (r_2^3 - r_1^3)$$

The result is:

$$r_g = \frac{2}{3} \frac{(r_2^3 - r_1^3)}{(r_2^2 - r_1^2)}$$

Finally, S is a function only of geometric dimensions:

$$S = \frac{1}{3} \frac{1}{r_2} \frac{r_2^3 - r_1^3}{r_2^2 - r_1^2} \tan \theta$$

and by saying:

$$\frac{r_2}{r_1} = a$$

S becomes:

$$S = \frac{1}{3} \frac{1 - a^3}{1 - a^2} \tan \theta$$

Figure 4.32 lists values of S for factors a and θ.

a \\ θ	30	45	60
0	0.19	0.34	0.58
0.2	0.20	0.35	0.60
0.5	0.23	0.39	0.68

Fig. 4.32. Swirl numbers, S, as a function of impeller dimension, $a = \dfrac{r_2}{r_1}$, and the angle, θ, of the impeller with respect to the axis.

These numbers show the dominant importance of the angle of inclination of the fins. However, these results are only as viable as the initial hypotheses, which is to say viable as long as the design of the fins permits modifying the air currents without incurring a measurable pressure drop. The results of current practice ordinarily justify these hypotheses, at least for values of θ that do not go over 60°.

4.3.8. Hypothesis for the mechanism of stabilization by an obstacle or by rotation

Imagine a turbulent stream coming out of a burner and revolving symmetrically around the axis of this burner (direction O'-x, in Fig. 4.33). We have seen above that either an obstacle centered on the axis or a rotation of fluids around the axis produces a pressure reduction Δp in a plane, O, across the outlet of the burner. Imagine that, in this same plane at the distance r_0 from the axis, a fluid stream leaves the burner. At point A (with $OA = r_0$) the current possesses an overall rate of flow that can be broken down into the three following terms:

$\vec{U}$ = axial velocity along the axis Ox.

$\vec{W}$ = tangential velocity, perpendicular to the axis and in a plane separated from the axis by distance r_0.

$\vec{V}$ = radial velocity perpendicular to the axis and in the plane of the axis.

The three terms being considered are the average components measured directly in a turbulent current by instruments such as Pitot tubes that measure impact pressure. Downstream of plane O, the constraints of the burner wall are

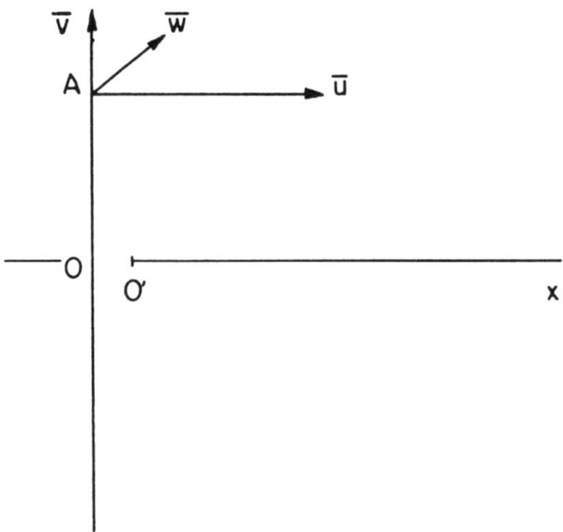

Fig. 4.33.   Diagram of vectoral speeds for analyzing rotating flows.

absent, and except for fluid friction, the motion of the current is governed only by inertia, by centrifugal force according to the rotation, and by any pressure gradient between the axis and ambience.

Let us consider the ring of fluid bounded by the radii r_0 and $r_0 + \Delta r$ and the length Δx. The volume of this ring, $\mathcal{V}$, is:

$$\Delta\mathcal{V} = 2\pi r\,\Delta r\,\Delta x$$

and the weight, ΔM, is:

$$\Delta M = 2\pi r \rho\,\Delta r\,\Delta x$$

where ρ is the density of the fluid.

If the current has a tangential velocity of rotation, W, the result is that for each elemental ring of fluid there is a centripetal acceleration, Γ_{cent}, equal to:

$$\Gamma_{cent} = \frac{W^2}{r_0} \tag{4.1}$$

At the same time, the existence of a pressure reduction, Δp, along the axis produces a pressure gradient equal to $\partial p/\partial r$ at distance r_0 from the axis. The resulting force, F, of this gradient is exerted on the fluid ring and is equal to:

$$F = 2\pi r_0\,\Delta x\,\frac{\partial p}{\partial r}\,\Delta r$$

The acceleration corresponding to this force is directed toward the axis and is equal to:

$$\Gamma_p = -\frac{F}{\Delta M} = -\frac{1}{\rho}\frac{\partial p}{\partial r} \tag{4.2}$$

The minus sign in Eq. 4.2 indicates that the induced gas flow is from zones of high pressure toward zones of low pressure.

In flowing from planes at O to O', x distance apart, the components of the fluid ring will see their distance to the axis changed according to the laws of uniformly accelerated movement, this acceleration being the sum of the two terms (4.1) and (4.2):

$$\Gamma = \frac{W^2}{r_0} - \frac{1}{\rho}\frac{\partial p}{\partial r} \tag{4.3}$$

Γ will be positive (centrifugal expansion of the jet) if the first term is the larger; it will be negative (directed toward the axis) if:

$$\frac{1}{\rho}\frac{\partial p}{\partial r} > \frac{W^2}{r}$$

In plane O' the radius of the ring is thus changed to become:

$$r = r_0 + Vt + \frac{1}{2}\Gamma t^2 \qquad (4.4)$$

where t is the time it takes to travel between the planes O and O'. This time is expressed directly as a function of x and of U:

$$t = \frac{x}{U}$$

from which:

$$r = r_0 + V\frac{x}{U} + \frac{1}{2}\Gamma\frac{x^2}{U^2}$$

This variation of the radius of the fluid ring brings a change in the enclosed volume equal to the difference between the volume at O:

$$d\mathcal{V}_1 = \pi r_0^2 \, \Delta x$$

and the volume at O':

$$d\mathcal{V}_1' = \pi r^2 \, \Delta x$$

If, as a first approximation, we consider that any transfer of matter between the fluid ring and the enclosed volume is small compared to the mass of the enclosed volume $(d\mathcal{V}_1 - d\mathcal{V}_1')$. Assume in addition that this expansion is adiabatic, since heat exchange between the nucleus and the surrounding fluid currents is limited, we can write:

$$\frac{\Delta p}{p} + \gamma\frac{\Delta V}{V} = 0$$

where γ is the polytropic coefficient of expansion for the gases in the nucleus.

In summary, when the acceleration is negative, the fluid currents converge toward the axis progressively, and all along the axis the pressure increases from upstream to downstream, thus explaining the recirculation of the gases in the center as shown in Fig. 4.16.

4.3.9. Characteristics of flames stabilized by rotation

Consider a burner design (Fig. 4.34) that rotates the air inside a cylindrical barrel that discharges through a flared outlet of length L, diameter D at the barrel, and an angle of flare, α. According to the intensity of the rotating motion identified by swirl factor, S, flames in the vicinity of the burner are observed to either have low axial recirculation (type I, Fig. 4.35) or a pronounced axial

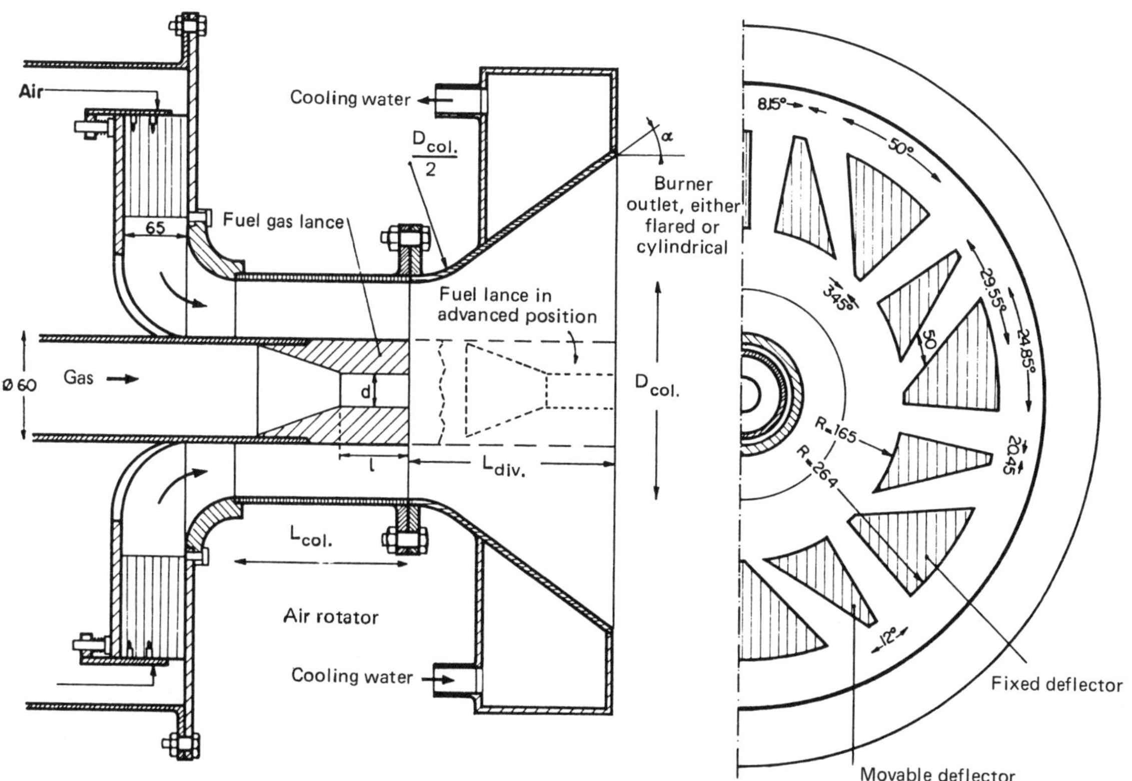

Fig. 4.34. Design of the experimental burner used for the photographs in Fig. 4.35.

$D_{col.}$ = Collar diameter.
$L_{col.}$ = Collar length.
$L_{div.}$ = Divergent length.

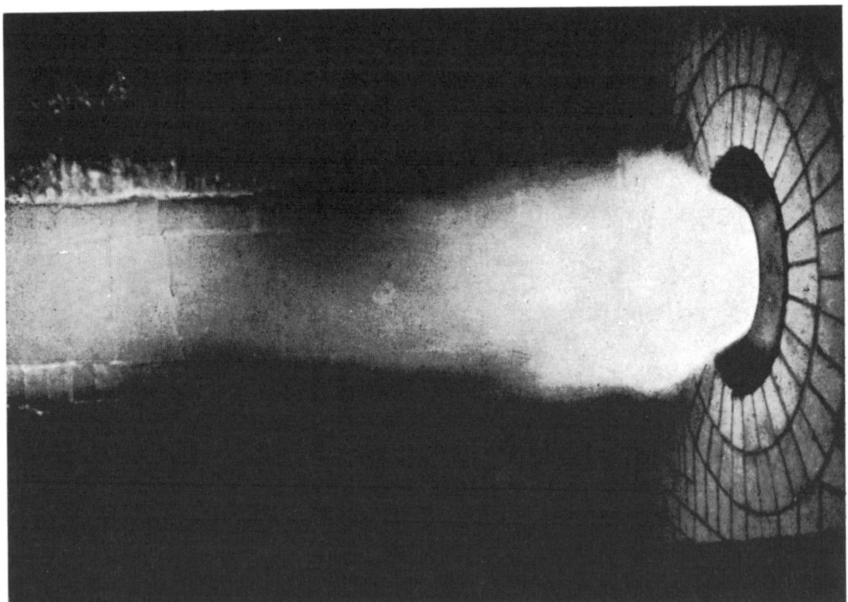

Type II flame: The burner's outlet bell is flared at 35°
and the combustion air is rotated at high speed.

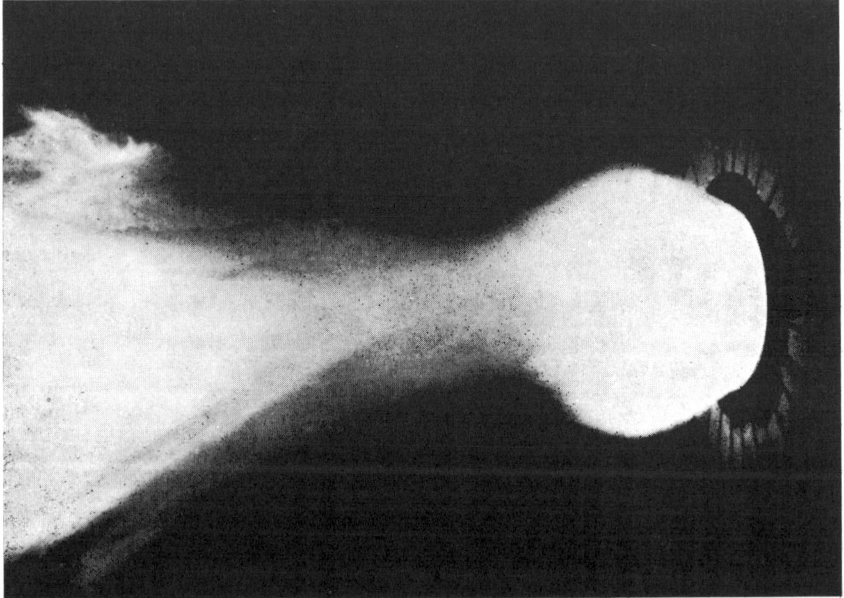

Type I flame : The burner's outlet bell is flared at 35° but
the combustion air is given only a mild rotation impulse.

Fig. 4.35.   The effects of rotating the combustion air on the flames
             from a given burner.
             (Photograph by *IFRF*).

recirculation (type II, Fig. 4.35). In both cases, however, the products of combustion formed in the firebox participate in a general recirculation motion described in Chapter 3.

Flames of type I, associated with a high velocity axial ejection of the fuel, are long and rather similar to flames obtained by a central fuel jet carrying along the peripheral air of combustion. The small fraction of fuel brought back to the burner is mixed with the combustion air on the periphery of the jet, where it burns, forming a stable combustion zone at the nose of the burner. The rest of the fuel burns in an elongated flame downstream of the nucleus of recirculation. This combustion is not noisy.

Increasing the angle α of the flared outlet bell increases the effect of the stabilizing nucleus. In a general way, the shape of type I flame is determined by mixing of the fuel and the air, and closely dependent on the fluid dynamics at the nose of the burner. The length of the flame does not depend on the rate of discharge of the burner, at least not for a gaseous fuel. (The experiment at Ymuiden was conducted over flow ratios of 12 to 1).

Flames of type II, associated with moderate axial velocities for the fuel, develop in the vicinity of the burner.

Almost all of the fuel is forced into the stabilizing nucleus, where it mixes intimately with the combustion air. This mixture is constantly ignited by contact with the hot combustion products present in the outlet bell of the burner. The sudden and incessant ignitions give birth to a noisy and intense combustion. These flames are not very luminous but do shine, however, because of a combustion temperature that is higher than that of flames of type I. The differences in temperature are due to the higher mixing rates of fuel and air for flames of type II. Depending on the fuel, the radiation of type II flames will be concentrated more in the vicinity of the burner than radiation from type I flames.

For type II flames from a burner with a bell whose angle of opening is $35°$, the flexibility of regulation is greatly improved if the gas injection rod discharges through an orifice ring, diverging orifices or a flared outlet (Fig. 4.36). The orifices and flared outlet have an angle of inclination from the axis of $35°$. The injector with a conical outlet exhibited excellent stability through an operating range of 8/1. At normal operating capacity, the flame remained stable even in the absence of rotation in the combustion air. However, rotating the air affords sensitive control of the length of flames of this type which collect into a ball in the combustion chamber right in front of the burner (Fig. 4.37).

Finally, we should keep in mind that type II flames are obtained when almost all the fuel is carried into the recirculation zone. If the fuel is liquid, fine atomization favors its transport into this zone; also the opening angle of a liquid jet can be played with to induce a similar flow of air and mist. If the fuel is gaseous, the recirculation zone is fed gas at a low velocity, so that the gas stream can be bent and carried along by the air currents. A gas injector with a circular opening

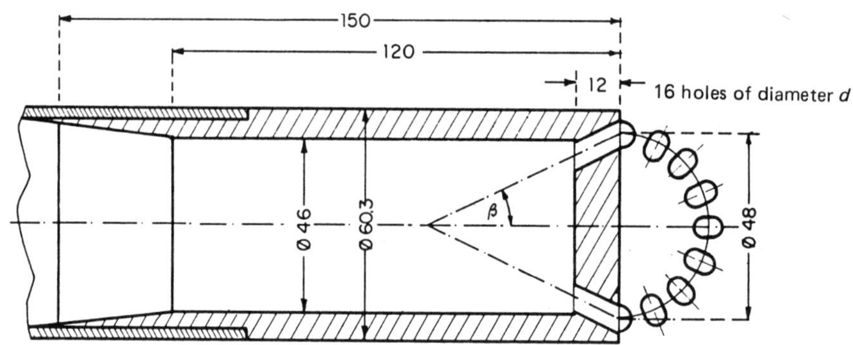

Fig. 4.36a.   Lance with a ring of diverging orifices.

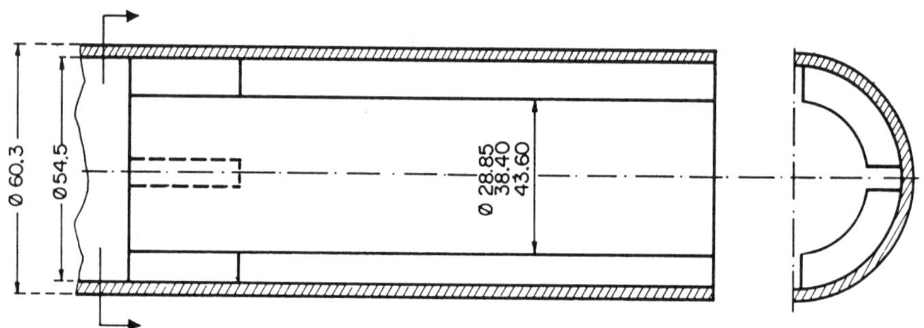

Fig. 4.36b.   Lance with axial flow.

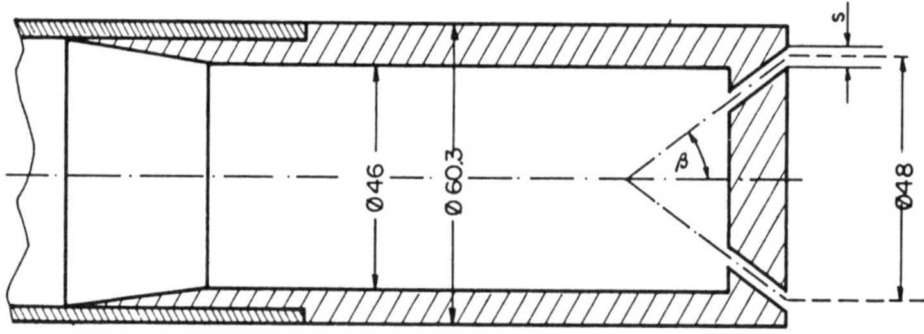

Fig. 4.36c.   Lance with a diverging conical outlet.

Fig. 4.36.   Three different types of fuel-gas lance used to feed type II
flames of Fig. 4.35.

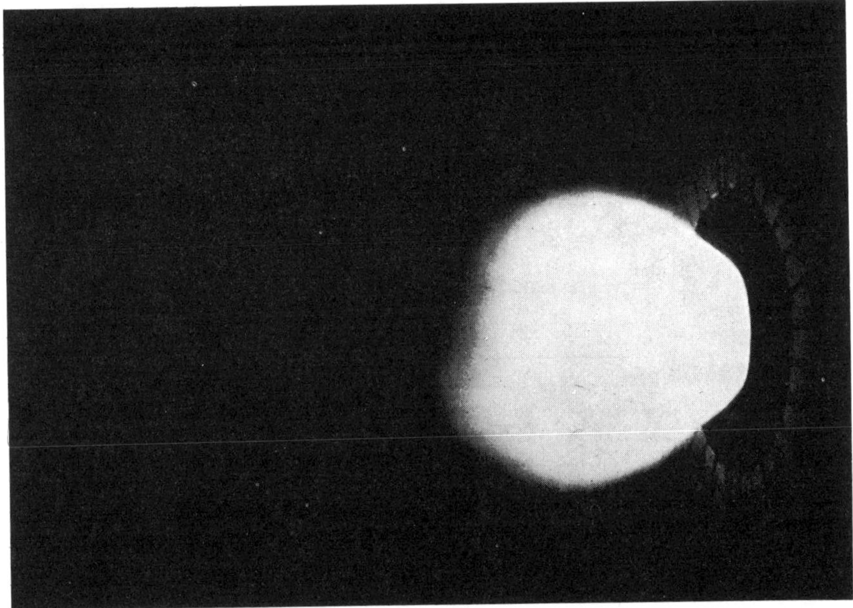

Fig. 4.37.   The ball of flame produced by rotating combustion air
to give a high swirl number of 1.4.
(Photograph by *IFRF*).

is particularly suitable for achieving this condition. The injection lance shown
in Fig. 4.30a achieves this condition when all the gas is furnished by a low
pressure circuit.

4.4. STABILIZING WITH A BURNER QUARL

The quarl of a burner can serve as a preliminary firebox to the firebox proper
(Fig. 4.38). The burner discharges into this quarl which is thus fed with all air
and fuel for combustion. The flame is established within the opening's volume
and thus serves to initiate the combustion that proceeds out into the firebox.

A cylindrical shape is the simplest for a burner quarl. Experiments, those
made at Toulouse and Ymuiden particularly, permitted analyzing the stabilizing
role of burner quarls of this sort. Placed in front of a high temperature furnace,
these experiments have shown that:

(a)   When its length is at least equal to its internal diameter, the quarl plays a
stabilizing role downstream of the fuel injector nozzle.

(b) The recirculation zone established in the quarl gets larger as the diameter of the opening gets larger.

(c) In practice, stabilization due to burner quarls, or burner blocks, does not depend on the temperature of the walls, which can be made either of refractory or sheet metal cooled by circulating water.

(d) Finally, any stabilization achieved with a cylindrical quarl gets more effective as the jet impulse gets greater.

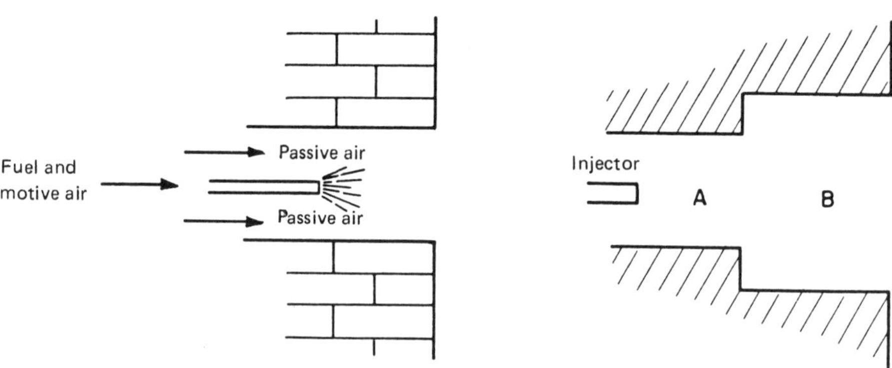

Fig. 4.38. Recessed burner with cylindrical quarl.

Fig. 4.39. Stepped-out burner quarl for better stabilization.

A cylindrical burner quarl can play the role of stabilizer only when the firebox that it feeds operates at a high temperature, on the order of 800° C or more, so that a recirculating zone within the quarl is enclosed with flame. This is easily understood when one considers that the quarl is a small combustion chamber. Its flows correspond to those analyzed by the Craya-Curtet theory (Chapter 3), which permits predicting how the expansion of a jet from a burner inside the quarl involves formation of reverse-flow currents directed toward the burner. A suitable position for the burner requires that Craya-Curtet recirculation begin in the outlet plane between the quarl and firebox and from there draws back into the quarl some of the flue gas present in the firebox.

This recirculated flue gas initiates the combustion in the quarl by its high temperature. Consequently, this mechanism does not work with a cold firebox, particularly during lighting. To overcome this difficulty, a two-step quarl is frequently incorporated in the burner-blocks used in practice (Fig. 4.39). The burner is situated in part A, which receives the reaspirated flue gas from part B. Taking into account the small volume of part B, it is filled with very hot burned gases as soon as the burner is lit and a fraction of these gases, going back up to A, maintains a flame nucleus in the vicinity of the burner. This two-stage quarl is thus able to play a stabilizing role when feeding a firebox with cold walls, such as occur in a boiler or tube-walled heater.

Also, the burner quarl can empty into the firebox through a truncated cone which permits amplifying the effects of recirculation.

4.4.1.   Calculating the dimensions for a burner quarl

Recirculation in a cylindrical burner-quarl is defined by the studies of Craya and Curtet. In Fig. 4.40 (see also Fig. 3.8) these authors have provided a correlation for the location and length of a zone of recirculation as a function of the dimensionless Craya-Curtet number, m (Section 3.2.3), as defined by:

$$m = \frac{G_0}{G_\infty} + \frac{G_a}{G_\infty} - \frac{1}{2}$$

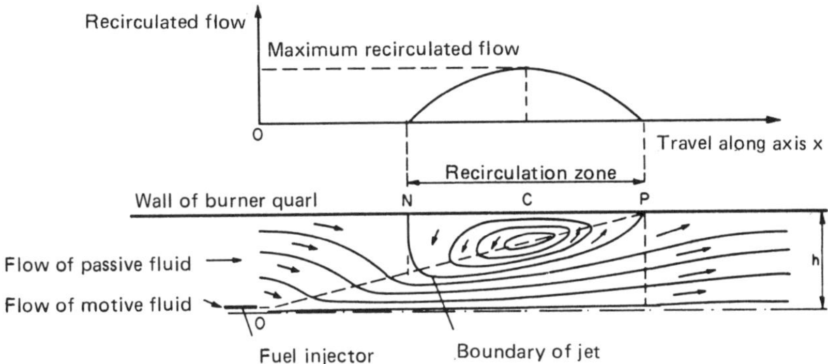

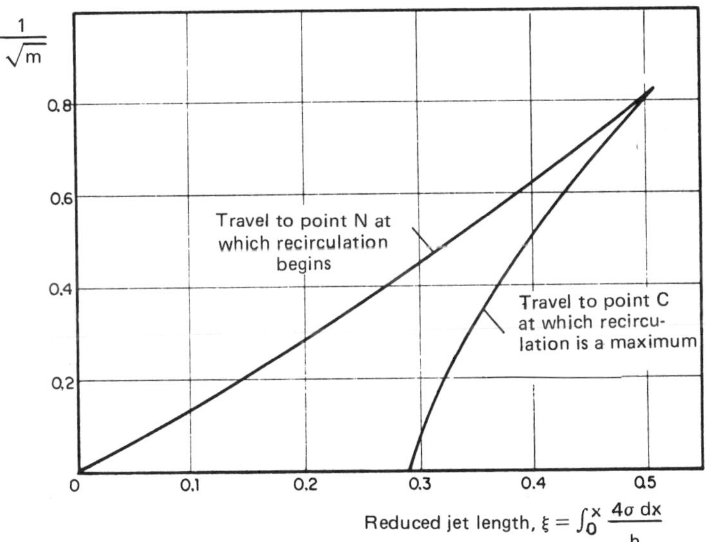

Fig. 4.40.   Chart for locating the recirculation around a jet discharging into a recessed burner quarl, by means of the Craya-Curtet number.

and of a reduced variable ξ such that:

$$\xi = \int_0^x \frac{4\sigma \, dx}{h}$$

with

x = downstream distance from the fuel injector,

h = radius of the opening,

σ = average value for the coefficient of tangential friction. In the case of a free jet in calm, infinite atmosphere, σ has a value of 0.0285.

The experimenters at Toulouse have verified that this same 0.0285 value of σ was applicable in their experiments with a central fuel jet penetrating a flux of combustion air (Ref. 4.2).

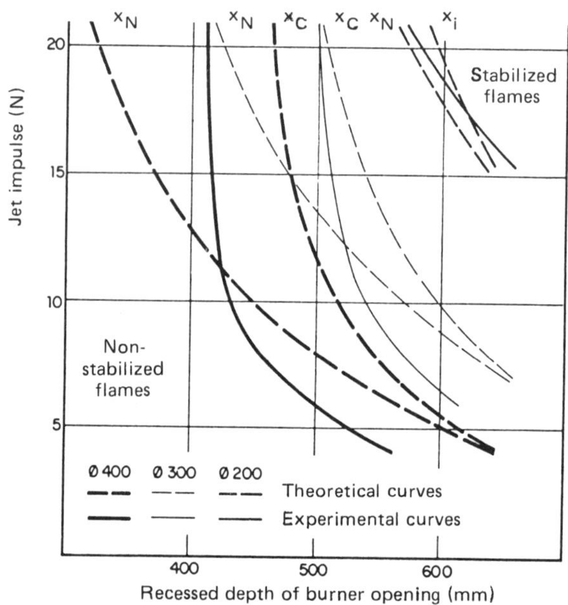

Fig. 4.41. Effects of burner-barrel diameter and fuel-jet impulse on the recirculation of hot combustion gases around a flame jet. (Theoretical jet travels to the beginning of recirculation, x_N and the point of maximum recirculation, x_C, are compared with experimental determinations of travel to point x_N for each diameter.)

For cylindrical burner quarls with diameters of 200, 300 and 400 mm, they calculated the theoretical locations of points N and C (Fig. 4.40) as a function of the impulse of the fluids introduced to the quarl. These locations have been compared to experimental observations. Figure 4.41, taken from their report, compares the calculated and observed values.

Good flame stability requires that the burner be located so that the abcissa of the center of the zone of recirculation, C, falls in the exit plane of the quarl to the firebox.

4.4.1.1. Sample calculation

Given a jet of natural gas with a specific gravity of 0.6 and an impulse of 20 N, a flow of combustion air of 1,800 m³/hr penetrating the quarl at a rate of 10 m/s, then:

$$G_0 = 20 \text{ N}$$

$$G_a = \frac{1,800}{3,600} \times 1.3 \times 10 = 6.5 \text{ N}$$

The velocity after mixing (V_∞) is around 110% the initial velocity of the air, taking into account the relatively small quantity of gas, from which $V_\infty = 11$ m/s and:

$$G_\infty = [1,800 + 180 \times 0.6] \frac{1.3}{3,600} \times 11 = 7.5 \text{ N}$$

The value of the parameter, m is then:

$$m = \frac{20}{7.5} + \frac{6.5}{7.5} - \frac{1}{2} = 2.7$$

and

$$\frac{1}{\sqrt{m}} = 0.61$$

In Fig. 4.40, the abscissa ξ of C corresponds to $\dfrac{1}{\sqrt{m}} = 0.61$:

$$\xi \cong 0.4$$

Taking into account the relation:

$$\xi = 4\sigma \frac{x}{h}$$

and the value $\sigma = 0.0285$, one obtains:

$$\frac{x}{h} = 3.5$$

Accordingly, in this example, the length of the cylindrical burner quarl that will stabilize the flame is equal to 3.5 times its radius.

4.4.2. Stabilized flames obtained with burner quarls

As stated above, flame stabilization is obtained with a burner quarl by means of the impulse in the discharged fuel and air, whose jet stream sets up a continuous reverse flow of hot combustion gases that ignite the air-fuel mixture inside the quarl. Figure 4.41 taken from the experiments at Toulouse brings out the fact that the area where the flame is stabilized depends less on the dimensions of the quarl as the jet impulse gets stronger.

4.5. OTHER METHODS FOR STABILIZING FLAMES

4.5.1. Pilot flames

Stabilization by means of a pilot flame, which is generally used in domestic gas burners, is little used in industrial applications that employ diffusion flames. The principle of operation of pilot flames consists of organizing at the base of the principal jet an auxiliary flame to heat the jet and feed it with combustion products that favor its ignition as soon as the mixture of fuel and oxidizer reaches the limits of inflammability. The auxiliary flame is conveniently made with a gaseous fuel by premixing a portion of the fuel with air and burning it in front of a flame-arrester screen that prevents flashback of the flames. Figure 4.42 illustrates a burner stabilized with a pilot flame such as was used in a series of tests at the experimental station of *GEFGN* at Toulouse. This burner has an air intake concentric to the injector. In the experiments, air leaves the burner at a low velocity on the order of 7-8 m/s while the gas, furnished under high pressure, supplies the impulse necessary to development of the flame within the enclosure. The premixed gas and air, discharging at low speed from the tube surrounding the gas tube, burn in the vicinity of the screen, and products of this combustion are aspirated into the high-speed central jet (Ref. 4.2).

4.5.2. Stabilization with an oxygen pilot

A similar method consists of replacing the pilot flame with a supply of pure oxygen. Figure 4.43 shows a gas injector consisting of an internal gas tube surrounded by a concentric tube carrying oxygen, such as was used at Toulouse in the tests described above. As a safety measure, the oxygen intake is cooled with water. With oxygen, it is recognized that stabilization is due to the presence of a methane-oxygen flame, which is carried from the root of the central jet along its periphery. The establishment of this oxy-gas flame is facilitated by the existence of a ring-shaped turbulent wake at the end of the injector.

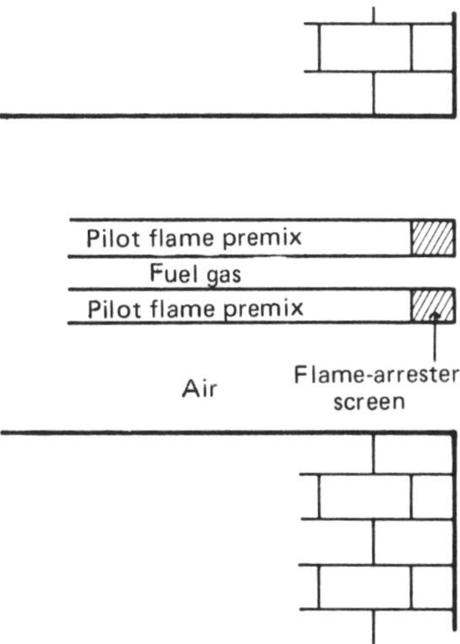

Fig. 4.42. Diagram of a gas burner with pilot stabilizer.

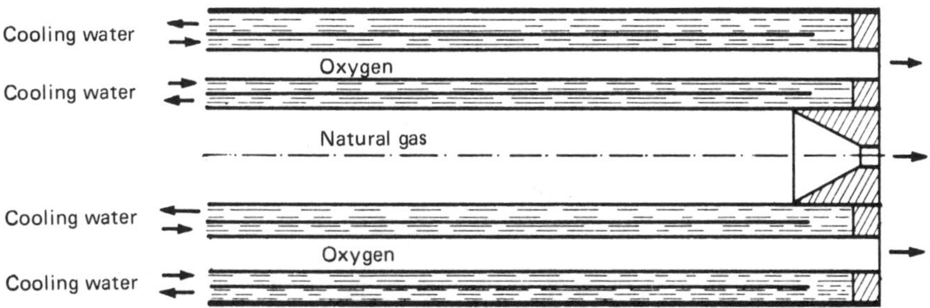

Fig. 4.43. Diagram of a natural gas burner stabilized with an oxygen pilot.

4.5.3. Effectiveness of pilot burners

Stabilization with a pilot burner depends on the rate of flow of gas or pure oxygen introduced in the pilot. As the specific impulse of the central jet increases, the flow from the pilot must also increase in order to keep the principal flame in place. The results of tests made at Toulouse are summarized in Figs. 4.44a and 4.44b. In these figures, the specific impulse, N, is expressed in newtons per 1,000 th/hr.

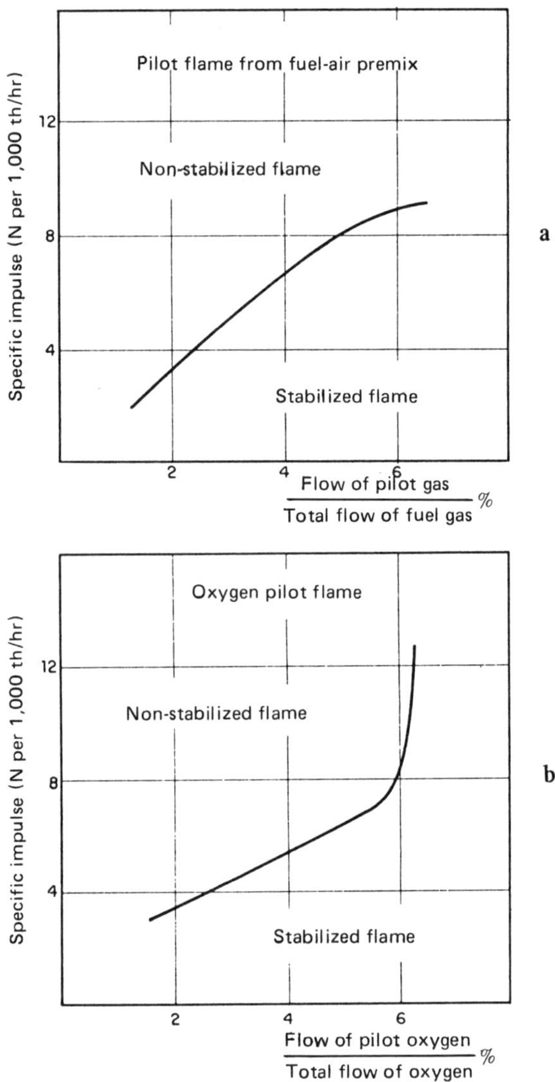

Fig. 4.44.   Stabile and non-stabile regions for flames with pilot lights.

The rate of flow of the pilot flame is the ratio of the rate of flow of pilot fluid to the total flow of fluid, whether gas or oxygen. It should be remembered that the maximum specific impulse possible for methane at ambient temperature is around 14.5 N/1,000 th/hr which corresponds to a discharge at the speed of sound. (This impulse can be increased by preheating the gas to increase the speed of sound, as for example, preheating methane from 0 to 250° C increases its speed of sound from 428 to 593 m/s. Corresponding to this variation is an increase of the specific impulse limit from 14.5 to around 20 N/1,000 th/hr).

The experiments at Toulouse demonstrate that for the gas pilot flame, the rate of flow of fuel in the pilot must increase as the specific impulse of the gas jet increases. Figure 4.44a also shows that the curve bends toward the horizontal beyond a rate of flow of 6% of fuel gas in the pilot flame; and this leads to the conclusion that this method of stabilizing is ineffective beyond a specific impulse of around 8 N/1,000 th/hr. By contrast, Fig. 4.44b shows that beyond a rate of flow a little over 6% of the stoichiometric flow of pure oxygen all flames are stabilized, no matter what the specific impulse of the natural gas jet. The effectiveness of this method of stabilizing is without doubt tied to the high temperature of the oxygen-gas pilot flame maintained at the root of the gas jet, preheating the jet and sowing it with combustion products that facilitate maintaining the flame downstream.

Also, combustion stabilized by oxygen pilot is much quieter than any other stabilized combustion—a phenomenon that the experimenters at Toulouse interpret as being caused by the decrease in turbulence through increased viscosity of the gases preheated with the pilot. The experimenters analyzed the mixing of fuel and oxidizer at the beginning of the flame for each stabilizing method. A non-stabilized flame obtained from a natural-gas nozzle centered in the combustion air intake pipe was used for purposes of comparison. In all the experiments referred to, the rate of the combustion air varied between 8 and 10 m/s at its entrance into the furnace.

The study compared the speed of mixing by taking samples of the fuel and oxidizer mixture downstream in the furnace along the axis of the burner. While allowing for possible bending of current lines, particularly during disc-stabilization, the compositions were related to the residence time of the gases in the furnace from the moment they left the burner to the moment of taking the sample. This time lapse was calculated from readings of the velocity of the gases along the axis.

A mixing factor, m, can be defined as being the ratio of oxides to fuel in the sample to oxides to fuel in stoichiometric proportion, as follows:

$$m = \frac{\left(\dfrac{\text{weight of oxidizer}}{\text{weight of fuel}}\right)\ \text{in the sample}}{\left(\dfrac{\text{weight of oxidizer}}{\text{weight of fuel}}\right)\ \text{stoichiometric}}$$

By this definition, the factor is zero for pure fuel, while it is infinitely large for air without fuel. For a stoichiometric mixture, the m factor is equal to 1.

Figure 4.45 shows that the mixture of gas with air is very rapid for the non-stabilized flame and that, by contrast, it is relatively slow during stabilization with pilot gas or oxygen. Mixing is also rapid for flames stabilized by disc.

There is a corresponding distribution of temperature from one flame to the other, as is shown in the curves in Fig. 4.46. The maximum temperature is observed along the axis of the stabilized flame. On the contrary, this maximum

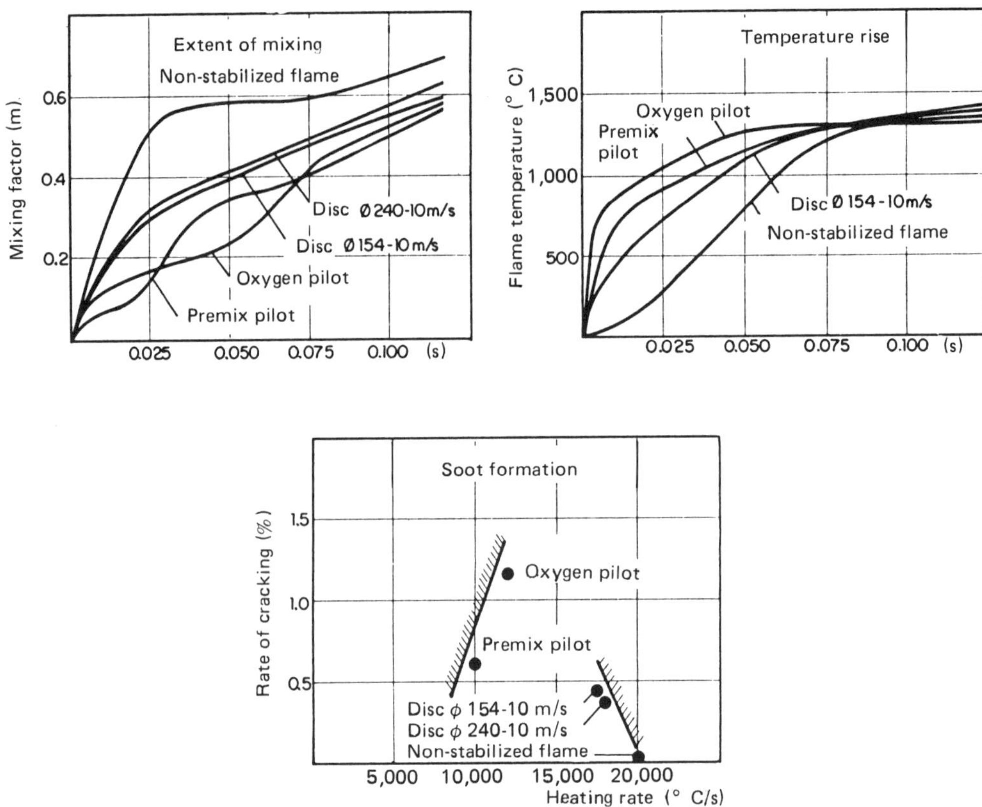

Fig. 4.45. Effects of mode of flame-stabilization on the combustion in front of a burner.

is observed in the vicinity of the burner on the circumference of the jet, both with disc stabilization and stabilization using oxygen or gas pilots. Heating is clearly more rapid in the case of a pilot.

The position of the flame front is naturally dependent on the mixing conditions. For a non-stabilized flame (but remembering that it is developed in a high-temperature furnace with walls at about 1,250° C)—the flame front appears at 60 to 70 cm downstream of the burner and presents strong fluctuations.

If the heat-release is related to the residence time of the gases after discharge from the burner (as the experimenters at Toulouse have done in Fig. 4.45) it is seen that a correlation appears between the rates of heating and the quantity of methane transformed into carbon particles by thermal cracking. The rate of heating is not the only factor intervening in the formation of the carbon, but its action is combined with the mixing factor, which characterizes the quantity of oxidizer present in the flame, and therefore, with the possibility of oxidizing

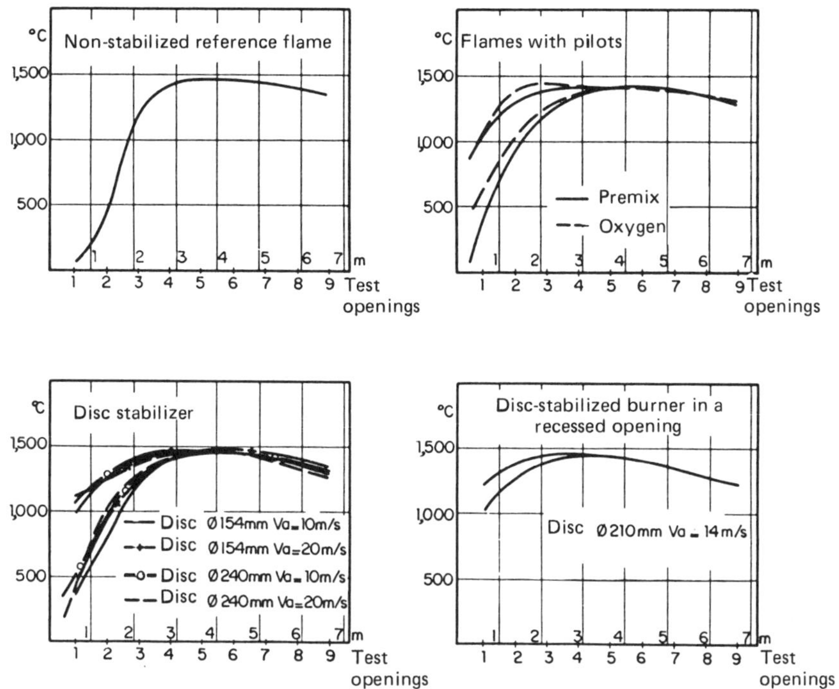

Fig. 4.46.   Effects of mode of stabilization on temperatures in the experimental furnace at Toulouse.

the carbon at the moment of its formation. The following table shows these data, such as they were observed in the experiments at Toulouse.

FORMATION OF CARBON PARTICLES IN NATURAL GAS FLAMES

Mode of stabilization	Mixing factor after about 0.04 s residence time	Soot concentration (g/m^3) (NTP)	Maximum rate of cracking (%)	Emissivity of the flame
Oxygen pilot	0.2	2.9	1.1	0.4
Premix pilot	0.3	2.5	0.6	0.36
Disc stabilization	0.4	0.8	0.4	0.32
Non-stabilized flame	0.6	traces	0	0.29

These brief remarks illustrate how essential to the properties of the flow is the role played by the flow conditions of the different fluids at the exit of the burner in the firebox.

REFERENCES

4.1   BEER, J.M., MONNOT, G. – *Ve Journée d'Etudes sur les Flammes*, Paris,21 Nov. 1963

4.2   ROGIER, J. – Report GN 6 of *Groupe d'Etudes des Flammes de Gaz Naturel*, Toulouse, Oct. 1972.

4.3   LEBLANC, B. – Report GN 10 of *Groupe d'Etudes des Flammes de Gaz Naturel*, Toulouse, Dec. 1972.

4.4   LEUCKEL, W. – Aérodynamique des écoulements tourbillons.

CHEDAILLE, J., LEUCKEL, W. – Flammes de mazout lourd et de charbon pulvérisé avec air de combustion en rotation. *VIIe Journée d'Etudes sur les Flammes*, Paris, 28 April 1968.

4.5   PERTHUIS, E. – Report of *Institut Français du Pétrole*, Ref. 17098.

5

high-temperature flames

5.1. INTRODUCTION

5.1.1. Preheating the air. Use of oxygen

Numerous industries carry out heating operations at very high temperatures, which require combustion at even higher temperatures. There are two fundamental methods of raising the temperature of combustion:

(a) Preheating the combustion air.
(b) Using oxygen-enriched air or pure oxygen.

Preheating the combustion air is ordinarily done by heat exchange with the hot flue gas. Technically, the extent of preheating is limited only by the temperature of the flue gas and by the materials of construction of the exchanger. A preheat to 500-600° C is easily achieved in metal exchangers, and heat exchange equipment using refractory elements permits preheat to the order of 1,200° C. However, refractory heat exchange equipment is usually regenerative, with alternating cycles of heat absorption and heat release, so that such equipment must be doubled.

It should be noted that preheating by recovering the heat from the flue gas saves energy, no matter what level of temperature. That is why this practice is so common.

Oxygen enrichment proportionately reduces the weight of nitrogen ballast found in the flue gas, so that the energy of combustion is applied to combustion products with a smaller thermal mass, and the temperature of those products is proportionately higher. The maximum effect is observed in combustions with pure oxygen.

The increase in the temperature resulting from the use of oxygen, or preheated air is limited by dissociation of the combustion products at high temperature, particularly carbon dioxide and steam; these dissociations involve a considerable quantity of energy.

5.1.2. Non-symmetrical flames

Non-symmetrical flames can achieve higher-than-theoretical temperatures. In certain arched furnaces heated by diffusion flames, oxygen is introduced underneath the flame to create a thermal gradient with the hot part of the flame nearer a charge on the floor of the furnace and the less-hot part nearer the roof, which because of this, receives less radiation, and thus can operate at a lower temperature to the advantage of its mechanical resistance.

Non-symmetrical flames are also useful for certain smelting furnaces, in which a flame is directed toward the charge so as to promote heat exchanges by forced convection. If a jet of oxygen is supplied within the area of the flame's impact on the charge, combustion in that area is strongly activated. Heat-dissociated products of the equilibrium reaction in the fuel/oxygen flame recombine on the surface of the charge, giving up the energy of recombination to the charge. This type of heat convection, called "live convection" by Veron (Ref. 5.1) permits a very high rate of heat exchange due to this particular use of oxygen.

5.1.3. Auxiliary burners; short flames

Certain applications, for example, an auxiliary burner on an electric smelting furnace or the injection of hydrocarbons in the tuyeres of a blast furnace, require extremely rapid combustion, either because the volume available for the flame is limited or because the flow rate limits the time available. In such cases, combustion with oxygen affords compact flames or very rapid combustion.

Rapid combustion can also be obtained by means of a burner with a pre-chamber, in which a first stage of combustion is organized in a rich mixture, leading to a cracking-conversion of the original fuel (liquid or gaseous) into a new gaseous fuel rich in hydrogen. A high hydrogen content perceptively increases the rate of combustion relative to that of the original fuel. Also, the cracked fuel is filled with carbon particles, which produce flames with high radiant emissivity. Two-stage combustions do not necessarily involve the use of oxygen; but if it is available, the design of the burner can be simplified. In the same way, use of air preheated to over 500° C facilitates primary combustion in a burner pre-chamber thus allowing good stabilization of the primary flame, even without the fluid-dynamic devices described in Chapter 4.

5.2   GENERATING HIGH COMBUSTION TEMPERATURES

The fundamental methods for raising flame temperature (i.e., preheat and oxygen enrichment) can be combined; one can enrich a previously preheated air with oxygen.

Preheating pure oxygen to some tens of degrees above ambient temperature is not used because of the danger of fire that hot oxygen presents even with metals. Preheating the fuel offers only limited possibilities, because the stoichiometric weight of fuel consumed in combustion is minor compared to the weight of air. Heavy hydrocarbon fuels should be heated only to the temperature needed for good atomization; and any superheating or even any prolonged heating should be avoided. The large molecules of the heavy hydrocarbons are not thermally stable and tend to crack long before reaching their boiling point at atmospheric pressure. In addition, heavy fuel oils are not Newtonian liquids, and prolonged heat can displace or rupture suspensions with asphaltenes through oxidation during the course of heating.

With natural gas, the chemical stability of the methane permits preheating to a temperature of around 600° C, if the gas is not in contact with metals that catalyze cracking. Since nickel is one such metal, its catalytic properties preclude the use of austenitic stainless steel as a construction material for fuel preheaters. Thus, only combustion air is preheated in industrial practice, usually through heat exchange with the flue gases, and as follows:

(a)   Using metallic exchangers up to maximum temperatures on the order of 600° C and above a minimum temperature set by the acidic dew point of the flue gases (see Section 7.3.1).

(b)   Using ceramic regenerative heat recovery units up to and beyond temperatures on the order of 1,200° C.

The combustion-temperature increase made possible by these procedures can be calculated. It is simply necessary to know the heat of combustion of the fuel and to set the ratio of air to fuel. Table 5.1 gives the adiabatic combustion temperatures for five fuels in stoichiometric mixture with air or oxygen.

TABLE 5.1

FLUE GAS TEMPERATURES FROM ADIABATIC COMBUSTION
OF STOICHIOMETRIC MIXTURES (°C)

Fuel	Air preheated to:			Oxygen at 27°C
	27°C	600°C	1,200°C	
Methane	1,955	2,190	2,380	2,775
Propane	1,990	2,210	2,410	2,815
Butane	1,995	2,210	2,410	2,825
Kerosene	2,000	2,215	2,410	2,830
No. 2 heavy fuel oil	1,975	2,200	2,395	2,830

5.2.1. Axial flames

5.2.1.1. Comparison of oxygen sources

A comparison of the effect of using oxygen or oxygen-enriched air was studied at Ymuiden (Ref. 5.2) using as fuels blast furnace gas with low heat value (LHV) = 1,100 kcal/m³ and natural gas from Groningue with LHV of 7,560 kcal/m³.

One burner was sufficient because the mass flow rates of blast furnace gas and oxygen, on the one hand, and of natural gas and air, on the other, were approximately the inverse (Fig. 5.1). In the first case, the oxygen was injected in the center of the stream of blast furnace gas, in the second, the natural gas was injected in the center of the air stream. The speeds of the oxygen and the natural gas at the nozzle were equal to 77 m/s and 84 m/s, respectively. In both cases, the same rotation defined by the number S, was given to the fluids, as follows:

$$S = \frac{\text{Flow rate of rotating movement about the axis}}{\text{Flow rate of axial movement x Radius of the throat of the nozzle}} = 0.8$$

The axial impulses were 13.5 N/10⁶ kcal/hr for the blast furnace gas/oxygen flame and 11.65 N/10⁶ kcal/hr for natural gas/air flame, respectively.

In addition two other flames were studied with blast furnace gas fuel, and an oxygen-enriched air (50% O_2 and 50% N_2) as the oxidizer on the one hand and atmospheric air, on the other. Figure 5.2 shows the burner that was used for these two flames. The same swirl number, $S = 0.8$, was attained for these two flames.

Table 5.2 summarizes the experimental conditions for each of the four flames. In these experiments the heating power was maintained practically constant. The excess of oxygen was also maintained constant (2% of O_2 in dry flue gas).

TABLE 5.2

OPERATING CONDITIONS FOR EXPERIMENTAL FLAMES STUDIED AT YMUIDEN

Flame No.	Fuel					Oxidizer				Impulsion		Heat Duty 10⁶ kcal/hr
	Nature	Flow (m³/hr)	Velocity (m/s)	Density (kg/m₀³)	LHV (kcal/m₀³)	Nature	Flow (m₀³/hr)	Velocity	Density	Gross (N)	Specific by LHV $\frac{N}{10^6 \text{ kcal/hr}}$	
1	Blast furnace gas	1,142	22.2	1.30	1,105	O_2	263	77.3	1.43	17.2	13.6	1.262
2	Blast furnace gas	1,142	22.2	1.30	1,105	Air O_2	343 197	14.3 58.1	1.293 1.43	15.8	12.5	1.262
3	Natural gas	171.1	84.1	0.833	7,560	Air	1,601	20.7	1.293	15.2	11.8	1.29
4	Blast furnace gas	1,134	36.4	1.31	1,129	Air	1,394	27.1	1.293	28.5	22.3	1.28

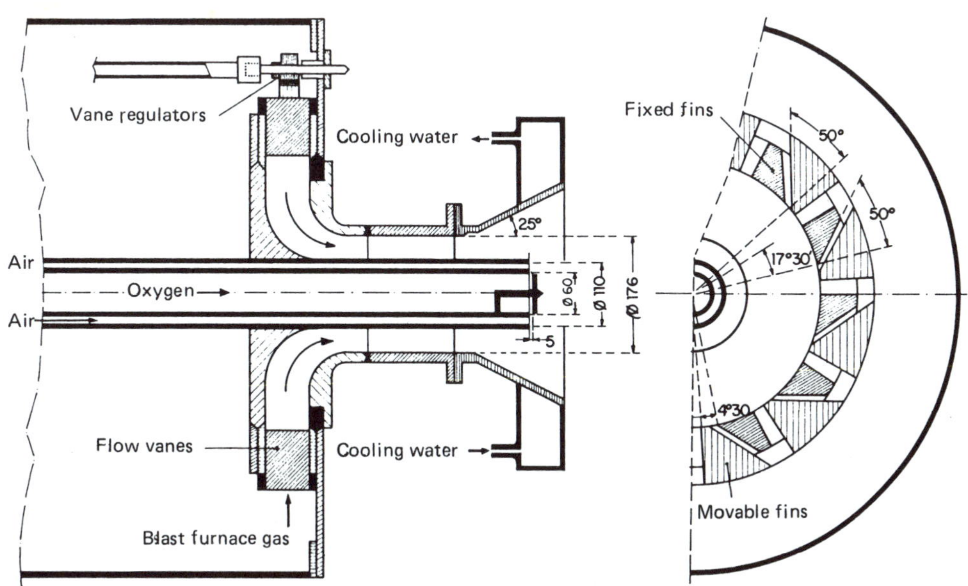

Fig. 5.1. Burner for blast furnace gas with oxygen.

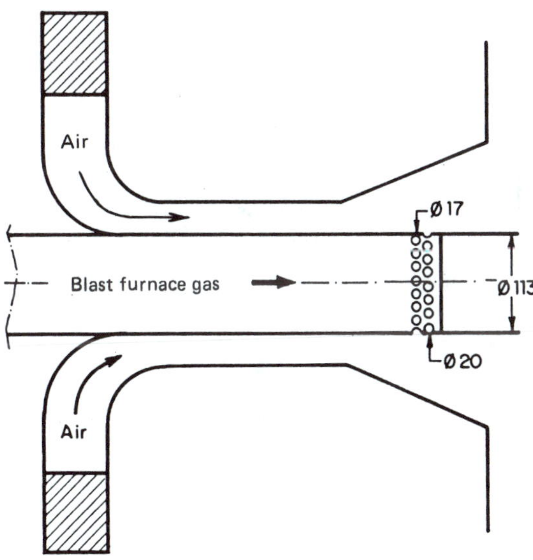

Fig. 5.2. Burner for blast furnace gas with enriched air.

As for the fluid dynamics, flame No. 4 was distinguished from the others by its very high impulse, which corresponds to the large quantity of inerts carried both by the blast furnace gas and the air, necessary to achieve the chosen heat rate of flow.

The cooling system of the experimental furnace permitted conducting these experiments in a hot enclosure. Figure 5.3 shows the temperatures of the roof, of

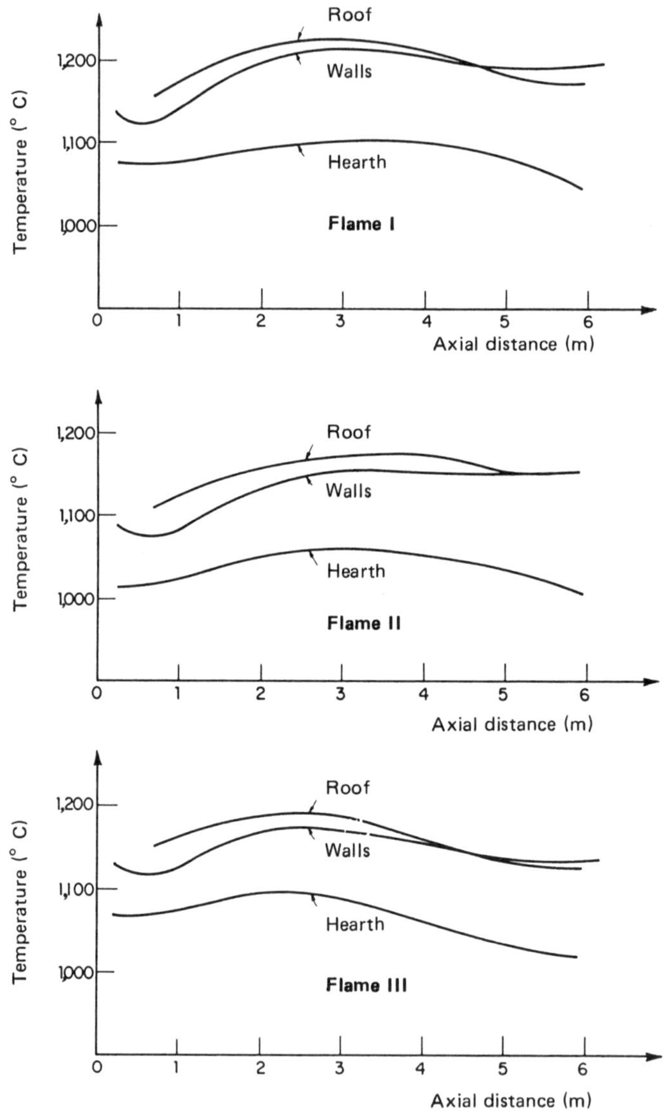

Fig. 5.3. Temperature profiles for burning natural gas and blast furnace gas, when taken along the axial travel of the flame in a hot-walled furnace.

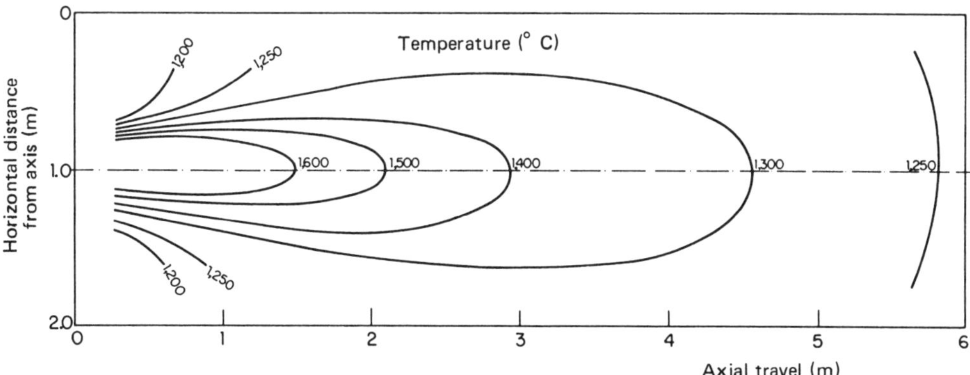

Fig. 5.4. Flame No. 1.

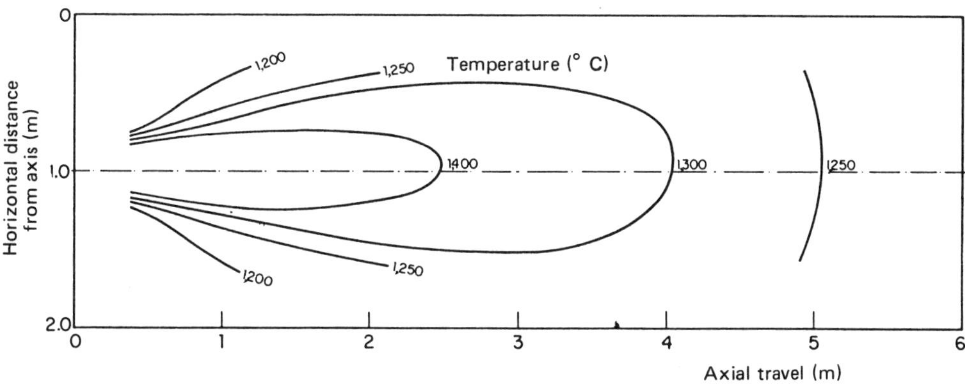

Fig. 5.5. Flame No. 2.

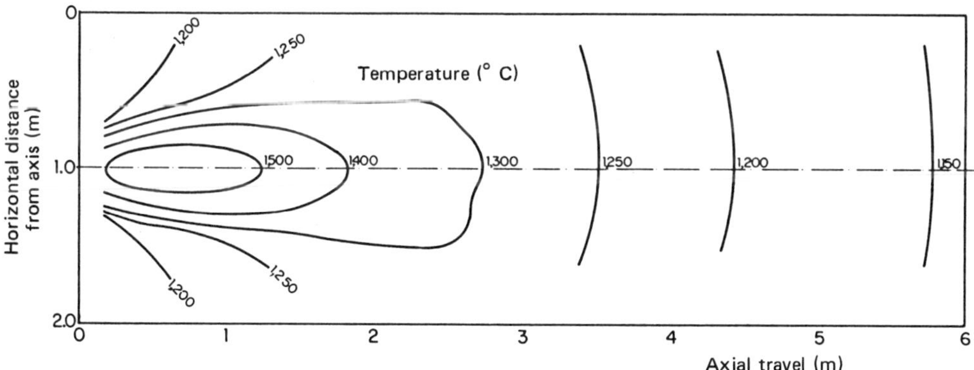

Fig. 5.6. Flame No. 3.

Figs. 5.4, 5.5, and 5.6. Isotherms for flames from natural gas and
blast furnace gas in a hot-walled furnace.

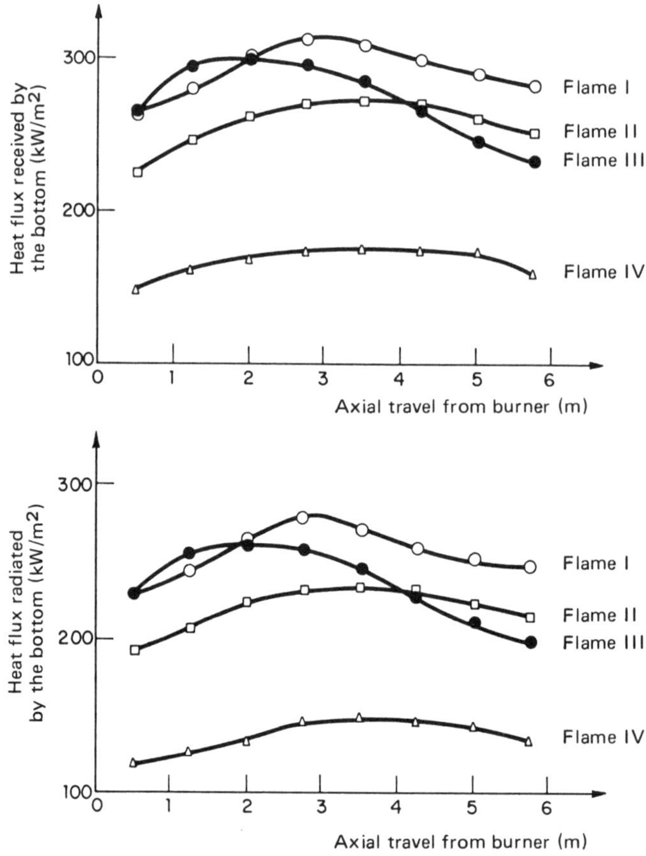

Fig. 5.7. Radiant heat exchange at the bottom of a hot-walled furnace fired with gas flames (see Table 5.2).

the wall and of the hearth, as a function of distance to the burner, for the first three flames. Also, Figs. 5.4, 5.5 and 5.6 show the distribution of temperatures in a horizontal plane passing through the axis of flame fronts 1,2 and 3. Finally, Fig. 5.7 shows the distribution of the radiant flux absorbed by the hearth for all four flames.

From these measurements, the experimenters were able to calculate the fractions of the released heat that were effectively absorbed by the furnace hearth; the numbers are the following:

	(%)
Flame No. 1	22.2
Flame No. 2	20.9
Flame No. 3	21.1
Flame No. 4	16.0

The heat exchange is thus more active with flame No. 1 from blast furnace gas and oxygen than with flame No. 3 from natural gas and cold air. In so far as the interpolations are valid, natural gas and blast furnace gas are equally effective when the latter is burned in an oxygen-enriched air of 75% oxygen.

Oxygen enrichment thus permits up-grading a very poor fuel to the point where it is comparable to the combustion of natural gas with air, and without preheating. However, this method is ordinarily without economic advantage, unless there is a breakdown and a temporary need to substitute a poor fuel for a rich one.

5.2.1.2. Non-symmetrical flames

In certain applications it is desirable to promote heat exchanges between the flame and the charge without perceptibly increasing the temperature of the refractory of the furnace, especially of the roof. In some open-hearth steel furnaces, this was attempted by squirting liquid fuel up into the gases recirculating back toward the burner along the roof. Atomized into these very hot gases that were poor in oxygen, the fuel oil cracked, forming a cloud of carbon particles, which protected the roof from radiation from the flame below. A correct application of this technique requires, however, that the smoke thus formed be recaptured in the flames coming out of the burners, so that the carbon particles are completely burned without accumulating in the dust collectors. This problem in fluid dynamics has to be resolved in both directions corresponding to cycles of heat absorption and heat release in the heat recovery units.

Another technique consists of passing the fuel and only a part of the oxidizer through the burner with the balance of the oxidizer introduced under the flame by means of an auxiliary lance. The balance of oxidizer is oxygen that promotes combustion in the lower part of the flame, causing a local temperature that is higher and permitting a more intense radiation towards the high temperature flame. This technique was studied as early as 1959 (Ref. 5.3), and the results of that study are reviewed below.

A.   *Non-symmetric axial flames parallel to the charge*

Figure 5.8 shows the experimental equipment used for this study. Under a normal burner placed in the axis of the enclosure to be heated, there is an auxiliary injector for introducing a part of the oxidizer (air or oxygen) so that the lower part of the flame is well supplied with oxygen. The result is that there is a more intense local combustion and a hotter flame.

To determine the recirculation in the enclosure with this equipment, the total impulse G_t is taken as the sum of the impulse G_f of the fluids coming out of the burner and the impulse G_a of the oxidizer leaving the auxiliary injector. Another important variable is the distance δ of the axis of the auxiliary injector from the axis of the burner. Finally, a third variable appears when oxygen is supplied

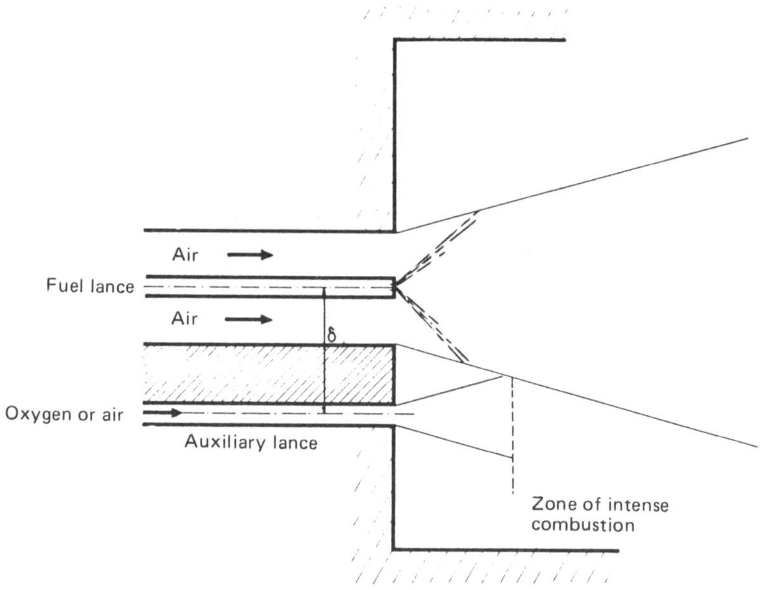

Fig. 5.8. Burner with auxiliary injection of oxidizer.

by the auxiliary injector, i.e., the average oxygen concentration, taking into account the rate of flow of air passing through the burner. The tests were made either in atmospheric air or in enriched air with an average oxygen concentration of 30% oxygen by volume.

Table 5.3 assembles some measurable results for the flames, some with air enriched to 30% oxygen and others with atmospheric air, for two flow levels.

TABLE 5.3

NON-SYMMETRICAL FLAMES OBTAINED WITH THE BURNER OF FIG. 5.8

Flame	G_t (kgf)	δ (cm)	Excess oxygen (% of oxidizer)	Average O_2 content of oxidizer (vol. %)	Flame temperatures (°C) Maximum	At 5 m downstream of burner	Emissive coefficient of the flame	Distance from point of measurement to burner (m)	Maximum radiation of the flame (cal/cm².s)	Distance from point of measurement to burner (m)	Comments
1	2.78	0	6.5	29.9	1,755	1,425	–		8.6	5	Oxygen. Symmetrical.
7	1.62	0	3.5	29.6	1,795	1,485	0.46		11	1.5	Oxygen. Symmetrical.
12	1.76	14.5	10	29.5	1,940	1,450	0.85	2.5	3.0	2	Oxygen. Non-symmetrical.
13	2.81	0	7	20.8	1,480	1,365	0.56	2	7.8	2	Air. Symmetrical.
20	1.67	14.5	9	20.8	1,420	1,330	0.88	1.5	8.3	3	Air. Non-symmetrical.
22	1.63	0	13	20.8	1,440	1,330	0.82	2.5	13	2	Air. Symmetrical.

Other flames permit comparing the effect of a non-symmetrical feed of oxidizer for the same average oxygen content and the same impulse level.

With conventional introduction of the oxidizer in the form of one single axial stream, a comparison of flames Nos. 1 and 13 indicates the effect of oxygen enrichment: the maximum temperature in the flame increases 270° C. With oxygenated air, the maximum radiation appears at the tail of the flame, while with air it appears at around 2 m from the burner. Flames Nos. 7 and 22 show that if the injector impulse is weaker, the oxygen enrichment raises the maximum temperature 350° C, while the maximum radiation in both flames is situated at 1.50 or 2 m from the burner. The influence of a jet impulse that creates a recirculation of flue gases inside the furnace is here found again; when this recirculation is intense, the oxidizer is polluted by the flue gases and the maximum temperature decreases.

As far as asymmetrical oxidizer injection is concerned, a comparison of flames Nos. 20 and 22 shows that the difference between the two is very slight when the oxidizer is air. On the other hand, if pure oxygen is asymmetrically introduced under the flame (compare flames Nos. 7 and 12) the non-symmetrical flame shows an increase of 150° C in maximum temperature and a considerable increase in its emissive power at 2.50 m downstream of the burner. Thus, it is better, when using oxygen, to use it directly without premixing with combustion air, in order to achieve the desired local increase in flame temperature.

B.   Non-symmetric flames impinging on a charge

Up-grading flames by adding undiluted oxygen led the experimenters at Ymuiden to study the reaction of flames plunging toward a cooled furnace bottom while extending an oxygen lance into the furnace to bring the oxygen close to the bottom (Ref. 5.4). Figure 5.9 shows the arrangement of the burner in these experiments.

Only one fuel was used: heavy fuel oil. It was atomized either pneumatically or mechanically. The distance between the burner and the point of impingement

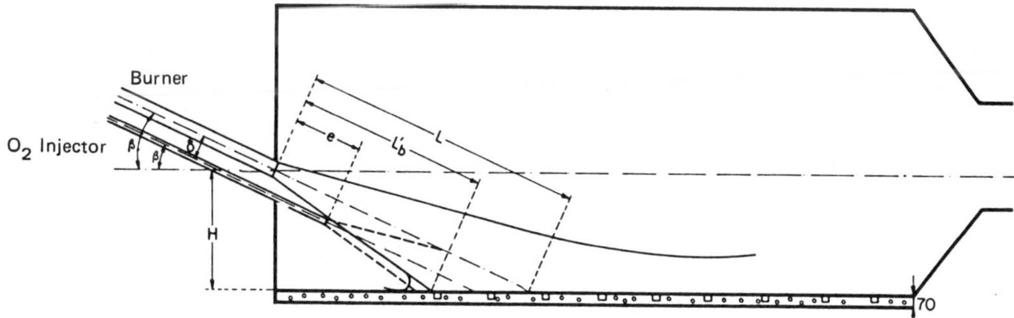

Fig. 5.9.   Diagram of a furnace hearth heated by a plunging flame
with localized oxygen supply via lance.

of its axis on the hearth was enough to keep droplets of fuel oil from reaching the bottom.

The oxidizer was comprised of air introduced to the burner (including any air of atomization) and oxygen (99.5% pure) introduced through a lance located under the burner and penetrating for distance e beyond the outlet of the burner. The combination produced an overall oxidizer oxygen content of 30%. Throughout the experiments the excess oxygen in the flue gas was maintained at 3% O_2 for air oxygenated at 30%, corresponding to 10% excess of air in similar combustion with pure air.

The furnace hearth was instrumented to permit measuring the heat fluxes. The position of the end of the oxygen lance, which was pushed more or less far into the furnace, played a principal role in heat transfer to the bottom. In the most effective cases, heat transfer to the hearth with plunging flame and oxygen supply reached 35-37% of the heat release compared to 18-20% in the case of horizontal non-symmetric flames with oxygen supply developed parallel to the hearth (see paragraph A and Fig. 5.8). The increase in heat transfer with a plunging flame is even more pronounced as the temperature of the hearth increases 150-200 °C due to the action of the flame. For some flames fed with oxygen a part of the heat transfer is accounted for by live convection.

Calculations made by the experimenters at Ymuiden show that the convection transfer can be very high because of "live convection" ($h_e = 0.38$ kW/m² . °C) in the case of a flame where the recombination of dissociated products enhances the effect of local impulse due to oxygen. We note nevertheless that the transfer calculations are done from experimental values that are difficult to measure (temperature and rate of gaseous currents in the vicinity of a wall, radiation received by the wall), so that some uncertainty is associated with the calculated values. This does not detract from the interest of combining the impingement of a flame with oxygen fed locally.

5.2.2. Swirling flame impinging on a target

In other tests (Ref. 5.5) the researchers at Ymuiden studied heat transfer between flames and a thermal charge placed opposite the axis of the burner at a distance of 0.75-1 m. The only fuel used was Groningue natural gas containing 81.3% methane and 14.4% nitrogen.

Different flames were compared, particularly flames that were produced by flows rotated with the combustion air at a moderate swirl number, S, of 0.90 sufficient for assuring flame stability. Figure 5.10 shows the burner used in these experiments. One of the flames was produced by air previously enriched with oxygen (30% O_2 by volume). In no case was the oxidizer preheated. The maximum temperature observed in the flames with air was 1,475° C, the temperature of the gases coming into contact with the target varying between 1,400 and 1,475° C. With an air enriched to 30% oxygen, the maximum observed

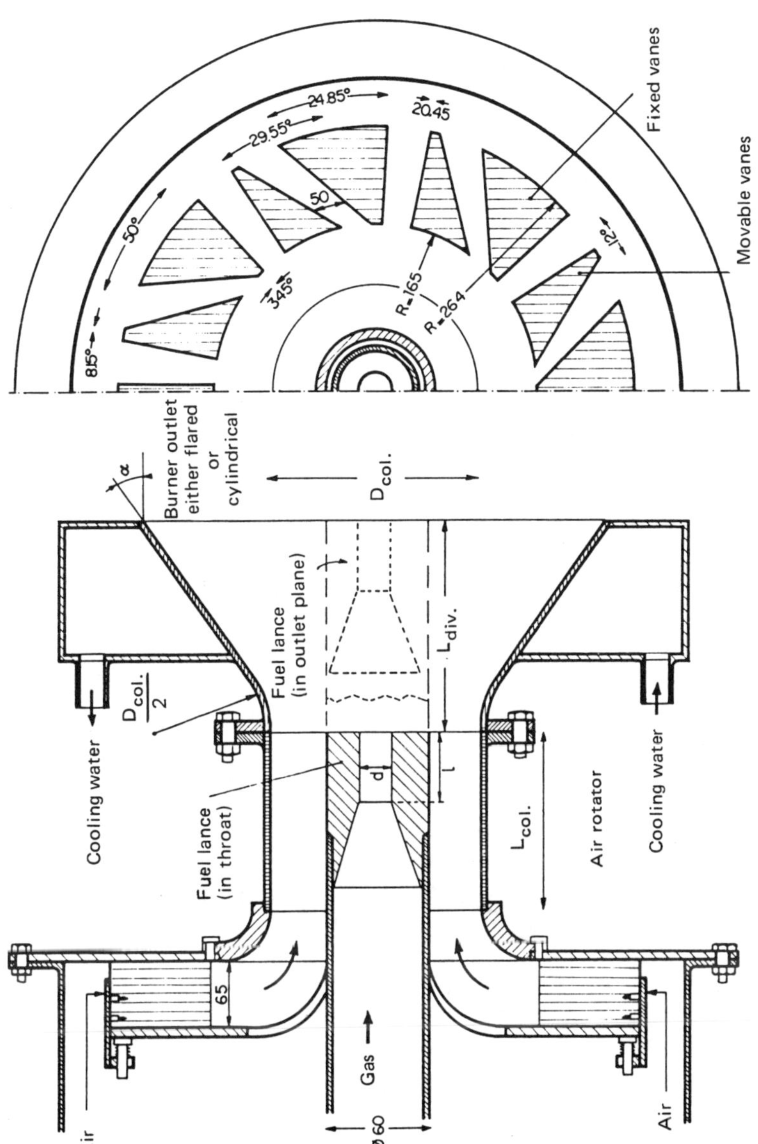

Fig. 5.10. Test burner used to heat a facing surface spaced at 0.75–1.0 m.

$D_{col.}$ = Collar diameter.
$L_{col.}$ = Collar length.
$L_{div.}$ = Divergent length.

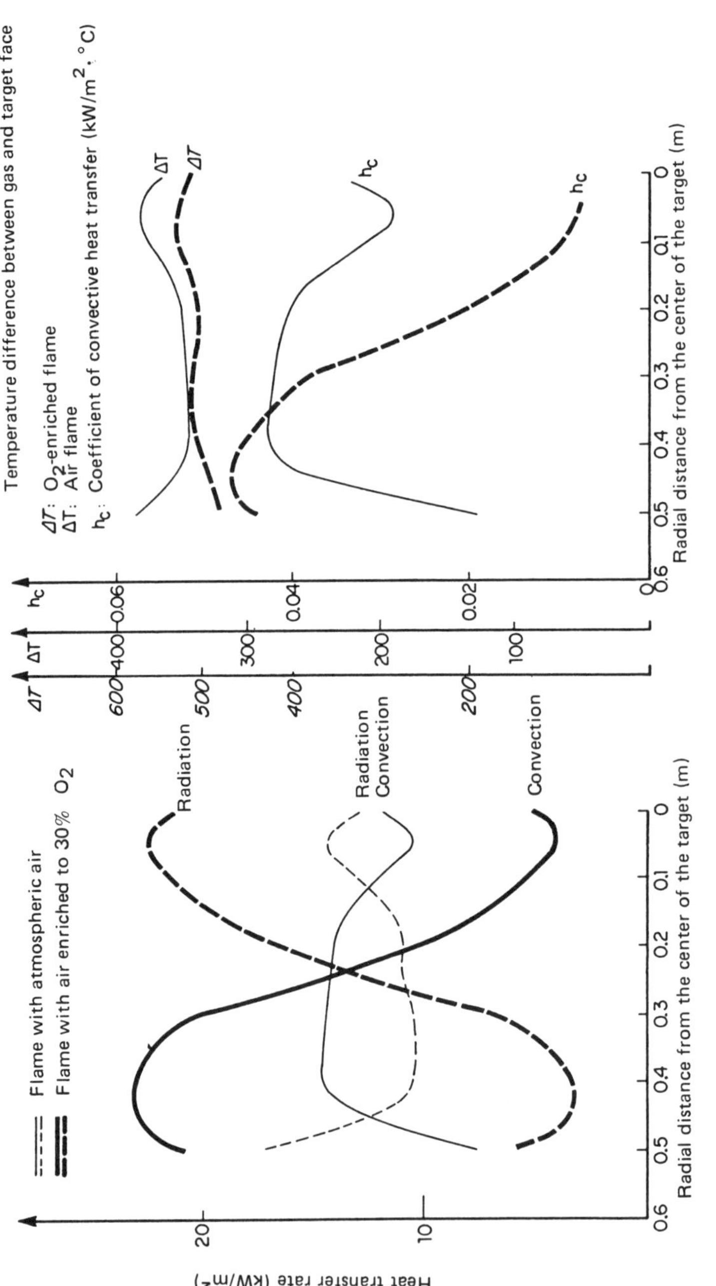

Fig. 5.11. Effects of oxygen enrichment on flame-to-target heat transfer rates, temperature-differences, and convective heat transfer coefficients.

temperature was 1,700° C and the gases in the vicinity of the target were at temperatures ranging from 1,550 to 1,650° C.

Figure 5.11 shows the difference between the flame-to-target heat transfer for a flame with air and a flame with oxygen-enriched air (30% oxygen). Transfer rates are higher with oxygen, due principally to an increase in the difference of temperature between the gas and the wall. On the other hand, the coefficient of heat transfer by convection reaches similar values in both cases.

5.2.3. Enriched or preheated air

A comparison of these two methods for raising flame temperatures was made at Toulouse as part of the research done by the *Groupe d'Etude des Flammes de Gaz Naturel* (*GEFGN*) (Ref. 5.6). The fuel used in this study was natural gas from Lacq (Table 5.4).

TABLE 5.4

COMBUSTION CHARACTERISTICS OF LACQ NATURAL GAS

Volume composition (%)	
CH_4	97.6
C_2H_6	1.9
C_3H_8	0.2
N_2	0.3
Specific gravity	0.567
High heating value (th/m³ NTP)	9.675
Low heating value (th/m³ NTP)	8.710
Stoichiometric ratio $\mathcal{V}_a$ (m³ of air/m³ of gas NTP)	9.76
Volume of flue gas from stoichiometric combustion (m³ of flue gas/m³ of gas NTP)	
Dry basis	8.75
Wet basis	10.77

Figure 5.12 shows the design of the burner used in these experiments. Specifically, the gas is introduced in the form of two concentric jets, the central jet with a great speed supplies almost all of the kinetic impulse, while a circular orifice introduces a part of the gas at a moderate speed of a few meters per second. Oxygen is introduced by a concentric ring of 22 holes, and the combustion air is introduced at a low speed through a circular orifice (Fig. 5.12d). The combustion air can be preheated to a temperature of 600° C.

Not only does the oxygen enrichment increase the temperatures of combustion but it simultaneously modifies the length, L_0, of the diffusion flames, which is defined in an infinite and calm atmosphere of air, as being:

$$L_0 = 6.65 \frac{M_0}{\sqrt{G\rho_f}} \left[1 + \frac{\mathcal{V}_a}{d_0} \right] \tag{5.1}$$

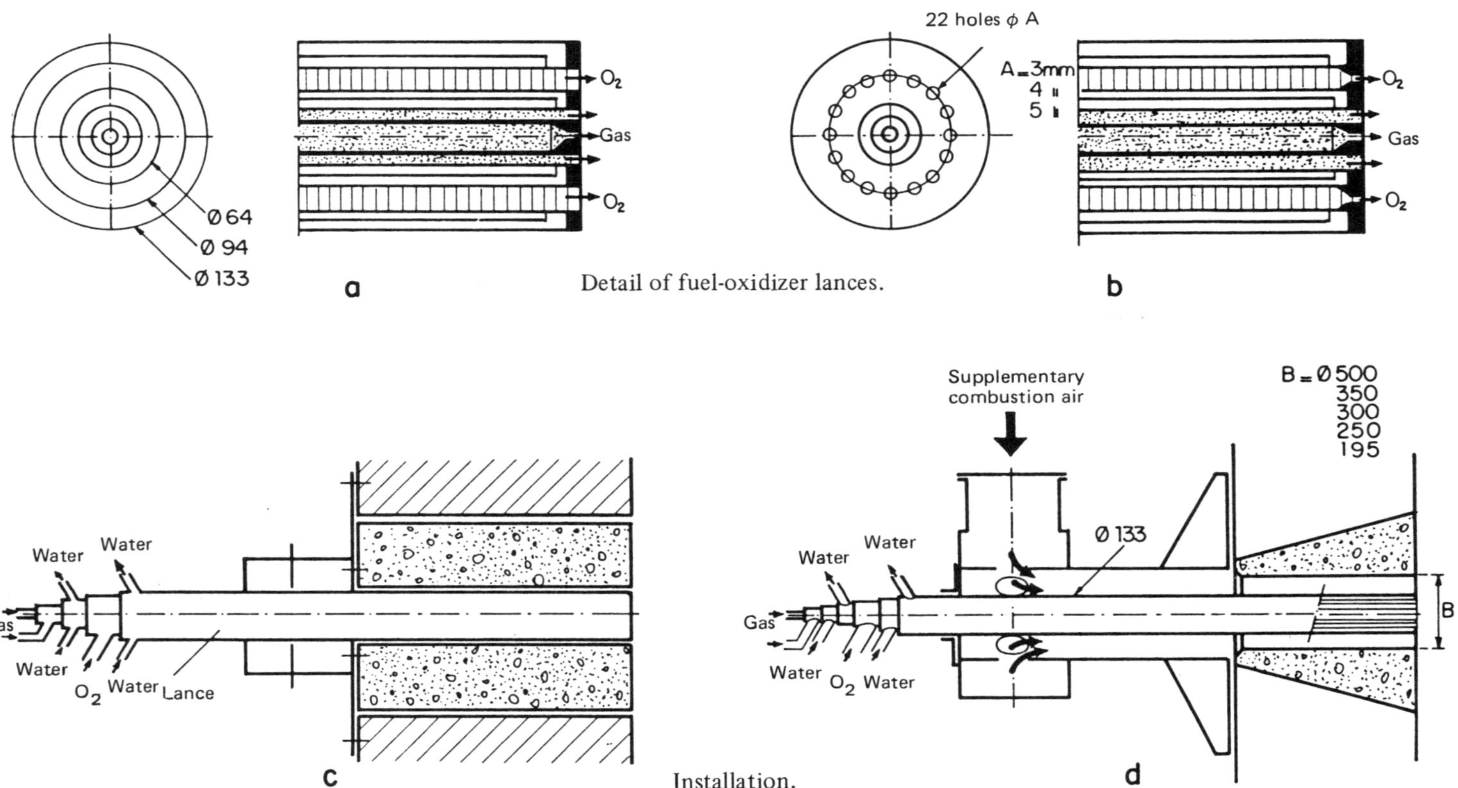

Fig. 5.12.   Experimental oxy-gas burners used for high-temperature flames.

where

M_0 = the mass rate of flow of gas (kg/hr),

G = the kinetic impulse carried by the gas (N, newtons),

ρ_f = average density of flue gas at the temperature at the end of the flame (kg/m³),

$\mathcal{V}_a$ = oxidizing power of the gas (m³ of air/m³ of gas in stoichiometric mixture),

d_0 = specific density of the gas at standard conditions. Numerical coefficient 6.65 for Lacq gas results from units chosen.

In an oxygen-enriched atmosphere, the length of the flame for the same fuel is, L_0', which is less than L_0 because the jet must carry along a smaller weight of oxidizer to complete the combustion. The length, L_0', of the flame under these conditions is given by the relation:

$$L_0' = 6.65 \frac{M_0}{\sqrt{G\rho_f}} \left[1 + \frac{\mathcal{V}_a}{d_0} (1 - 0.77\,k) \right] \tag{5.2}$$

where

$$k = \frac{\mathcal{V}_{O_2 \text{ injected}}}{\mathcal{V}_{O_2 \text{ stoichiometric}}}$$

Formulas (5.1) and (5.2) assume that the impulse G is supplied by the fuel and that the oxidizer is introduced symmetrically around the gas jet at low speed (Fig. 5.12).

When the speed of the oxygen at a constant fuel impulse increases, the flame lengthens about 20% as the speed increases from 1 m/s to the speed of sound (Fig. 5.13). Similar relations were observed inside the experimental furnace, which was shaped like a box 1.5 x 1.5 m, by 6.5 m long.

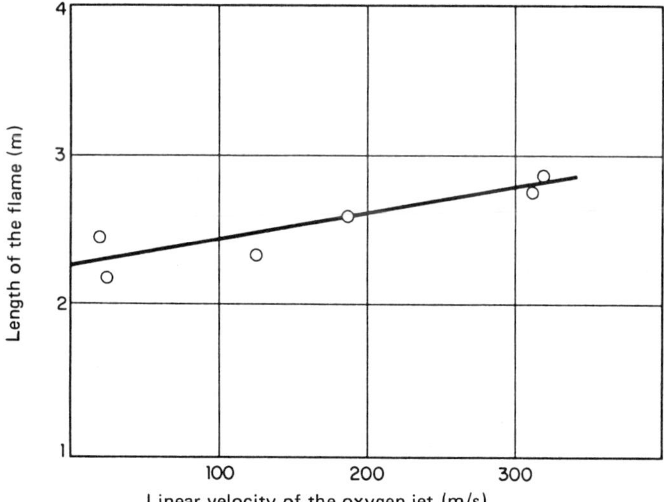

Fig. 5.13. Influence of an oxygen jet on the length of a flame.

If L is the length of flame observed in a furnace using atmospheric air as oxidizer the experiment shows that:

$$\frac{L}{L_0} = 0.32\sqrt{m} + 0.48 \tag{5.3}$$

where m is the Craya-Curtet number (Chapter 3).

If the oxidizer is enriched with oxygen, the same empirical relationship is confirmed for moderate speeds of introduction of the oxidizer:

$$\frac{L'}{L'_0} = 0.32\sqrt{m} + 0.48$$

Figure 5.14 summarizes the observations made at Toulouse by the experimenters who measured the savings of fuel that could result for the same heat transfer to the bottom of the experimental furnace either by preheating the air to 600° C, or by enriching the cold air with oxygen. Transposing these results to an industrial furnace cannot be done without precaution; in particular, the thermal concentration (ratios of the heat duty to the cross-section of the furnace) figures in the advantages that can be credited to using oxygen.

The gain in heat transfer for oxygen-enrichment can be defined by:

$$\text{gain} = \frac{(\text{transfer for the enriched flame}) - (\text{transfer for the air/gas flame})}{(\text{transfer for the air/gas flame})}$$

The gain thus defined varies as a function of the relative oxidizing power, k, on which depends the rate of flow of gas entering in the furnace.

Figure 5.15 summarizes the observations of the experimenters at Toulouse.

5.2.3.1. Effects of oxygen source on heat transfer

Two flames were created with the same rate of flow of natural gas, the same specific impulse, the same oxygen enrichment and the same overall equivalence ratio of fuel to oxidizer, and the heat transfers to the bottom of the furnace were compared for two different ways of introducing oxygen:

(a) Diluted oxygen in the combustion air.
(b) Pure oxygen at the base of the gas jet coming out of the burner.

The introduction of pure oxygen leads to a more active transfer toward the bottom and the walls of the furnace. These same tests were repeated for other pairs of flames, while varying characteristic, k, of the oxygen. Figure 5.16 summarizes the observations of the experimenters at Toulouse. These experiments confirm the interest that is attached to the use of undiluted oxygen not only in the case of flames striking a target (paragr. 5.2.1.2.b) but also in the case of axial flames extending through a furnace.

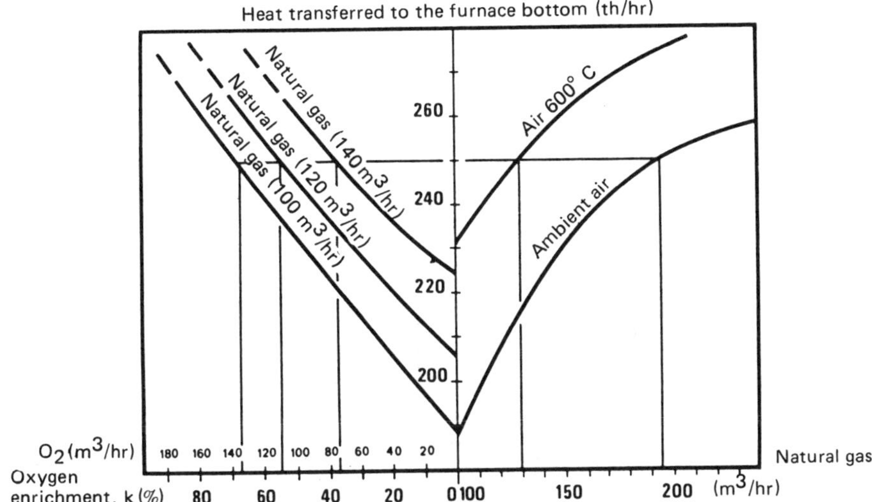

Fig. 5.14.   Relations between the heating capacity for axial flames with oxidizers of enriched oxygen, preheated air, or air.

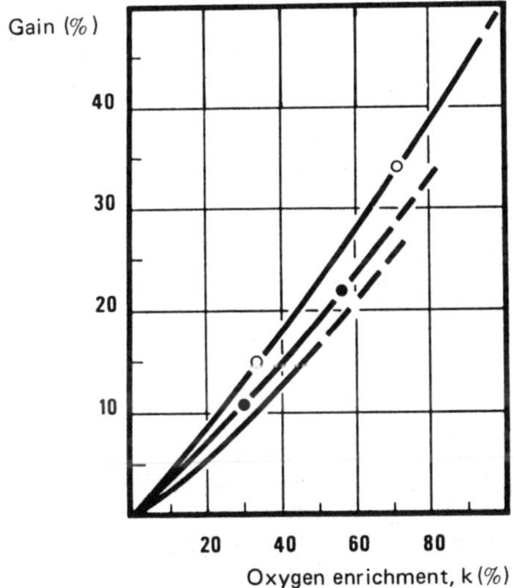

Fig. 5.15.   The gain in heat transfer as a function of oxygen enrichment, k.

——— o ——— Natural gas (100 m³/hr).

——— • ——— Natural gas (120 m³/hr).

——————— Natural gas (140 m³/hr).

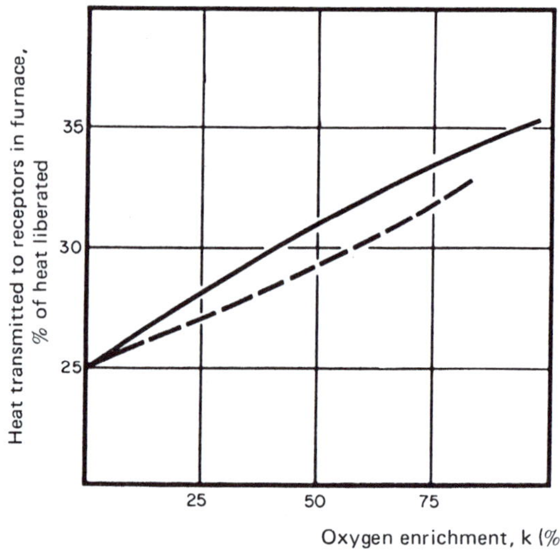

Fig. 5.16. The effects of oxygen additions to heat transmission within a furnace.

———— O_2 injected at the base of a natural gas jet.

- - - - O_2 added to the combustion air.

5.2.3.2. Wall flames

The burner in Fig. 5.17 can produce flat flames through radial injection of the natural gas perpendicular to the air flow and axial introduction of the oxygen. The flared refractory burner opening promotes the spread of gas flow and flame along the burner wall. A charge to be heated was placed parallel to the burner wall in a plane 1.90 m from the wall.

Different flames were studied for the same rate of flow of natural gas with relative oxygen supply, k, equal to 0, 20 and 40% and with preheated air at

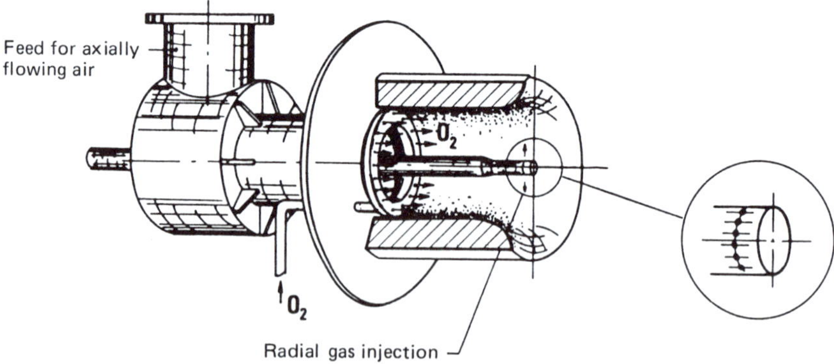

Fig. 5.17. Burner design for creating flat wall-flames.

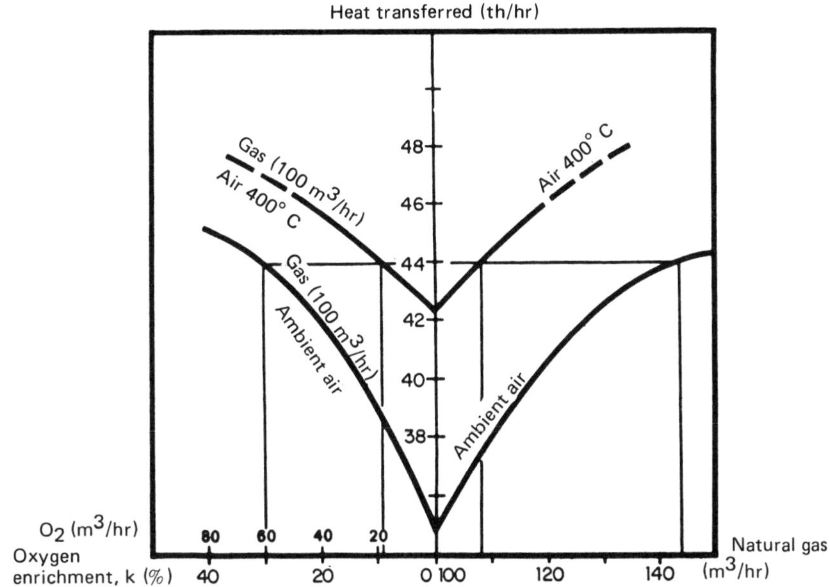

Fig. 5.18.   Heat transfer from a flat wall flame to a target, as a function of O_2 enrichment and combustion-air preheat.

400° C. A comparison of oxygen enrichment to preheating with respect to heat transfers to the charge led to the curves in Fig. 5.18, where the increases of heat transfers are due partly to the rise in the flame temperature which brings on a rise in the temperature of the refractory wall of the burner. Consequently, with this type of burner, the wall radiation appreciably contributes to the heat transfer. The following maximum temperatures were taken at the surface of the recessed burner opening during these tests (Table 5.5).

TABLE 5.5

EFFECTS OF AIR PREHEAT AND OXYGEN ENRICHMENT ON
TEMPERATURES AROUND A RECESSED BURNER OPENING

Oxidizer		Maximum temperature of wall surface (°C)
Temperature (°C)	Oxygen enrichment (k %)	
20	0	1,300
20	20	1,390
20	40	1,405
400	0	—
400	10	1,410
400	25	1,500

5.3. COMBUSTION IN TWO STAGES

The goal of this technique has been to modify the chemical nature of the fuel by means of a previous combustion in very rich mixtures. The research we discuss here used natural gas, although the oldest research (which has been applied industrially) used heavy liquid fuel and was carried out by the *Office Central de Chauffe Rationnelle (OCCR)* and the *Institut Français du Pétrole (IFP)* (Ref. 5.7).

Partial combustion with a natural gas rich in methane forms a hot flue gas containing, in addition to unreacted methane, CO_2, CO, hydrogen, higher homologues of methane, possibly unsaturated hydrocarbons, and particles of carbon. The reactions being exothermic overall, this flue gas constitutes new fuel existing at a high temperature. This new fuel is subsequently burned in the firebox. The high proportion of hydrogen plus the high temperature give it a high rate of combustion with an easily stabilized flame and thus possibilities for high flow rates at the exit of the burner, with consequent possibilities for creating strong kinetic impulse to the fluids entering into the combustion chamber. Thus, precombustion permits "hard" flames that are difficult to obtain from direct combustion of a natural gas principally made of methane.

Furthermore, precombustion gas is loaded with particles of soot that are only gradually burned in the principal firebox, so that they confer radiating properties to the firebox flame — radiation such as is not usually achieved with natural gas flames under a strong impulse.

This staged combustion can be easily used in a burner made up of a primary combustion chamber discharging either directly or through a restricted outlet into the combustion chamber of the furnace. The primary combustion is carried out with a high fuel-to-oxygen equivalence ratio of the order of between 2 and 4. The primary flame should be perfectly stable, and this stability can be obtained by fluid dynamic devices, such as the recessed burner opening, obstacle or swirl or by using preheated air or oxygen as primary oxidizer.

5.3.1. Two-stage combustion with air

In a study made at Toulouse by *GEFGN* (Ref. 5.8) the researchers used the experimental equipment shown in Figs. 5.19 and 5.20. The primary mixture of air and gas corresponds to an equivalence ratio of between 2 and 2.5. The primary air was introduced, either at ambient temperature, or preheated to 600° C. When the primary air was cold, stabilization was necessary. Fig. 5.20 shows stabilization by premixed flame. Once the primary air is preheated to more than 500° C, however, the primary flame is stable even in the absence of any device.

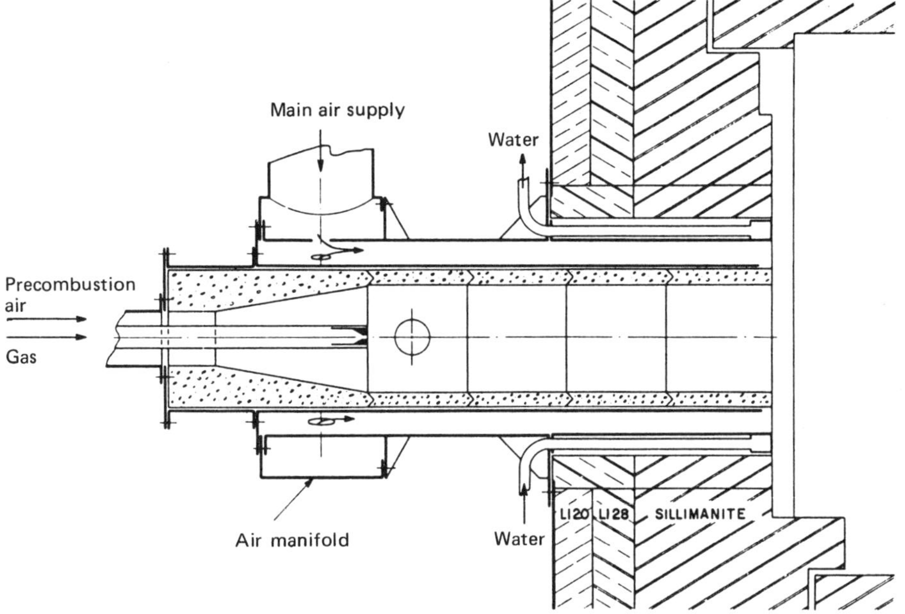

Fig. 5.19.   Burner for creating high-temperature flames through precombustion of fuel gas.

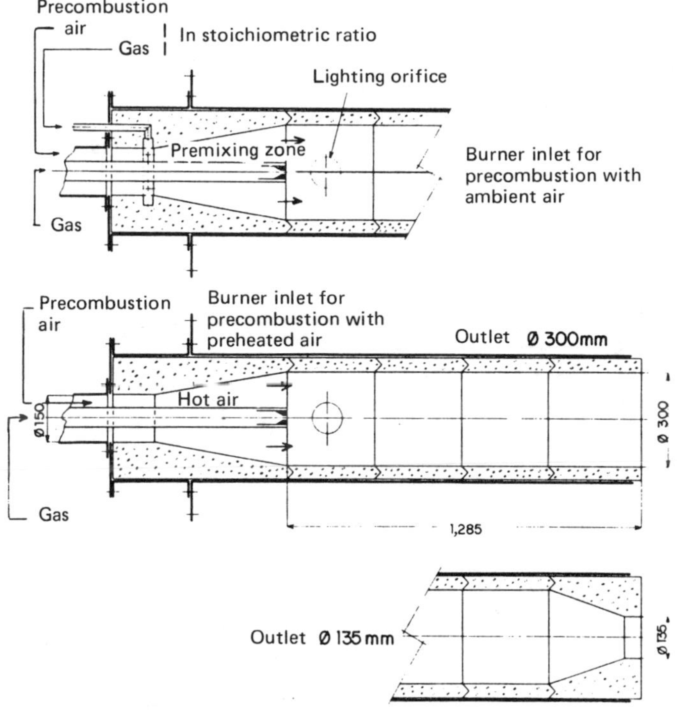

Fig. 5.20.   Dimensions and ignition arrangements for precombustion burner.

The composition of the gases resulting from the primary combustion varies with the equivalence ratio of the primary mixture and with the fluid-dynamic mixing conditions between air and gas in the precombustion chamber. Table 5.6 gives some results from Toulouse.

TABLE 5.6

PARTIALLY OXIDIZED FUEL GAS COMPOSITION
FROM PRECOMBUSTION WITH AIR

Gas composition. (vol. %)	CO_2	$C_n H_m$	O_2	CO	H_2	CH_4	H_2O (vol. % of dry gas)	N_2
Ambient air, equivalence ratio = 2.1	5	~ 1	~ 0.5	~ 6	8 - 12	6 - 10	0.14	By difference to 100%
Air at 600°C, equivalence ratio = 2.5	3 - 4	2	0.5	9 - 10	17 - 19	5 - 6	0.12	

The curves in Fig. 5.21 show the composition of the gases as a function of the rate of aeration (inverse of equivalence ratio of the primary mixture; also shown is the influence of preheating the primary air on the formation of soot. The extent of cracking is expressed in terms of the carbon found in the soot as a percentage of the total carbon supplied by the natural gas. These curves clearly

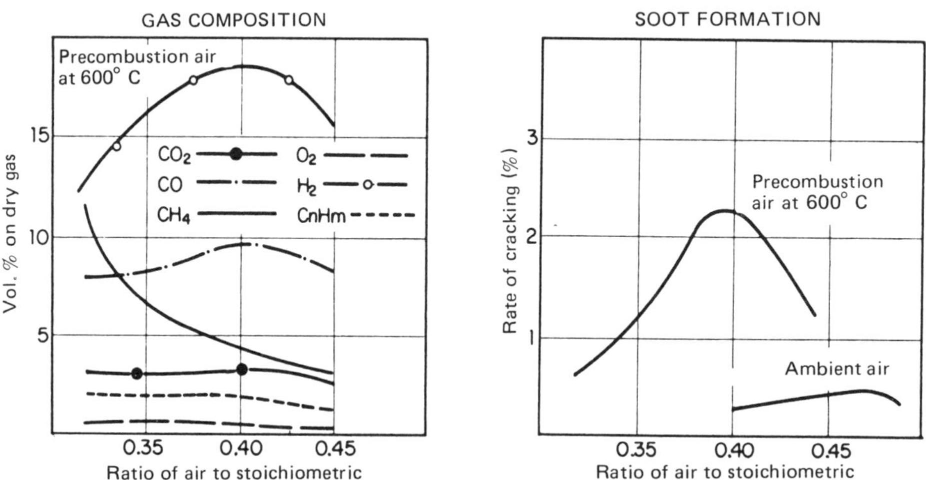

Fig. 5.21. Experimental composition curves to identify the optimum ratio of precombustion air to gas in a precombustion burner.

show that with the primary air preheated to $600°$ C, an optimum equivalence ratio of 2.5 leads simultaneously to a maximum formation of both hydrogen and soot. With a cold primary air, the formation of soot is about 1/8 that for air at $600°$ C at an equivalence ratio of about 2.2.

Secondary flames from combustion of the flue gas from partial oxidation of natural gas are shown to differ from a flame from the direct combustion of methane. The fuel for the secondary flames differs from natural gas in three respects:

(a) Its heat of combustion is low (on the order of 1.4 th/m³, compared to 9.67 th/m³ for gas from Lacq).

(b) Its hydrogen content is high (8-12% with cold primary air; 17-19% with primary air at $600°$ C).

(c) Its temperature is high (1,200 to $1,300°$ C) so that the precombustion gas supplies sensible heat to the firebox in an amount equivalent to 15 to 25 % of the total heat of the fuel.

The high temperature and high hydrogen content confer a high speed of combustion in air, and this permits easily obtained stable flames, even at the high flow rates characteristic of burners with a strong kinetic impulse. Moreover, the carbon particles carried by the partial-oxidation flue gas become incandescent in the secondary flame and are at the source of a continuous spectrum of radiation waves ranging from the visible to infrared.

The result of these two properties is that it is possible through two-stage combustion of natural gas to produce within an enclosure radiant flames with strong impulse such as are generally produced only with diffusion flames or by heating the combustion air to a temperature of more than $1,000°$ C.

5.3.2. Use of oxygen for precombustion

The use of pure oxygen as oxidizer for the first stage of two-stage combustion achieves a higher overall equivalence ratio than with air. Work done at *Air Liquide* and the *Institut Français du Pétrole* (Rcf. 5.9) has shown that an equivalence ratio equal to 6 could be used, that is, 3 volumes of methane for 1 volume of oxygen. The temperature resulting from this partial oxidation is $1,550°$ C. The burner shown in Fig. 5.22 permits introducing oxygen and gas at will along the axis of the burner barrel to feed a primary flame. By shutting off the axial intake of primary gas, a primary combustion can still be maintained at the equivalence ratio of the mixture of oxygen and gas introduced through the concentric nozzle. For the same overall rates of flow (at standard conditions) of oxygen (20 m³/hr) and of gas (60 m³/hr) the products resulting from such a primary combustion are considerably different as is shown by the numbers in Table 5.7.

Dividing the natural gas feed into axial and radial components thus affords a secondary gas that is much richer in hydrogen and in carbon monoxide, by

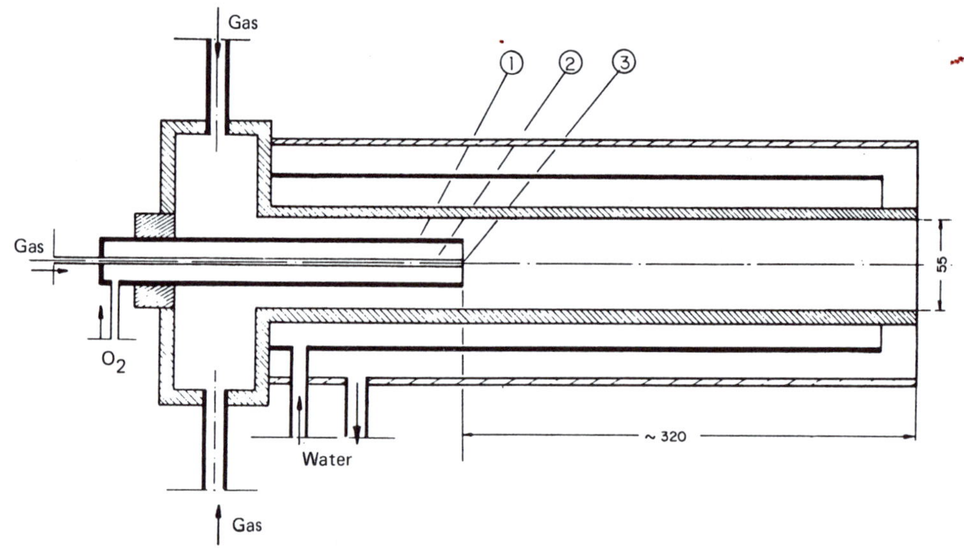

Fig. 5.22. Precombustion burner using oxygen as partial-oxidizer. (Courtesy l'*Air Liquide*).

TABLE 5.7

PARTIAL-OXIDATION PRODUCTS OF THE PRECOMBUSTION
BURNER IN FIG. 5.22

	CH_4 (%)	CO_2 (%)	O_2 (%)	H_2 (%)	CO (%)	$C_n H_m$ (%)
Without axial gas jet	40	14	19	12.5	8.5	6
With axial gas jet	31.5	3.5	1	35.5	23.5	5

improving the mixture of the gas and the oxygen. The effectiveness of the methane conversion through combustion with oxygen depends closely on the method of mixing and the time available for the reactions. The results of these tests were confirmed at *Shell International Petroleum Co., Ltd.* on two-stage burners with the purpose of producing luminous flames from natural gas (Ref. 5.10). In the Shell experiments, the primary combustion was developed in a chamber where the flame from an almost stoichiometric mixture of gas and oxidizer was struck perpendicularly by jets of natural gas. The oxidizer of the primary flame was either preheated air, or oxygen-enriched air. By this double combustion, the Shell researchers achieved secondary flames that were comparable to heavy fuel oil flames from the point of view of radiation.

It should be noted that, in certain equipment, the formation of carbon leads to soot deposits on the walls of the primary reactor, but this deposit can be dissipated by a temporary air purge, which momentarily modifies the primary combustion and consequently the properties of the secondary flame inside the heating enclosure.

In summary, two-stage combustion of natural gas permits modifying the nature of the fuel introduced into the firebox in such a way as to produce flames with qualities appropriate to various desired heat applications. This procedure requires that the modified fuel thus created be used immediately before losing part of its sensible heat, which represents a large proportion of the chemical energy supplied by the original fuel.

5.4. CONCLUSION

Taking into account the relative weight of oxidizer to fuel used in combustion, any modification either to the temperature or the oxygen-content of the oxidizer will be reflected in the nature of the flames and their temperature, and consequently in the heat transfer between flame and charge.

To a certain point, preheating and oxygen enrichment have comparable effects, even though the preheating air temperatures in the described experiments were insufficient for general conclusions. Nevertheless, the data show clearly that if pure oxygen is available it is much more effective to use it in that form, without premixing with air. Oxygen supplied to a flame at a point still rich with fuel creates a local high temperature that leads to more active heat exchange, either by radiation or by convection.

Furthermore, the supply of oxygen can even be delivered in the immediate vicinity of the charge to be heated, at least if the thermal stability of this charge permits it, and the consequent reaction on the surface of the charge of ions from the flame leads to the phenomenon of live convection described by Veron. This is a very interesting method of heat transfer offering an added feature related to the use of oxygen — a feature that is intermediate between flame and plasma.

REFERENCES

5.1   VERON, M., ROCARD, Y. — "La convection vive." *Bulletin technique Babcock et Wilcox*, Paris, 1948.

5.2   *IFRF*. — Doc. F17/a/18, 14 July 1975.

5.3   *IFRF*. — Doc. F31/a/32 by KISSEL, R.R. (Oct. 1959).

5.4   *IFRF*. — Doc. F31/a/41[2] by CHEDAILLE, J., MINEUR, J.M. (May 1967).

5.5   *IFRF*. — Doc. F35/a/6 by VIZIOZ, J.P., LOWES, T.M. (Dec. 1971).

5.6 *GEFGN.* – Report GN 13 by MARQUE, D. (Dec. 1973).

5.7 *OCCR.* – "Gazogène à fuel". Instruction, June 1950. *IFP*, Report 24 June 1949 by GI-VAUDON.

5.8 *GEFGN.* – Report GN 7 by ROGIER, J. (Nov. 1968).

5.9 IVERNEL, A.B., PERTHUIS, E. – *Congrès International de Chimie Industrielle*, Brussels, Sept. 1966.

5.10 *Shell International Petroleum Co. Ltd.* – Reports NG/69/3 and NG/69/8.

6

premixed combustion

6.1. GENERAL NATURE OF PREMIXED COMBUSTION

The preceding chapters have been devoted to flames resulting from a progressive mixing of fuel and oxidizer. In certain applications the mixing could be achieved wholly or in part ahead of the enclosure in which the combustion takes place.

A volume filled with a fuel mixture presents a certain danger. The flashback of flames to automotive carburetors is well known to people who have tried to prime an engine through the carburetor for cold starts, and if engine flashbacks do not usually involve unhappy consequences, they illustrate an important problem related to handling such mixtures. Another example well known in laboratories is the flashback of fire to the nozzles of Bunsen burners fed with town gas rich in hydrogen.

Industrial heaters, which involve very large rates of flow, generally reduce the risk of uncontrolled pre-ignition through equipment designs that restrict the volume of premixed fuel and oxidizer.

We have just talked about combustion mixtures that assume the fuel to be in the form of gas or vapor, including mixtures of a mist of very fine fuel droplets in suspension in the combustion air. Experimental evidence has shown that the combustion of a well homogenized mixture of air and fuel does not depend on the fluid movements that bring about in diffusion flames contact between the air and the fuel. Nevertheless, the fluid movements that supply the burner and the combustion chamber with fresh feed are usually turbulent, and this turbulence, pre-existing ahead of the flame, modifies its propagation. Because of the turbulence, therefore, premixed combustion depends appreciably on the movement of the fluids.

Premixed combustion also depends on chemical variables: the nature of the fuel, its concentration in the mixture, the nature of the oxidizer, the concentration of oxygen in the mixture, the nature of the accompanying inerts that can be not only nitrogen of the air but also combustion products recycled to the fresh feed or compounds introduced to provoke reactions favorable to certain applications.

Combustion is also sensitive to the temperature of the feed mixture and to a lesser degree on its pressure.

Depending on the nature of the fuel-air mixture, an increase in pressure can either increase or decrease the speed of propagation of the flame (Ref. 6.1). In the majority of the practical applications, combustion takes place at atmospheric pressure or close to it, consequently this variable is of little importance.

The temperature of the mixture acts directly on the speed of combustion. For a mixture at rest (Ref. 6.1) the rate of combustion is approximately proportional to the absolute temperature of the mixed gas. Here, then, is a variable whose importance can be great, but preheating a fuel-oxidizer mixture is rarely used because excessive preheating might favor slow pre-combustion oxidation reactions or lead to autoignition and dangerous live combustion upstream of the heating unit. These considerations usually preclude an interesting method for recovering the heat lost in the flue gas. The flue gas can, for example, be used to preheat the combustion air before its mixture with the fuel, but such preheating must be limited to avoid any premature initiation of live combustion.

6.2. STABILITY OF FLAMES FROM PREMIXED FUELS

Flames from premixed fuels can be stabilized as diffusion flames through a correct organization of flow for the mixture.

Nevertheless, stabilization can often be obtained by acting on the chemical or thermal factors on which the rate of combustion depends, understanding that precautions must always be taken to have the combustion of a fraction of the flowing mixture occur near the burner in order to stabilize the flame. The speed of combustion must be equal and opposite to the speed of flow all along the perpendicular to the flame front. This condition attained in a Bunsen burner permitted Gouy, in 1879, to calculate the speed of fundamental laminar combustion by measuring the surface of the flame front outlined by the blue cone and the overall measurement of the gaseous rate of flow in the burner.

We shall examine in succession the influence of the chemistry of the fuel mixture achieved upstream of the combustion chamber, the influence of the temperature of that mixture, and the influence of fluid dynamic factors, such as rate of flow and type of flow established.

6.2.1. Influence of equivalence ratio of the fuel-oxidizer mixture on laminar speed of combustion

Table 2.5 of Chapter 2 gives values of the fundamental speed of combustion as a function of the equivalence ratio of the mixture. For most fuels, the maximum speed occurs when the equivalence ratio is about 1. Only hydrogen and to a certain extent acetylene behave differently by presenting a maximum combustion speed for very rich mixtures.

At the burner outlet friction between the gas and the walls creates a velocity gradient through the fluid, and the flame is stabilized in the vicinity of the burner, if there is a place in this velocity gradient where the speed of deflagration is equal to the speed of flow. Because the flow is turbulent in industrial burners, the stable speed of deflagration corresponds to a velocity of flow in the turbulent regime (see Chapter 2). The point at which the two speeds are equal is displaced as a function of the rate of flow, but the wall friction (by imposing a gradient of speeds ranging from 0 at the wall to a maximum in the flow stream) permits flame stabilization at various rates of flow within a given section.

For low rates of flow, the speed of turbulent deflagration can become higher than the maximum speed in the flow stream, and the flame travels back upstream into the burner, as happens when the flame of a Bunsen burner flashes back to the nozzle.

For high rates of flow, the velocities of turbulent deflagration and of flow can be equal only in the immediate vicinity of the wall. But in this case the wall promotes a recombination of free radicals occurring in the chain reaction, and the combustion is found to be inhibited. The flame detaches itself from the burner and is blown out.

These two limits of flashback and blowout depend on the chemical nature of the premixture, particularly, the concentrations of fuel and oxygen; between them, a stable flame attaches itself to the nozzle of the burner (Fig. 6.1).

The proximity of the flame to the walls often heats them up considerably unless they are cooled, and, when hot, the walls can participate in heating the premix before its arrival at the flame front. Such a preheat increases the speed of turbulent deflagration and hence enlarges somewhat the limits of stability. The result is a progressive development of flame front during the start-up of a burner before a stable regime can be established.

If the wall of a burner is made of refractory, the rise in the observed temperature can be considerable, so that the wall radiates in front of it. This effect is systematically used in a wide range of industrial burners that promote radiant heat transfer from the refractory.

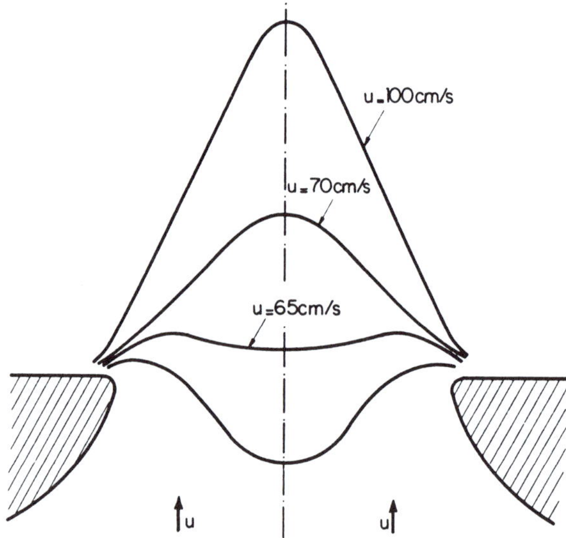

Fig. 6.1. Blue flames at the nose of a burner : their shapes reflect
           the fluid flowing velocities across the nozzle in relation to
           the speed of turbulent deflagration.
           (From : *La Combustion des gaz et les flammes. ATG*).

6.2.2. Influence of preheating the premix

First of all, preheating increases the velocity of flow in the outlet section of
the burner, due to the expansion of the gases. Also, preheating introduces energy
that raises the temperature of combustion, and the concentration of free radicals
formed in the flame depends on this temperature. Thus, any increase in this
free-radical concentration, even slight, tends to increase the speed of the
oxidation reactions and hence the fundamental speed of combustion.

There are not many direct measurements of the effect of preheating because
of the difficulties due to the occurrence of reactions in the premixture as soon
as the temperature gets high. Table 6.1 (Ref. 6.1) shows some data taken from
different publications.

TABLE 6.1

THE EFFECTS OF TEMPERATURE ON THE STANDARD VELOCITY
OF FLAMES FROM TYPICAL FUEL GASES
(Stoichiometric mixtures)

Fuel mixture	Temperature (K)		Velocity (cm/s)
	Initial	Final	
CH_4/O_2 [1] [2] containing 62.5% N_2	346	2,422	85
	383	2,448	89
	463	2,466	97
	539	2,492	106
C_2H_2/O_2 [2] containing 79% N_2	322	2,174	92.5
	357	2,193	97.5
	413	2,211	105.5
	509	2,238	119
CH_4/N_2O [3] containing 40% N_2	338	2,512	49
	414	2,542	54
	499	2,573	59.5
	575	2,589	64.5
	633	2,612	68

Fuel/oxygen ratios are stoichiometric. The velocities were measured by a Schlieren optical system (Section 2.3.2.4). The percentage nitrogen is overall. Flame temperatures were determined by the observation of sodium rays (5,890 Å).

[1] Vandenabeele, H., Corbeels, R., Van Tiggelen, A. – *Combustion and Flame*, 1960, 4, 253.
[2] Van Tiggelen, A., Vandenabeele, H. – "Kinetical parameters in premixed laminar flames". Techn. Note No. 2, Contract AF 61 (514)–1245; Monitoring Agency, doc. No. AFOSR-TN-59-1149; Astia doc. No. 234-580, 1960, p. 25.
[3] Van Tiggelen, A., Vandenabeele, H., d'Olieslager, J. – "New development of a kinetical study of flames". Tech. Note No. 2; Contract AF 61 (514)–1245; Monitoring Agency, doc. No. AFOSR-TN-60-1210; Astia doc. No. 254-500, 1961, p. 19.

6.2.3. Influence of pressure of the premix

The use of fireboxes operating under high pressures of 25 to 30 atm has been suggested by different researchers, and although these conditions are exceptional in practice, it is interesting to note how the fundamental speed of combustion varies with pressure.

The published data generally agree, relating the speed of fundamental combustion V_{n_p}, at pressure p, to the speed V_{n_a}, at atmospheric pressure p_a, by the equation:

$$\frac{V_{n_a}}{V_{n_p}} = \left(\frac{p_a}{p}\right)^n \tag{6.1}$$

Exponent *n* varies according to the nature of the flames. It can have values between − 0.3 and + 0.3 for the flames of hydrocarbons. Figure 6.2 (Ref. 6.2) shows that in an important area of speed V_n, attainable with hydrocarbons, the influence of pressure is very slight. This conclusion is satisfactory even though

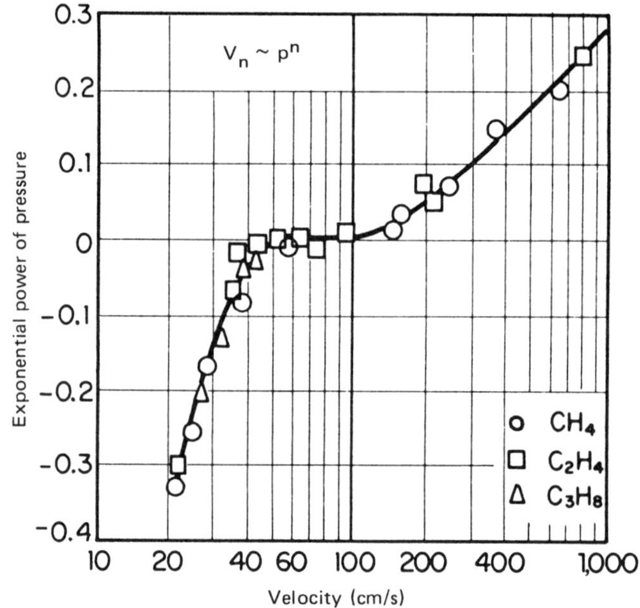

Fig. 6.2. Effect of pressure on the standard speed of propagation of combustion.
(From: Lewis, B. *Selected Combustion Problems. AGARD Combustion Colloq., Butterworths,* London, 1954).

the experiments from which it is taken have some uncertainty. A more recent study (Ref. 6.3) uses a rapid compression machine (Fig. 6.3) to compress and preheat known mixtures. As compression raises both pressure and temperature of the fuel mixture, the speed of combustion is measured during the delay of autoignition of the mixture. The authors were able to separate the influence of temperature and pressure for stoichiometric mixtures of propane and air (Fig. 6.4), as well as normal heptane and air (Fig. 6.5). Thus they obtained results that agree well with the formula given by Van Tiggelen (Ref. 6.1):

$$\frac{\overline{V}_{n_i}}{\overline{V}_{n_0}} = \frac{T_i \, T_{m_0}^{1/2}}{T_0 \, T_{m_i}^{1/2}} \, \exp\left(\frac{E}{2R}\left(\frac{1}{T_{m_0}} - \frac{1}{T_{m_i}}\right)\right) \tag{6.2}$$

with:

$$T_{m_0} = T_0 + \sigma_{m_0}(T_{f_0} - T_0)$$

$$T_{m_i} = T_i + \sigma_{m_i}(T_{f_i} - T_i)$$

where

V_{n_i} = fundamental speed for an initial temperature, T_i, of the mixture,

V_{n_0} = standard fundamental speed ($T_0 = 298$ K, $p_0 = 76$ cmHg)

σ_m = relative speed increase for the reaction corresponding to an average reaction speed, subscript i is for T_i the initial temperature; subscript 0 is for $T_0 = 298$ K,

E = overall energy of activation found to be close to 38 kcal/mol (Figs. 6.4 and 6.5),

T_{f_0} = temperature of the flame that is propagated in a mixture with the initial temperature T_0,

T_{m_0} = average temperature in the mixture with a flame that is propagated from initial temperature T_0,

$T_{f_i,\ m_i}$ = subscripts indicating an initial mixture with temperature T_i.

In summary, Eqs. (6.1) and (6.2) proposed by A. Van Tiggelen satisfactorily represent the influence of the pressure and initial temperature of the mixture on the normal speed of the flame. Stability depends on this flame characteristic and thus on these conditions.

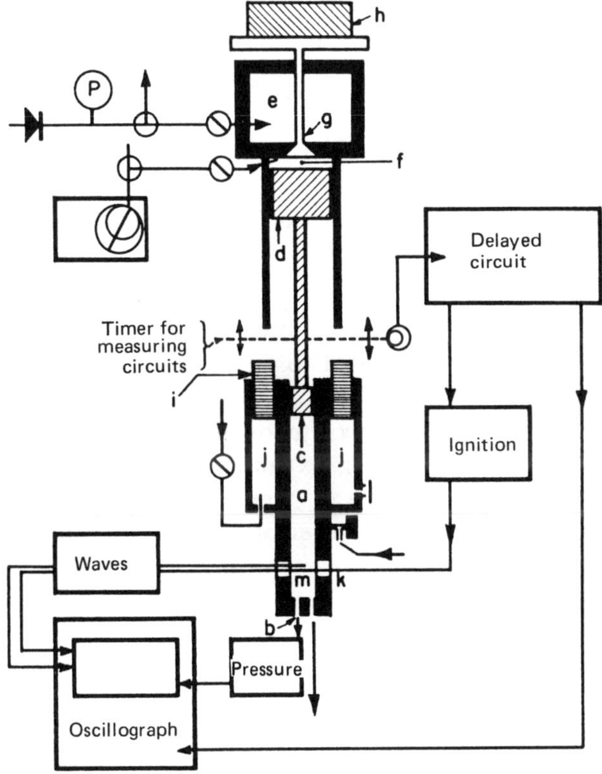

Fig. 6.3.   Diagram for rapid compression machine to measure speeds of combustion.

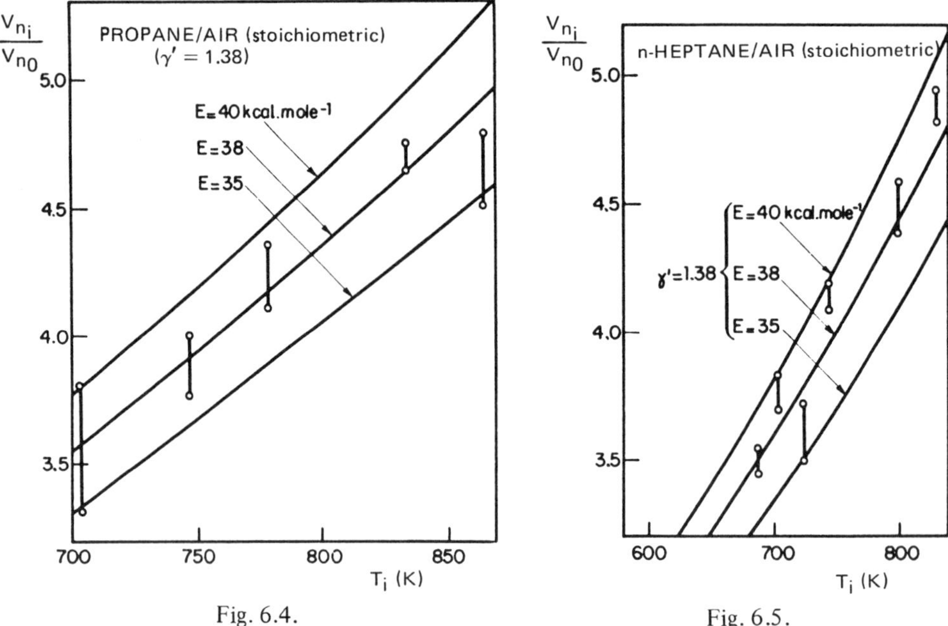

Fig. 6.4.   Fig. 6.5.

Figs. 6.4 and 6.5. Effects of temperature on the normal velocity of
propagation of flame (see Eq. 6.2. for terms and
relations).

6.3. USE OF DOMESTIC BURNERS WITH PREMIXED FUEL

Domestic gas burners, which are fed premixed fuel for the most part, are also
called blue-flame burners, as opposed to diffusion-flame burners, which are also
called white-flame burners.

In domestic appliances, the mixture of gas and oxidizing air is obtained
through aspiration of the air by an expansion of the gas which arrives at the
burner under a light pressure. The gas draws only a part of the necessary air,
and combustion terminates at the level of the flame with the addition of air
from the atmosphere.

6.3.1. Operating limits of domestic burners with premix

For a given burner fed with a gas of constant composition and pressure, a
region of operating stability can be established similar to Fig. 6.6. This diagram
has as its abscissa the rate of primary aeration, n defined by:

$$n = \frac{\text{volume of primary air}}{\text{volume of stoichiometric air}} = \frac{\mathcal{V}_A}{\mathcal{V}_a}$$

where $\mathcal{V}_a$, represents the oxygen requirements of the fuel.

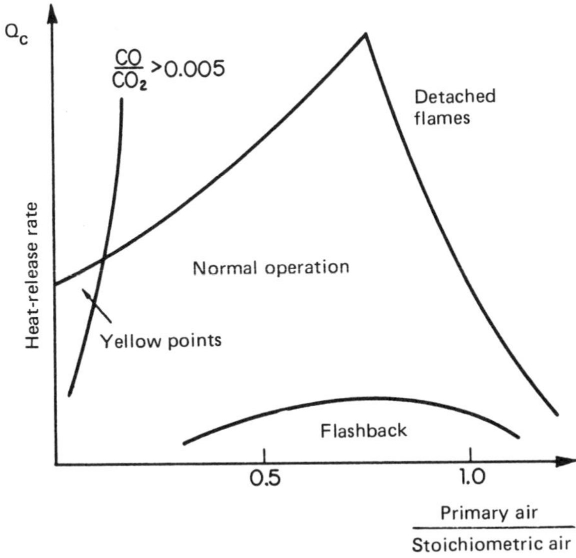

Fig. 6.6. Operating regions for domestic gas burners.

The ordinate is the rate of heat flow, Q_c, the product of the rate of flow of gas multiplied by its heat of combustion. For any given burner and outlet section, Q_c is proportional to the outlet speed of the gas mixture.

In diagram 6.6, a first curve on the right separates the zone of normal operation from the zone where the flames become detached because of high outlet velocities at a rate of aeration close to one. A second curve along the lower portion corresponds to low outlet velocities that permit the flame to travel back against the current to the gas injector. Two other limits (Fig. 6.6) correspond to, first, low n-values for primary aeration and the consequent appearance of yellow flashes in the flame due to particles of soot that can cause clogging, and second, a limit where the ratio of carbon monoxide to carbon dioxide $\frac{(CO)}{(CO_2)}$ in the fumes goes over a threshold of toxicity fixed by law in France at 0.005.

Figure 6.6 shows in addition that the rate of primary aeration of 0.7 permits great variations of rate of flow from the burner without leaving the stable zone. Actually, for many domestic gas appliances the rate of primary aeration is regulated close to 0.7, in order to give the maximum flexibility to the burner.

6.3.2. Methods for stabilizing flames of premix gas burners in domestic appliances

The geometry of a burner plays an important role in the flexibility of its operation, and the choice of an appropriate geometry has been the subject of much research in the gas industry.

Without attempting to enumerate all the equipment tested let us note by way of summary that most methods act on the orifices: their number, their outlet arrangements, their depth, and their spacing. The tendency to blowout is reduced by an increase in size of the orifice cluster, by an increase in the thickness of the wall, and by a reduction in the spacing between orifices (at the limit, slot burners are most stable). The tendency to flashback is reduced by increasing the thickness of the wall and by reducing the size of the orifice cluster. Figure 6.7 shows that the speeds at which flashback occurs increase with

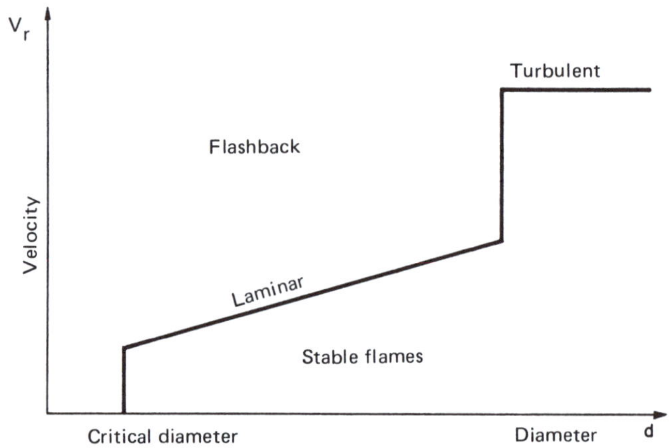

Fig. 6.7. The related effects of orifice diameter and velocity through the orifice on flashback in a domestic burner. Critical diameter depends on the burning velocity of the fuel-oxidizer mixture.

the diameter of the orifice when the flow of the gaseous mixture is laminar but is independent of diameter when the flow is turbulent. The critical diameter depends directly on the speed of combustion of the fuel-oxidizer mixture, which is to say, on the chemical nature of the fuel, on the equivalence ratio of the mixture, and on the temperature. It is accepted that for domestic use the critical diameter is 3 mm with natural gas and 1 mm for manufactured gas with a high hydrogen content.

6.3.3.   Domestic burners with pilot flames

Figure 6.8 represents a burner in which the throttling of a part of the gaseous mixture permits feeding the slots at a low rate while maintaining a constant flame. This flame, a source of heat and of active combustion intermediates, serves as a continuous source of ignition for the principal jets. Such a pilot flame affords a stable flame over a wide range of flow rates for a great variety of fuel gases. Such burners are used in a large number of domestic gas appliances. The flexibility of regulating is satisfactory but these burners do tend to be noisy; the constant re-ignition of the jets of fuel mixture by the pilot flame occurs with a series of small deflagrations, the source of emission of sound.

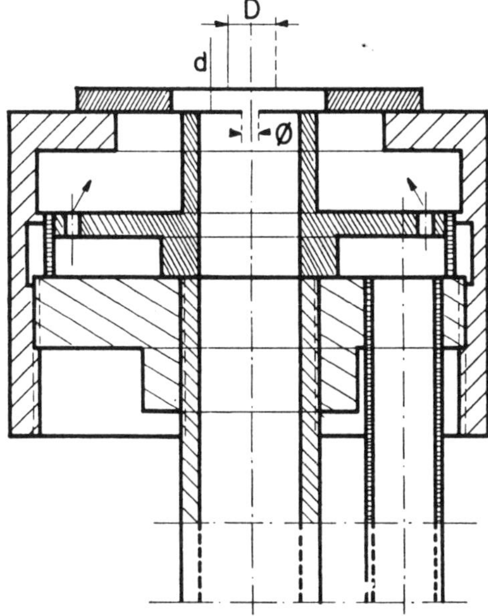

Fig. 6.8.   Arrangement for a burner with a pilot flame.

6.4.   INTERCHANGEABILITY OF FUEL GASES

Many gases are offered for domestic consumption, including manufactured gases whose composition varies with the manufacture, natural gases which are principally methane with some heavier hydrocarbons and inerts (CO_2, N_2, etc.) and liquefied petroleum gas which may be principally propane or butane.

The suppliers of gas appliances must therefore furnish equipment that can accommodate different conditions of use imposed by a specific consumer. They must know the conditions under which a given gas can be substituted for a different one without unduly upsetting the operation of the appliance. Numerous studies have been made on this subject, particularly those by Delbourg and his colleagues (Refs. 6.5, 6.6 and 6.7).

6.4.1. Definition of interchangeability

Two gases are completely interchangeable if they give identical flames from a given burner without modifying the way it is regulated or changing its geometry. This would be the case for a burner with air induced by the gas; the substitution of one gas for another completely interchangeable gas maintains:

(a)　The same rate of heat flow.

(b)　The same rate of primary aeration.

(c)　The same concentration-ratio of carbon oxides $\dfrac{(CO)}{(CO_2)}$ in the combustion products plus the same occurrence of yellow points.

(d)　The same stability of the flame with respect to blowout or flashback.

(e)　The same facility for mutual lighting by various parts of the burner.

(f)　The same efficiency of combustion.

(g)　The same oxidizing power and the same dew point of the combustion products.

Actually, all of these conditions are never completely filled. In practice, a certain tolerance is allowed and two gases may be considered interchangeable if they have the same Wobbe index and the same combustion index.

6.4.2. Wobbe index (W)

The rate of heat flow from a burner, Q_c, can be expressed as:

$$Q_c = P_c m s \sqrt{\frac{2(h-h_0)}{a\delta} \frac{B}{760} \frac{273}{t+273}} \tag{6.4}$$

where

P_c = heat of combustion per unit volume of gas,

m = coefficient, determined as the ratio of the effective cross-section of the injector to the actual geometric section,

s = cross-sectional area of the injector outlet,

h = pressure upstream of the orifice,

h_0 = pressure downstream of the orifice,
a = density of the air,
δ = specific gravity of the gas relative to air,
B = absolute pressure of the gas (mmHg),
t = temperature (° C) upstream of the orifice.

If for a given orifice cluster, the pressure differential $(h - h_0)$, B, and t are all constant, formula (6.4) can be reduced to:

$$Q_c = K \frac{P_c}{\sqrt{\delta}} \qquad (6.4 \text{ bis})$$

From this formula, the heat flux of the burner is proportional to the ratio:

$$\frac{P_c}{\sqrt{\delta}} = W \qquad (6.5)$$

which is known as the Wobbe index.

Accordingly, two Wobbe indexes — one for the high heating value (HHV) and one for the low heating value (LHV) — can be defined as W_{high} and W_{low} respectively.

In a burner to which air is aspirated by the gas, the induced air flow is calculated by the equation:

$$\left(1 + \frac{Q_A}{Q_g}\right) \left(1 + \frac{1}{\delta} \frac{Q_A}{Q_g}\right) = k_1 \frac{S_m}{S} \qquad (6.6)$$

where
Q_A = volumetric flow rate of induced air,
Q_g = volumetric flow rate of the gas,
δ = density of the gas,
k_1 = constant, characteristic of the burner,
S_m = cross-sectional area of the orifice admitting the induced air,
S = cross-sectional area of the injector.

If the flow rate of induced air is expressed per unit (1 m³) of the gas, or as Q'_A, and if the quantity of 1 is dropped from $\left(1 + \frac{Q'_A}{\delta}\right)$ and $(1 + Q'_A)$, then expression (6.6) reduces to:

$$Q'_A = k_2 \sqrt{\delta} \qquad (6.7)$$

Now, the unit rate of aeration, n, can be described as:

$$n = \frac{Q'_a}{\mathcal{V}_a} \qquad (6.8)$$

where $\mathcal{V}_a$ = oxidizing requirement of the gas (in m³ of air/m³ of gas).

Accordingly, Eqs. (6.7) and (6.8) can be combined to give:

$$n \mathcal{V}_a = k_2 \sqrt{\delta} \tag{6.9}$$

Typical fuel gases, except for gases with very low heating values, require about 1 m^3 of oxidizing air for each 1,000 kcal of heat released. Assuming this typical value, formula (6.4 bis) can be written as:

$$Q_c = K \, W \cong 1,000 \, Q'_a$$

and allowing for Eq. (6.8) $K \, W$ can be expressed as:

$$K \, W \cong 1,000 \, Q'_A = 1,000 \, n \mathcal{V}_a$$

which leads to an empirical correlation between the Wobbe number, W, and rate of aeration, n as:

$$n = k_3 \, W \tag{6.10}$$

The Wobbe index thus determines the limits in the rate of aeration; it affects the height of the blue flame cones, which actually depend on the rate of combustion and therefore, the equivalence ratio of the air-gas mixture.

If a burner is fed with a gas having a heat of combustion radically different from that of the initial gas such as, say, $W = 2$ instead of $W = 1$, while conserving the same rate of heat flow, the rate of primary aeration, n, is maintained while the burner's orifice is changed in accordance with the relationship:

$$\frac{h_1}{h_2} = \frac{\left(P_{c_2}\right)^2}{\left(P_{c_1}\right)^2} \, 2\left(\frac{\mathcal{V}_{a_1}}{\mathcal{V}_{a_2}}\right)^4 \frac{\delta_2}{\delta_1} \tag{6.11}$$

where subscripts 1 and 2 indicate conditions while firing the initial and the alternate gases respectively.

Applying the empirical law of 1,000 kcal, this ratio becomes:

$$\frac{h_1}{h_2} \cong \left(\frac{W_1}{W_2}\right)^2 \tag{6.12}$$

which can be put in the form:

$$\frac{\sqrt{h}}{W} = \text{Const.}$$

called the "gas module" introduced by Van der Linden.

This Van der Linden module predicts that a burner fed under a pressure of 15 to 25 mbars with natural gas (P_c of 9,000 to 10,000 kcal/m³) must be fed at 5 to 7 mbars with town gas ($P_c = 4,500$ kcal/m³) and at 40 to 60 mbars with liquefied petroleum gas (P_c of 22,000 to 28,000 kcal/m³) in order to maintain a constant rate of heat flow.

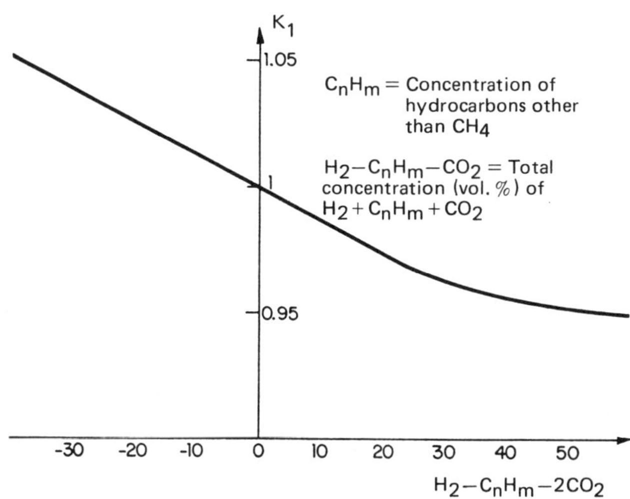

Fig. 6.9. Values of K_1 for Eq. (6.13) applied to manufactured fuel gases, as a function of hydrocarbon content of the gas.

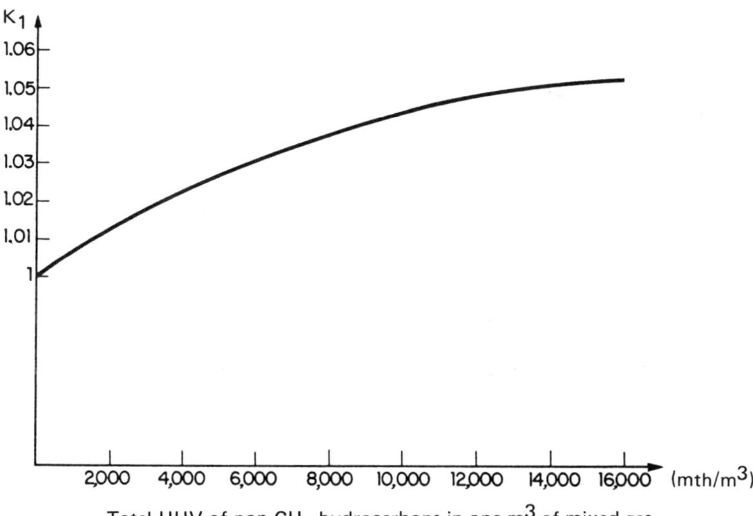

Fig. 6.10. Values for K_1 for Eq. (6.13) applied to natural gases, as a function of the non-CH$_4$ hydrocarbon content of the gas.

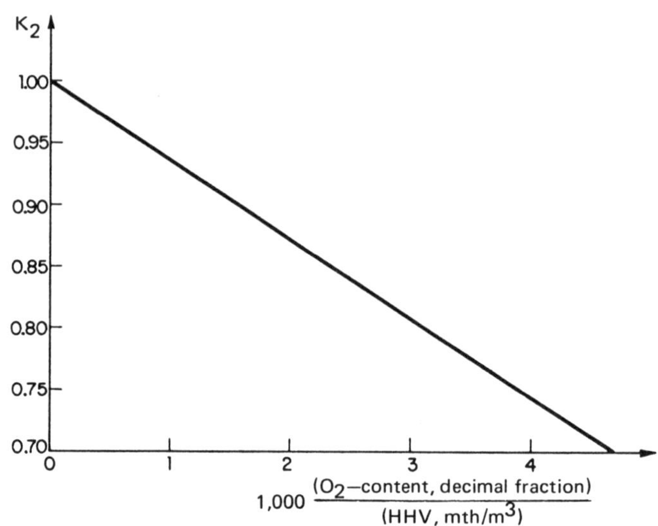

Fig. 6.11.   Values of K_2 for Eq. (6.13) applied to manufactured fuel gases, as a function of the O_2-content and HHV of the gas.

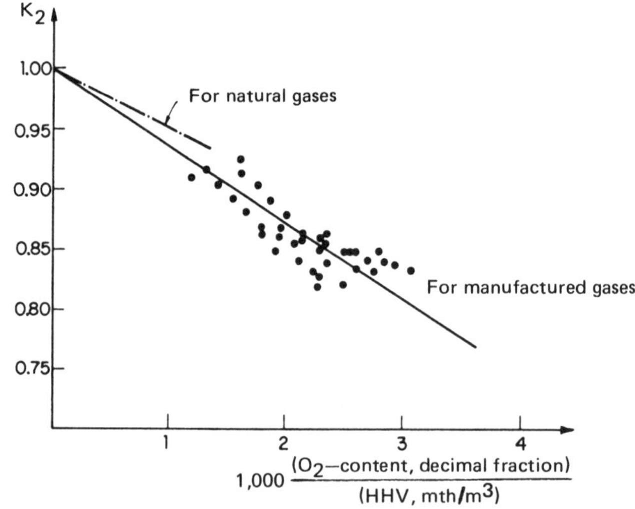

Fig. 6.12.   Values of K_2 for Eq. (6.13) applied to manufactured and natural gases, as a function of O_2-content and HHV of the gas.

It can be seen from these relations that only gases with relatively close heating values can be considered interchangeable. For this reason gas suppliers and appliance manufacturers generally group fuel gases into three large families according to their changeability:

(1) Town gas with a high heating value of combustion between 4,000 and 6,500 kcal/m^3.

(2) Natural gases with a high methane content and a heat of combustion around 10,000 kcal/m^3.

(3) Liquefied petroleum gas (butane and propane) with a heat of combustion ranging from 22,000 to 28,000 kcal/m^3.

Refining the analysis of these phenomena, Delbourg has shown that, for dealing with interchangeability of gases, a corrected Wobbe index, W' is of interest:

$$W' = K_1 K_2 W \qquad (6.13)$$

where

K_1 = correction for the rate of flow, taking into account the viscosity of the gases as that affects the value of the coefficient, m, in Eq. 6.4,

K_2 = correction coefficient for the concentrations of O_2, CO and CO_2 in the gas: it takes into account, principally, the variations in the actual rates of aeration of the flames from oxygen-containing gas, and it affects the height of the blue cone.

Values of the parameter K_1 can be found for the first family of gases in Fig. 6.9 and for the second family in Fig. 6.10. Values of K_2 for gases of the first and second family are found in Figs. 6.11 and 6.12, respectively.

It is suitable to note that the heating capacity of the appliances depends only on W'', which Delbourg calls the half-corrected Wobbe index:

$$W'' = K_1 W$$

6.4.3. Monitoring burners

Given the number of variables to watch out for, the quality of gases delivered to consumers must be monitored with a test burner. The apparatus, usually of the Bunsen burner type, furnishes reference points directly related to the characteristics of the gas being fed.

Delbourg designed a control burner illustrated in Fig. 6.13. It consists essentially of a burner with two laminar diffusion flames, a burner with premix, valves and a manometer. In operating this burner, one seeks the gas pressure h_n necessary for the two laminar flames to attain a given height on a graduated scale (A in Fig. 6.13). As a first approximation, this flame height depends only on the rate of heat flow achieved. In the final adjustment, it also depends on the diffusion of the gas in air; by means of a tube above its orifice (A in Fig. 6.13) one of the two laminar flames is transformed into a premix flame, and the necessary time is measured for it to evaporate 20 cm³ of water. An empirical curve that goes with the apparatus gives h_n/h as a function of this time. A statistical study has shown that, if the white flames were stable under pressure h_n and if $h_n/h = 1.5$, the operation of domestic appliances with white flames is satisfactory. The premix burner (B in Fig. 6.13) is then fed under pressure h_n. This burner consists essentially of an air intake whose opening is calibrated, a water jacket to remove heat from the metallic parts, and 30 orifices of 2.5 mm diameter distributed in a crown-form about a central tube 120 mm long and 8 mm in diameter whose blue cone of flame is measured by a graduated scale. A chart gives the opening position of the air as a function of h_n so that the rate of primary aeration is 0.8. A tested gas will not disturb the operation of domestic appliances from the instant that the premix flames are stable with

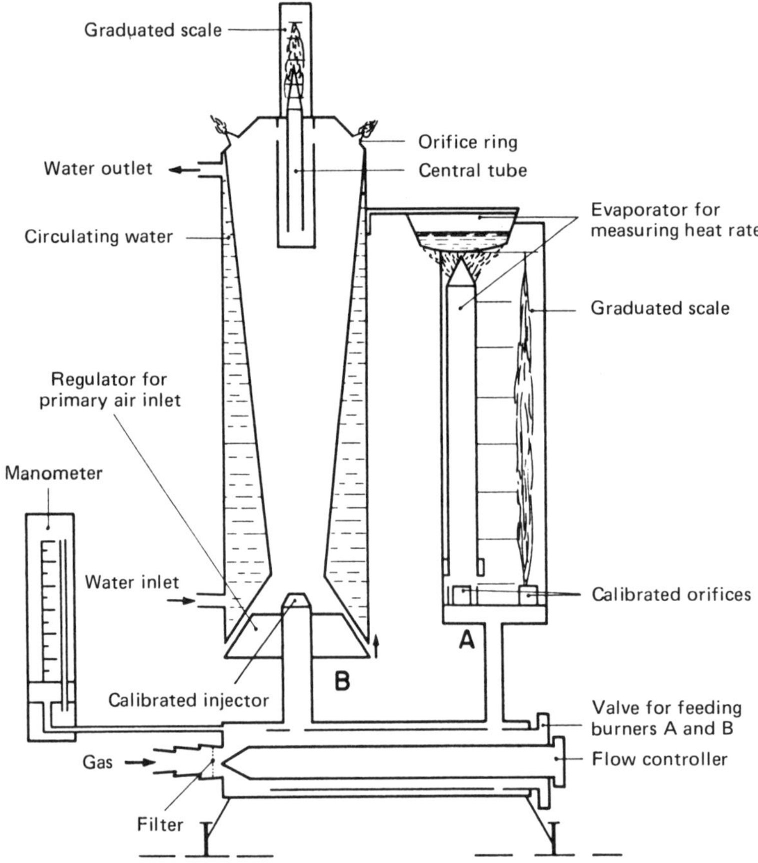

Fig. 6.13. Test burner for monitoring fuel gas supplied to domestic consumers.

h_n less than a maximum value h_m and that the blue cone of the central flame is less than a limiting height marked on the scale.

Experiments have shown that the simple separation of their flames from the orifices in the crown is equivalent to a blowing out, while a slight recession of the central flame is equivalent to a flashback to the injector in domestic appliances. The advantage of the control burner is that it is more sensitive to the quality of the gas than the commercial appliances. Observations have been made on hundreds of commercial appliances.

6.4.4. Combustion potential

The development of chemical analysis with chromatography plus computer calculation make possible the use of an additional index, called the combustion coefficient, C, which is calculated from the chemical composition of the gas. The combustion coefficient indicates the heat capacity of a burner of given dimensions and is a useful complement to the information about the interchangeability between gases. It is expressed as the effect of each constituent relative to that of an equal proportion of hydrogen, rather than as a function of the speed of combustion of each one of the constituents in the gas. The combustion potential is expressed by the equation :

$$C = \frac{H_2 + 0.3\,CH_4 + 0.7\,CO + \nu\,C_nH_m}{\sqrt{d}} \times u$$

where
H_2, CH_4, C_nH_m, CO represent the percent concentration of those constituents in the gas,
0.3, 0.7, etc, are empirical coefficients,
ν is an empirical coefficient given by the experimental curve in Fig. 6.14 for gases in the first family and by the curve in Fig. 6.15 for gases in the second family,
u is a coefficient to account for the possible presence of oxygen in the gas and is given by the curves in Figs. 6.16 and 6.17 for gases of the first and second families, respectively.

Besides the Wobbe index and the combustion potential, combustion can be evaluated in terms of:

(a) The "yellow points" index, which is related to the minimum rate of aeration necessary for avoiding the yellow points that appear at the extremity of the flames from domestic burners.

(b) The carbonizing index, which is the tendency of gas to form carbon at the pilot light with non-aerated flames.

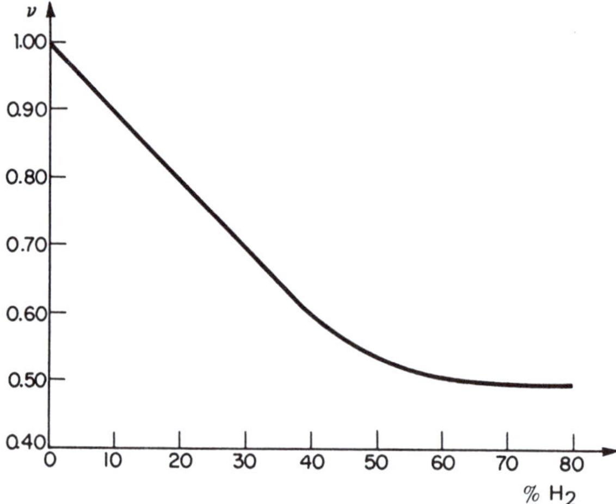

Fig. 6.14. Values of coefficient ν for calculating the combustion potential of manufactured fuel gases, as a function of the H_2-content of the gas.

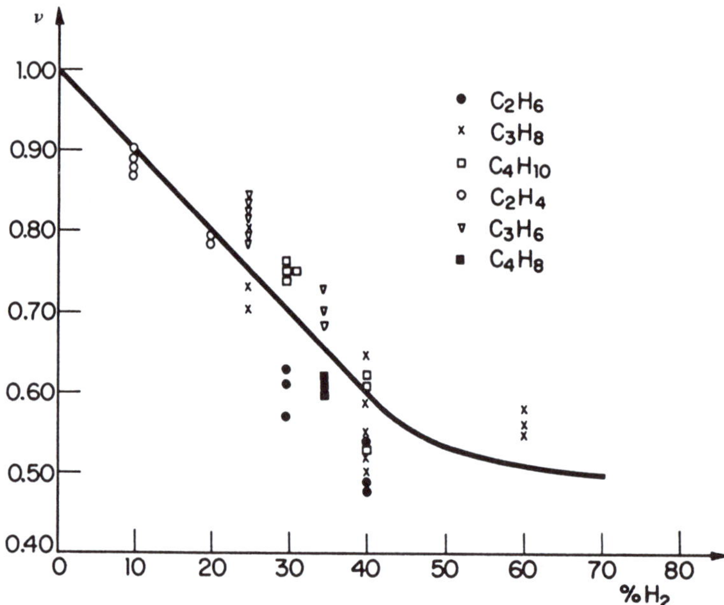

Fig. 6.15. Values of coefficient ν for calculating the combustion potential of natural gases, as a function of the H_2-content of the gas.

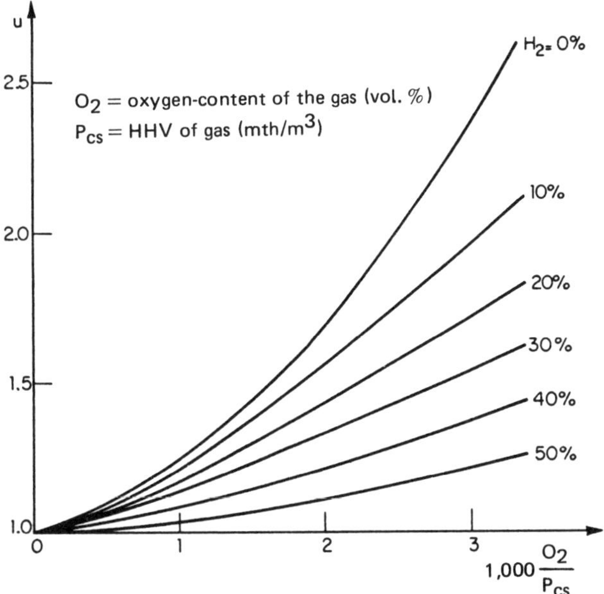

Fig. 6.16. Values of coefficient u for calculating the combustion potential of fuel gases containing oxygen, as a function of the H_2-content, O_2-content and heating value of the gas.

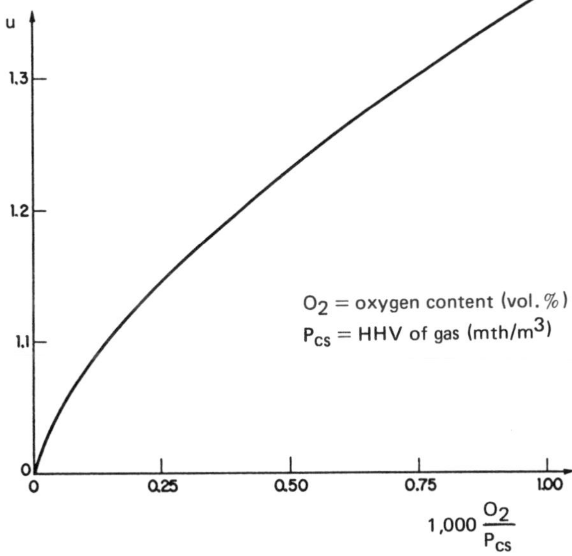

Fig. 6.17. Values of coefficient u for calculating the combustion potential of natural gases, as a function of oxygen content and HHV of the gas.

6.4.5. Conclusions about interchangeability

Without going further into the definition and the manipulation of the different indexes defined in gas research, it can be concluded that the analysis of a gas, its heat of combustion, and its density allow a confident prediction of the possibilities for substituting one gas fuel for another in currently manufactured domestic appliances. These studies have as a corollary good information for manufacturers of gas appliances, which are currently furnished with stabilizing pilot flames and regulation of the primary air by simply changing a nozzle. Such appliances are completely adaptable either for manufactured town gas with a high hydrogen content, for most natural gases composed principally of methane, or for liquefied petroleum gas (butane or propane).

6.5. PREMIX BURNERS

In order to avoid handling large volumes of air and gases capable of detonating, mixing is done only in the body of the burner itself. Combustion air will be carried along by the gas when the gas is furnished under a high enough pressure, as occurs in air induction burners. For other burners, the air is supplied independently of the gas, possibly through a ventilator incorporated in the apparatus.

6.5.1. Industrial induction burners

Figure 6.18 illustrates an induction burner. Fluids 1 and 2 reach cross-sections s_1 and s_2 at pressures p_1 and p_2, respectively. In the constricting section of Fig. 6.18, fluids 1 and 2 are mixed and pass through the throat at a pressure lower than the upstream pressure by an amount equivalent to the conversion of pressure energy to kinetic energy of motion, as predicted by Bernouilli's law. The purpose of the flared section downstream is to reconvert the kinetic energy

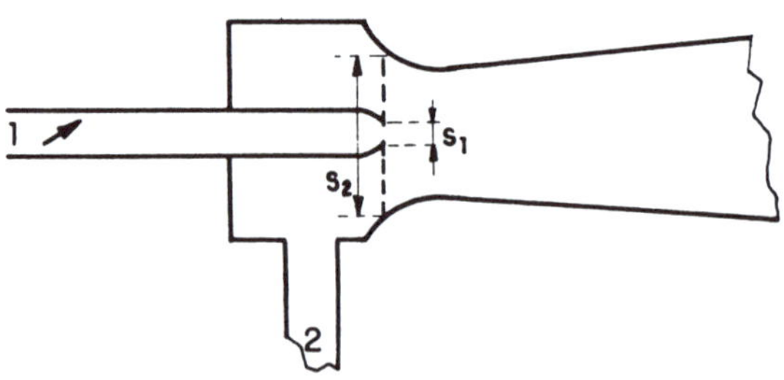

Fig. 6.18.   A typical industrial air-induction burner.

back to pressure energy at the pressure reigning in the combustion chamber. In addition, this reconversion improves the homogeneity of the air-gas mixture. A flame stabilizing device may be added to the outlet of the burner (not shown in Fig. 6.18).

Experimental evidence confirms the theory that the ratio of the flowing velocities of fluids 1 and 2 depends only on the ratio of the pressures p_1 and p_2, as determined by the dimensions of each section of passage through the burner, according to a coefficient for energy conversion that depends on the geometry and the machining of the pieces (Ref. 6.8).

6.5.1.1. Inducting a fluid at atmospheric pressure

This situation can exist in two practical cases:

(a) The gas is the inductor fluid; it carries along, through its expansion, air taken directly from the atmosphere.

(b) The air is the inductor fluid; it has been previously compressed, while the gas is supplied at a pressure equal to atmospheric pressure through a pressure-reducing valve.

The ratio of the mass rate of flow of induced fluid to the rate of flow of motive fluid can be written:

$$R = \frac{\dot{m}_2}{\dot{m}_1} = K_2 s_2 \sqrt{\frac{p_1}{p_2} \frac{1}{K_1 s_1 (\mu S - K_2 s_2)}}$$

where

$\dot{m}_2$ = mass rate of flow of the induced fluid,
$\dot{m}_1$ = mass rate of flow of the motive fluid,
K_2 = coefficient of the rate of flow for cross-section, s_2,
K_1 = coefficient of the rate of flow for cross-section, s_1,
S = cross-sectional area of the throat,
μ = ratio of the pressure in the combustion chamber to the pressure at the throat.

At subsonic velocities, μ depends only on the pressures. It follows from this that the ratio of the mass rate of flow of the two fluids, R, is independent of the motive pressure, p_1; hence the heating power of the burner can be regulated by controlling p_1 to give the desired rate of flow m_1 and proportional flow, m_2, without changing the chemical composition of the fuel/air mixture.

6.5.1.2. Burners with induced air

The simplest case is represented by the burner shown in Fig. 6.19. The gas under pressure is injected into the burner through a contracting fitting from which it leaves at great speed so that it carries along the necessary atmospheric air through turbulent exchanges. The air and the gas mix as they pass through the expanding cone and the flame is maintained at the outlet of the burner.

All this can be more easily done if there is a refractory quarl with a high wall temperature in front of the burner.

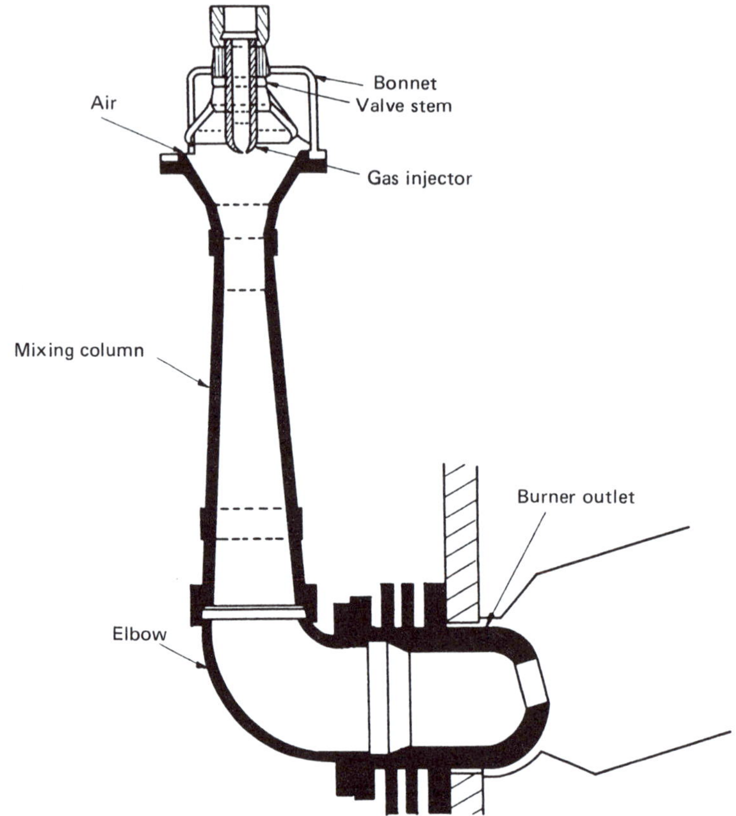

Fig. 6.19. Induction burner for pressurized manufactured gas. (From: *Heurtey and Co.*).

The rate of air flow is controlled in the air entrance by means of restrictions consisting of a moveable ring or sleeve (Fig. 6.20a), by means of a moving disc (Fig. 6.20c) or by means of a bonnet opposite the extremity of the injector (Fig. 6.20b). Control of heat release by the burner is achieved for a given family of gases at the same pressure by a flow-regulating valve. The heating power can thus be varied by a ratio up to 3 or 4 times the minimum power of the burner.

Figure 6.21 shows a *Gaz de France* research burner, in which the flame cannot flashback to the injector because of the diverging-converging shape. This burner permits inducting up to 5 m^3 of air per m^3 of gas with a low pressure gas (20 mbars). That amount of air corresponds to all that is needed with coke-oven gas but about half needed with natural gas. In this latter case, the pipeline pressure can be used to induct the required combustion air by means of two injectors arranged in cascade (Fig. 6.22).

a. By means of a moving ring.

b. By means of a cap eductor.

c. By means of a disc eductor.

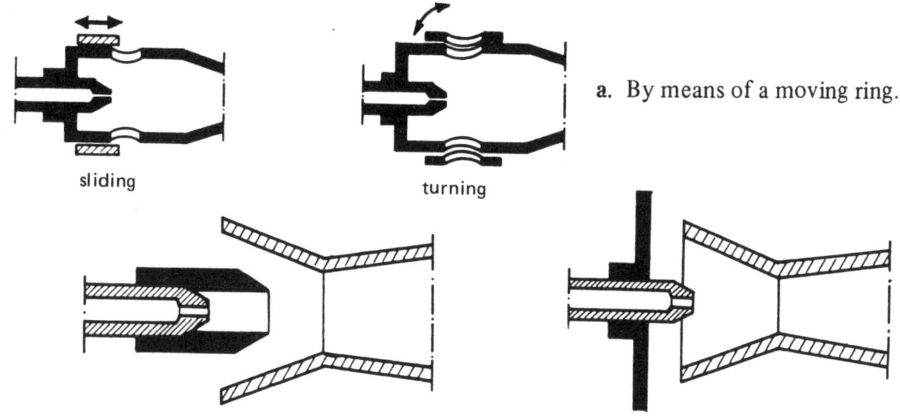

Fig. 6.20. Devices for controlling the air induced to domestic gas burners.

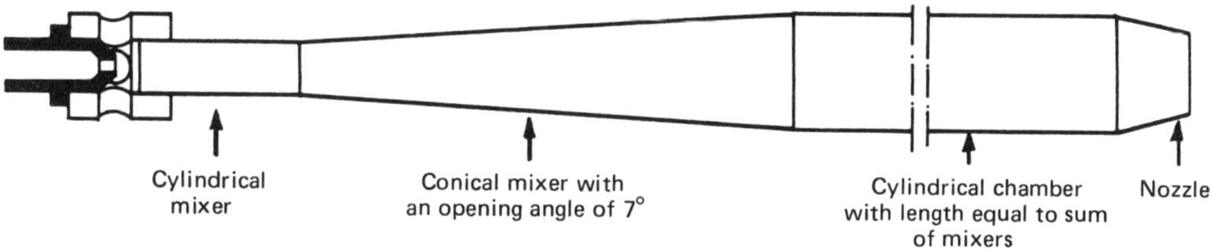

Fig. 6.21. An experimental burner to stop flashback by means of kinetic energy alone, as studied by *Gaz de France*.

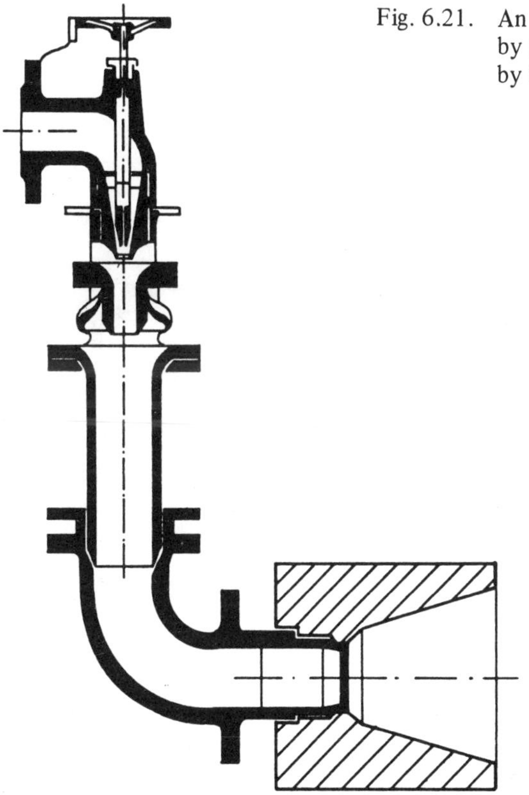

Fig. 6.22. Double air-induction burner for high pressure pipeline gas.

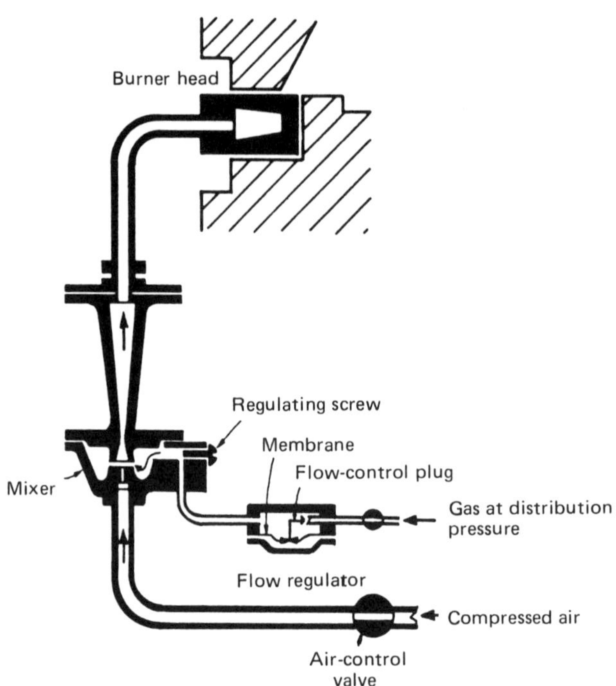

Fig. 6.23.   Gas-induction burner using pressurized air.

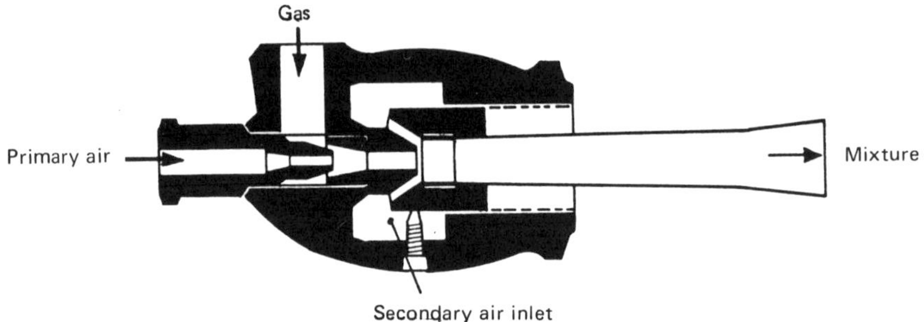

Fig. 6.24.   Gas-induction burner using pressurized air and pipeline gas.

When using gas having a low speed of flame propagation (natural gas, propane, butane) it is advisable to furnish the burner outlet with a stabilizing device such as an obstacle or pilot flame.

6.5.1.3. Burners with air-powered induction

If air pressure is available for feeding the burner, it can be used for inducting the fuel gas. The relatively larger mass of air-to-fuel gas used in combustion makes this operation easy, and the variations of rate of flow can go from 1 to 5 for a constant ratio of air to gas. But this induction system does not make use of the potential energy of pipeline gas under pressure, and usually requires an expenditure of energy for compressing the air. On the other hand, the gas can be induced from slightly below atmospheric pressure so as to avoid any accidental leak during a shut-down of the burner. The method thus presents an element of safety that is not available with burners using gas as motive fluid.

Figure 6.23 shows a model of an industrial burner with inducting air and depressurized gas.

Finally, there are burners with induction air under pressure and gas introduced at the pressure of the pipeline (Fig. 6.24). This type of burner is extremely simple, but the ratio of flow rates of air and gas is not strictly constant over variations in the rate of flow of primary air or the pressure of the gas pipeline. The inherent advantage of induction burners is therefore missing.

6.6. CONCLUSIONS

The use of burners with premix is particularly widespread in all domestic appliances, the majority of which operate according to the flow schemes of motive gas and induced air. The principal advantage of these burners is their ease of control, which is obtained simply by varying the rate of flow of gas with a cock-valve placed upstream of the burner.

From the standpoint of the manufacturer, it is possible to sell appliances that can use any gas by simply changing the nozzle; fuels as different as coal gas or liquefied petroleum gas can be used. These adaptations are only possible, however, because of the effectiveness of pilot-flame devices for stabilizing flames.

Induction burners can be useful in industry for giving off intense heat in an enclosure. But the unit power of these burners remains limited by the possibilities for accumulating upstream of the flame a quantity of premix whose accidental ignition could be dangerous. These safety considerations become exigent as the speed of combustion of the air-gas mixture becomes rapid.

REFERENCES

6.1 VAN TIGGELEN et al. — *Oxydations et Combustions.* Editions Technip, Paris, 1968.

6.2 AGARD. — *Selected combustion problems.* Londres, 1954.

6.3 DE SOETE, G., BRASSELET, J. — "La vitesse normale de propagation en pré-mélange dans des conditions d'explosivité". *Rev. Inst. Franç. du Pétrole*, XXIV-12, 1969.

6.4 DELBOURG, P., LIARD, R. — *9ᵉ Congrès International de l'Industrie du Gaz*, the Hague, 1964.

6.5 DELBOURG, P. — *Congrès de l'Association Technique du Gaz (ATG)*, Aix-les-Bains, 1959.

6.6 SCHNECK, H., BRUNET, G. — *Congrès de l'Association Technique du Gaz (ATG)*, 1965.

6.7 BIARD, E. — *Techniques de l'utilisation du gaz*, Paris, 1948.

6.8 ROUX, A. — *Journal des Usines à gaz*, T 74 (August 1950).

7

pollution caused by combustion

This chapter cannot entirely cover a subject as vast as pollution; however, it seems necessary to try to present some data on the pollution that can result from the combustion of hydrocarbons in industrial fireboxes.

Pollution from combustion products affects the atmosphere principally by filling it with solid particles or noxious gases. Unfortunately, when this atmospheric pollution is removed by washing the flue gases, the waters from the gas scrubber can pollute surface water or the water table. Also industrial processes using combustion often cause heating and de-oxygenation of surface waters, with disastrous biological effects.

In a work published in 1933, R. Humery (Ref. 7.1) quotes a report in old French of an investigation made in Rouen in 1510 regarding the problem of pollution:

"... Master Guillaume Moullet, doctor, says that the smoke that comes from the carbon of the earth is bad and dangerous, and that it smells of sulfur and that there are no persons who can bear said smell, and because of this people could fall into a coma or die suddenly; and it seems to him that one should make those who burn carbon live in distant places and outside of the good streets. And it seems good to him that all other things be done so that they cannot harm the human body..."

"... Jehan Mustel said that one should be allowed to use said carbon provided that those that use it raise their chimneys up above their houses..."

The use of this same "carbon from earth", or coal, was forbidden in London as early as 1273 by a law of Edward I, because the smoke was considered noxious to human health. These reports show that the need to fight against atmospheric pollution has been a social problem for a long time.

Because the development of a social recognition of pollution, as well as an articulation of that recognition, is better documented in France, a history of the French social attitudes toward pollution might be useful to English-speaking readers.

7.1. EVOLUTION OF FRENCH LEGISLATION

In addition to a few royal edicts, legislation designed to limit the harmful effects of smoke first appeared in France in 1790 and 1791, and it was taken up again and developed during the 19th century. Originally the legislation was principally concerned with the emissions from industrial "establishments and shops", particularly those using "machines with fire" for burning coal.

As late as the beginning of the 20th century, noxious smoke was detected only by its appearance, and lacking effective smoke suppressors, the administration of Paris forced industry into using coke instead of coal, thus obtaining appreciable qualitative results. In 1907 and 1912, the gas companies of Paris and its suburbs carried out an investigation on the change-over and found the following:

	In 1907	In 1912
Chimneys with a lot of smoke	326	172
Chimneys with little smoke	32	98
Chimneys with no smoke	68	156

A law appeared in 1917 applying to "dangerous, insalubrious, or unsuitable" installations and this gave the heads of the Departments of France the power to prohibit the production of noxious smoke through inspection of classified installations. The heads of each *arrondissement* of Paris and mayors of communities had been able to continue assuring "the public health" in their community through a law of 1884 which extended their authority over each fireplace. In the following fifteen years, different resolutions were made either on the local level in large cities, or on the level of the larger departments.

In 1932, a law, called the Morizet law, was promulgated with the aim of suppressing industrial smoke. This law forbids the emission of toxic or corrosive soot, or dust or gases. In spite of its imperfections (it is too absolute, ignoring technical difficulties, and it applies only to industrial fireboxes but not domestic ones) the Morizet law gave public bodies enough authority to intervene and remove the most bothersome cases of pollution. It had the merit of getting industrialists used to the idea of responsibility for what their smokestacks were pouring out into the atmosphere. But the concern was still limited to the visible aspect of smoke; and only rarely perhaps did the users make a connection between pollution and the imperfect control of combustion (the latter also being wasteful).

Then, sixteen years after the Morizet law, on March 10, 1948, a law was promulgated on the use of energy. This law, at first applicable to the fireboxes of large industrial consumers of energy, had as its objective the saving of fuel by reducing losses through smoke and unburned particles. Indirectly this law contributed to reducing atmospheric pollution from fireboxes. A decree of April 18, 1957 extended the application of the 1948 law to include small domestic apparatus using combustion.

Subsequently, the battle against pollution and the concern for protecting the quality of the environment led the French legislature to promulgate the law of August 2, 1961, which concerns all sources of pollution. For this to be applied, decrees of application are necessary. This law is connected to the 1917 law for classified establishments. It differentiates precisely when it applies to private buildings, industrial installations, automobiles, or radioactive substances. One of the first decrees of application was a decree of November 7, 1962; it concerns the rules for use of petroleum products and the characteristics of those products.

A subsequent decree of September 17, 1963 creates special-protection regions and fixed threshold limits for the pollutants. Later decrees define the conditions relative to each one of the regions. In this way, four decrees of August 11, 1964, were taken together by the *Ministères de la Santé publique et de la Population, de l'Intérieur, de l'Industrie et de la Construction*, in order to:

(a) Create two special-protection zones in Paris.

(b) Regulate the use of fuel.

(c) Organize the control of emissions from combustion.

(d) Extend application of the rules of the preceding decrees to classified establishments situated in the special-protection zones in Paris.

A decree of June 22, 1967 prescribed apparatus for measuring an index of blackening. A decree of June 10, 1969, created an obligation to keep heating records for units of more than 1,000 th/hr. The texts of these decrees are supplemented by a certain number of pamphlets concerning their application.

Finally, an alert mechanism is being studied. Through a departmental decree, this would permit reducing the operation or even the shut-down of an installation generating excessive pollution.

In summary, it can be said that French legislation is first of all concerned with reducing or suppressing visible smoke. In 1917, establishments that are dangerous, insalubrious or unsuitable were set in a class apart. Even though still maintained, this distinction is being phased out little by little.

The scarcity of energy that followed the war of 1939-1945 led to putting the accent on saving fuel in 1948, and led consequently to a battle against pollution related to wastefulness. The 1960s saw the battle with pollution directed against different aspects, as is seen in successive texts. The battle continues today in the latest texts with a concern for coordinating measures taken in a European context.

7.1.1. French regulations since the end of 1975

At the moment, French regulations applicable to heating units are concerned with emission of SO_2 and solid particles. This regulation comes out of the law of August 2, 1961. There is no regulation yet for pollution from nitrogen oxides although such a regulation is foreseeable in the future.

7.1.1.1. Sulfur oxides

The regulation is not directly concerned with the amount of sulfur oxides in smoke, but it limits the sulfur content of fuels and treats the dispersion of fallout.

For liquid fuels, the maximum sulfur contents are the following:

	(%)	
No. 2 heavy fuel oil.......	4	
No. 1 fuel oil...........	2	
Special low sulfur light fuel.	1	
Domestic fuel oil........	0.5	This value was taken to 0.3 in October, 1980.

In 1976, two new categories of No. 2 heavy fuel oil were defined:

(a)   No. 2 heavy fuel oil (BTS) (low sulfur content) S < 2%.
(b)   No. 2 heavy fuel oil (TBTS) (very low sulfur content) S < 1%.

The departments of Nord and Rhone declared protection areas similarly to the regulations in Paris, by a decree of August 11, 1974. This allows the administration to determine "special-protection areas" by decree of May 13, 1974 where the sulfur content of heavy fuels is limited, as well as "alert zones" where fuels with low sulfur content must be used during a maximum of 48 hours when weather conditions are unfavorable or when excessive pollution is observed.

As for the dispersion of pollutants, the heights of chimneys and the exit speeds of smoke were the subject of a circular issued November 24, 1970 (*Journal Officiel* December 13, 1970).

7.1.1.2. Solid particles

Those instructions of November 24, 1970 were taken up and reinforced by a decree of June 20, 1975, with the purpose of assuring a wide dispersal of solid particle fall-out from heating units. The emissions are limited by color (standard AFNOR NF X 43 002). The mass of particles emitted, as expressed in milligrams per thermie without indicating the nature of the solid particles, is determined for large units by the standard NF X 43 003.

This decree of June 20, 1975, (*Journal Officiel* July 3, 1975) limits emissions to 200 mg/th for new units of less than 8,000 th/hr, starting from 1976. A limit of 250 mg/th is imposed on older units, starting from 1978.

In addition, this decree imposes:

(a) Control and regulation devices for heating equipment (draft gauges, smoke temperature recorders, gas analyzers, viscometer for heavy fuel oils, etc).

(b) Maintenance of a heating record-book that permits following the operation of the unit.

(c) Visits by experts.

Finally, it should be mentioned that qualified personnel have been recognized as able to fight pollution (creation of a "Certificat d'Aptitude Professionnelle de conducteur de chaufferie", for the operator of a boiler room, (*Journal Officiel* November 4, 1972).

7.2 SOLID PARTICLE POLLUTION FORMED DURING COMBUSTION

Combustion of hydrocarbons, liquids or gaseous, can cause the formation of the solid particles found in smoke:

(a) Hydrocarbon molecules in gas phase diffusion flames can be heated to a high temperature without oxygen, and this causes them to crack.

(b) Liquid fuels that burn as a mist can crack either in vapor phase or in liquid phase.

(c) Solid residues from sediments and metals are present in heavy fuels (No. 1 and No. 2) in the form of organometallic compounds. These residues become associated with carbon particles from cracking. The total mass of particles thus formed with heavy fuels can be relatively important, on the order of several 10^{-4} s of the weight of the fuel.

These sources of pollution are combated in several ways. Physical separation, such as double centrifuging, can eliminate part of the sediment and water that may be associated with the fuel. Careful atomization of the fuel, primarily by preheating to a well-regulated temperature at the burner so as to reduce the viscosity at the moment of atomizing, can be particularly effective for mechanical atomizers. Figure 7.1 shows that with heavy fuel oils an increase of $10°C$ can correspond to as much as 160 cSt reduction in viscosity. Under a constant feed pressure, the rate of flow of the burner's spray nozzle will accordingly vary more than 10%. However, this can represent an important defect in the burner operation because the increase in flow of liquid fuel is not accompanied by a corresponding increase in flow of oxidizer. Consequently, liquid fuel is typically

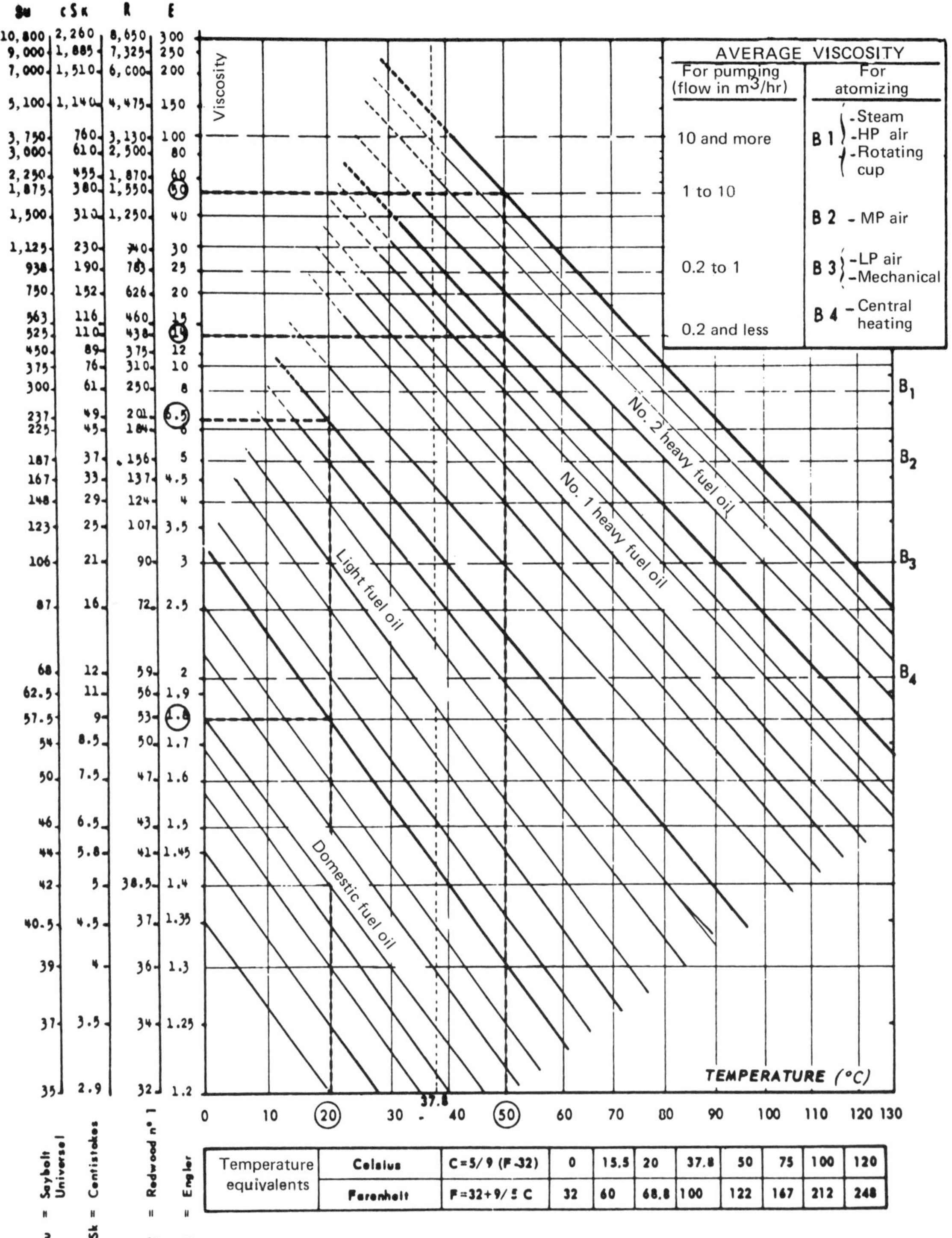

Temperature equivalents	Celsius	C = 5/9 (F -32)	0	15.5	20	37.8	50	75	100	120
	Farenheit	F = 32 + 9/5 C	32	60	68.8	100	122	167	212	248

Fig. 7.1. Nomograph for determining preheat temperatures needed to pump and atomize liquid fuels.

fed by a controlled positive-displacement pump, and the reduction in viscosity is reflected in the pressure at the nozzle, in the atomization, and in turn by the combustion.

7.2.1. Formation of solid particles

Two kinds of solid particles can result from burning hydrocarbons, depending on whether the solid derives from vapor-phase cracking or liquid-phase cracking. Particles coming from the vapor-phase cracking are small, those from the liquid-phase cracking are much larger.

7.2.1.1. Formation of small soot particles

These particles form a suspension in the flame and are carried to a high temperature at which they radiate in the visible and infrared spectrum. The flames that contain them are called white and luminous. Such flames have long been used as a source of light in oil lamps, candles, gas lights, etc.

Despite much research, the mechanism of the formation of these soot particles is still not completely defined; what follows is a summary of contemporary thinking, as a result of a large number of experiments reported in the excellent summaries, References 7.3 and 7.4.

First of all, a fundamental characteristic of these particles is their nearly spherical shape, with diameters ranging broadly between 200 and 500 Å no matter from which hydrocarbon fuel. Secondly, these spheres can agglomerate into chains that can be seen with an electron microscope (Fig. 7.2).

These particles do not form directly but appear during combustion of the hydrocarbons under certain conditions of temperature, concentration, and residence time, after a sequence of chemical reactions involving intermediates made up of higher acetylenes, as well as of polycyclic aromatic hydrocarbons. The condensation and simultaneous dehydrogenation of these intermediates produces compounds with molecular weights on the order of 500-600 g/mol. The intermediates are often ionic radicals; this facilitates their condensation and is the cause of their short lifetime. They have, however, been identified in certain laboratory flames; and it has been assumed that they also show a transitory existence in large luminous industrial flames.

Nucleation, the transition from the gaseous to the solid phase, is accomplished by some as yet undetermined mechanism. It leads to particles that are at first smaller than 20 Å but that grow to several hundred Å by condensation and coagulation on their surfaces.

A corollary of these hypotheses is that particles of carbon can form only as long as the intermediates appear and have a long enough lifetime. As a matter of fact, the condensation reactions in which the intermediates participate do compete with the oxidation reactions of the combustion. Furthermore, different

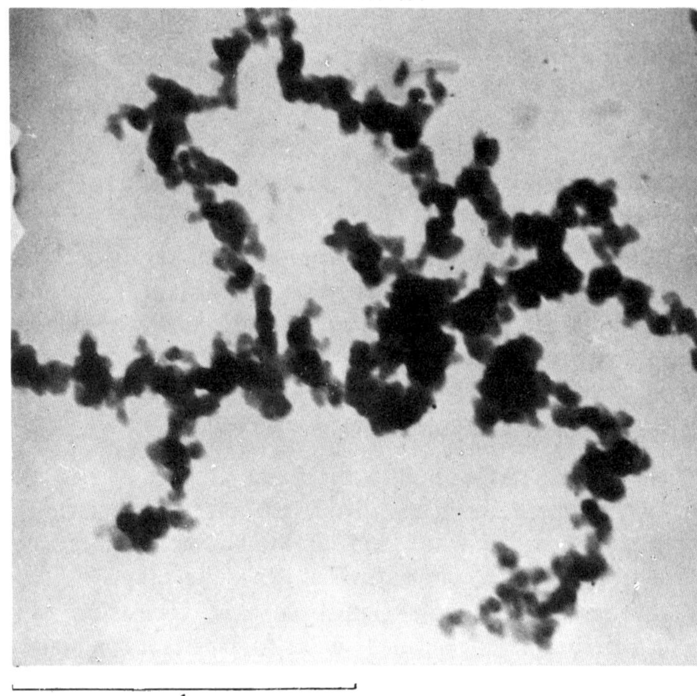

Fig. 7.2.   A typical chain of small soot particles.
(Photograph by *IFRF*).

authors (Refs. 7.5 and 7.6) think that the oxidation reactions mostly involve
ȮH radicals; any combustion reaction that decreases the concentration of these
ȮH radicals in the flame must consequently favor the appearance of carbon.
This interpretation agrees with observations of laboratory flames in flows that are
most often laminar; it is also the basis for explaining the effects of certain
metallic additives (Refs. 7.7 and 7.8).

7.2.1.2.   Formation of large soot particles

In addition to particles of soot, larger particles called "cenospheres" appear in
flames resulting from the combustion of fuel oils, the more so as the fuel gets
heavier. Resulting from liquid-phase cracking of droplets of fuel before the drops
have completely burned, these cenospheres have the same dimensions as the fuel
oil droplets from which they derive, some tens of microns. Their aspect is generally spherical and spongy (Fig. 7.3). Although infinitely fewer in number than
soot particles, they generally exhibit much greater total weight. It is principally
the presence of cenospheres that makes application of pollution regulations for
heaters fired with heavy fuel so difficult.

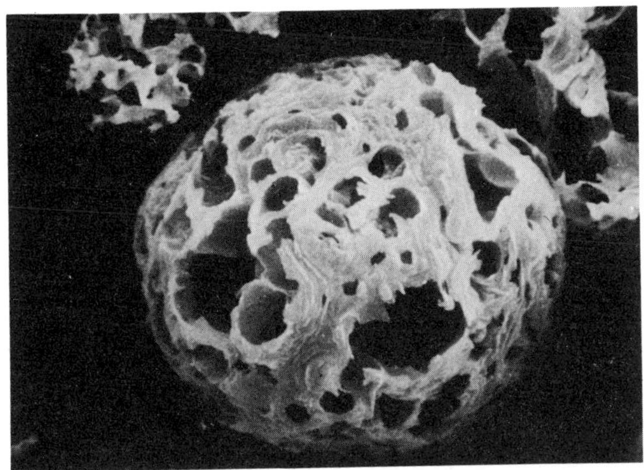

Fig. 7.3. Cenosphere (G x 1,000).

7.2.2. Combustion of solid particles in suspension in the flame

The combustion of solid particles is a surface phenomenon; its rate is governed by two variables:

(a) The activation energy corresponding to the temperature level of the particles.

(b) The diffusion of oxygen toward the solid surface and that of the combustion products away from that surface.

7.2.2.1. Combustion of soot

It now seems established that the semi-empirical mechanism proposed for the oxidation of pyrolytic graphite by Nagle, Strickland and Constable (Ref. 7.9) can be used for the oxidation of soot. The mathematical expression is a follows:

$$\frac{w}{12} = x \, \frac{K_a \, p_{O_2}}{1 + K_z \, p_{O_2}} + K_b \, p_{O_2} \, (1 - x)$$

where

w = the oxidation rate (g . atoms/cm^2 . s)

$$x = \left[1 + \frac{K_t}{K_b \, p_{O_2}} \right]^{-1}$$

p_{O_2} = the partial pressure of oxygen,

$$K_a = 10^{6.30} \exp \left(- \frac{15,100}{T} \right) \qquad \text{(g/cm}^2 \text{ . s . Pa)}$$

$$K_b = 10^{2.65} \exp\left(-\frac{7,650}{T}\right) \qquad (\text{g/cm}^2 \cdot \text{s} \cdot \text{Pa})$$

$$K_t = 10^{5.18} \exp\left(-\frac{48,820}{T}\right) \qquad (\text{g/cm}^2 \cdot \text{s})$$

$$K_z = 10^{6.33} \exp\left(+\frac{2,063}{T}\right) \qquad (\text{Pa}^{-1})$$

According to this mechanism, the activation energy for the reaction is close to 35 kcal/mol up to temperatures approaching 2,000 K, as determined in a number of experiments (Refs. 7.10 and 7.11). Above 2,000 K, it diminishes a little with temperature and then increases again. (It is assumed that the temperature of the soot particles is that of the surrounding gases (Ref. 7.12).) The reaction is of the first order relative to oxygen at low partial pressures, and the activation energy approaches zero for higher oxygen partial pressures.

In large, very turbulent industrial flames burning heavy fuel oil, the combustion of each particle of soot is very rapid, and studying the flame from upstream to downstream shows a decrease in the concentration of particles without much change in their average diameter.

We should remember that the presence of soot particles confers on flame a high radiant emissivity, which aids heat exchange with a charge. Radiating flames are very desirable in certain operations like manufacturing glass in open-hearth furnaces.

7.2.2.2. Combustion of cenospheres

Because of their size, cenospheres burn more slowly than soot particles, and this combustion is controlled principally by diffusion phenomena. Moreover, cenospheres contain in addition to a hydrocarbon skeleton the minerals that were present in the fuel. These minerals can be objectionable. In contact with refractory walls, an open-hearth bath, or tubes, they can lead to contamination and corrosion. For example, vanadium salts attack steels above 600° C.

The importance attached to particle pollution has brought on a number of recent studies (Refs. 7.13, 7.14, and 7.15). From these, the following general conclusions can be drawn concerning means for reducing this pollution:

(a) The characteristics of the fuel — its viscosity at the time of combustion and its composition (Conradson carbon index, ratio of light fractions, etc.) are very important.

(b) At the burner, anything is favorable that contributes to the fineness of atomization and the speed of evaporation, including higher wall temperature, nozzle design, higher fuel oil temperature at the moment of injection. All else being equal, conversion from mechanical to air or (better yet) steam atomizing reduces pollution.

(c) With respect to combustion, the use of preheated air and a moderate excess air (less than 50%) is favorable.

(d) The amount of unburned matter can be noticeably reduced through such techniques as emulsifying water in the fuel and adding certain metals to the fuel.

7.3. SULFUR POLLUTION

Crude petroleum contains varying amounts of sulfur, depending on its origin. The following Table gives the weight percent of sulfur in some crude oils:

SULFUR CONTENT (WEIGHT %) OF TYPICAL CRUDE OILS

African	Algeria Hassi-Messaoud	Libyan	Algeria Zarzaïtine	Nigeria Bomu	Angola Tobias
% S total	0.14	0.4	0.07	0.14	1.51
Mid-East	Saudi-Arabian		Kuwait	Irak Kirkuk	
	Safaniya	Khursaniyak			
% S total	2.95	2.67	2.50	2	
Iran	Gach-Saran (heavy)	Agha Jari (light)			
% S total	1.6	1.34			
Others	Venezuela		Indonesia		
	Boscan	Centa-Zubia	Minas	Duri	
% S total	5.5	1.36	0.08	0.18	

During refinery operations, sulfur tends to concentrate in residual fractions that make up an important portion of industrial liquid fuels. The light distillate fractions can be treated with hydrogen to eliminate all or part of the sulfur. The sulfur compounds are converted to hydrogen sulfide, which can be absorbed and through controlled oxidation converted to elementary sulfur. Hydrotreating also lowers the organic nitrogen content of the petroleum fraction, as it increases stability and improves odor and color. The average straight-run distillates (gas oils or domestic fuel oils) can thus have their sulfur content reduced by a factor of ten (from 1.5% for a distillate of Near East crude to 0.15% for the hydro-refined products). A slightly smaller reduction is obtained for straight-run gas oils by catalytic cracking. The cost of hydrotreating and its hydrogen consumption depend on the contents of sulfur and olefins as well as on the boiling range of the treated product.

Large industrial units have not yet been built for treating crude oils and residues. The difficulty of treating these products is due principally to the presence of asphaltenes and metals which deactivate the hydrotreating catalysts. Studies in this direction have been pursued by *Institut Français du Pétrole* (*IFP*). The following Table gives some examples for Kuwait crudes:

TYPICAL RESULTS FROM HYDROFINING KUWAIT CRUDE

	Charge	Product	Charge	Product
Density	0.868	0.852	0.990	0.914
Sulfur (wt.%)	2.4	0.75	6.7	1.8
Asphalt (wt.%)	2.3	1.0	18.3	8.3
Engler viscosity 35° C	–	–	900	9.7
Engler viscosity 50° C	–	–	260	5.1
Freeze point (° C)	–	–	+ 12	– 30
Gasoline				
Distillation (° C)	48-186	51-184	30-200	30-200
Wt. % on crude	20.0	19.9	5.1	9.2
Vol. % on crude.	23.8	23.1	6.6	10.6
Density	0.731	0.735	0.745	0.746
Sulfur (wt. %)	0.09	0.008	0.91	0.15
Gasoil				
Distillation (° C)	197-350	208-347	200-350	200-350
Wt. % on crude.	29	31	21.1	33.4
Vol. % on crude	29.8	32.4	24.2	36.0
Density	0.844	0.839	0.857	0.846
Sulfur (wt.%)	1.30	0.20	3.3	0.40
Diesel index	56.3	58.6	45.5	53.9
Residue (> 350° C)				
Wt. % on crude.	48	45.4	73.8	57.4
Vol. % crude	42.4	–	69	53.4
Sulfur (wt. %)	4.15	–	8.0	2.9

From Wuithier P. *Le pétrole. Raffinage et génie chimique.* 2nd ed. entirely revised. Editions Technip, Paris, 1972.

During combustion, sulfur contained in the fuel is oxidized mostly into sulfur dioxide, SO_2, and in small part to sulfur trioxide, SO_3.

In the presence of oxygen, these sulfur oxides tend toward an equilibrium:

$$SO_2 + \frac{1}{2} O_2 \rightleftarrows SO_3$$

The equilibrium constant K_p for this reaction is defined by the relation:

$$K_p = \frac{p_{SO_3}}{p_{SO_2}(p_{O_2})^{1/2}}$$

K_p varies with temperature, according to $K_p = f(T)$, a relation proposed by Lowrison and Heppenstall as:

$$\ln K_p = \frac{22{,}600}{RT} - 10.68$$

where
$\quad T$ = temperature (K),
$\quad R$ = perfect-gas constant.

From this relation the proportion of sulfur dioxide at equilibrium with sulfur trioxide in the flue gases can be estimated as a function of the temperature and the excess air for combustion. Figure 7.4 shows the calculated fraction of sulfur transformed into SO_3, as Y %. The curves indicate that the oxidation of SO_2 to SO_3 is favored by a moderate temperature. The speeds of reaction are slow at those temperatures but the transformation can be catalyzed by certain metallic oxides in the fly-ash. Also, there is possibility of formation of sulfur trioxide at high temperature through oxidation of sulfur dioxide by nascent oxygen:

$$SO_2 + O \longrightarrow SO_3$$

This reaction leads to the formation of SO_3 in the flame, so that its concentration is larger than equilibrium (Fig. 7.5).

At high temperatures the preceding reaction is accompanied by a decomposition reaction:

$$SO_3 \rightarrow SO_2 + \frac{1}{2} O_2$$

The observed overall result depends on the relative speeds of these two reactions; and since the reaction with nascent oxygen is 8 to 9 times as fast as the decomposition of sulfur trioxide, the formation of SO_3 in the flame varies with temperature as shown in Fig. 7.5.

Because of these reactions, sulfur present in the fuel eventually appears as oxides of sulfur, SO_2 and SO_3 in the flue gas (Fig. 7.4), where those oxides combine with steam from the combustion of hydrogen to form the acids H_2SO_3

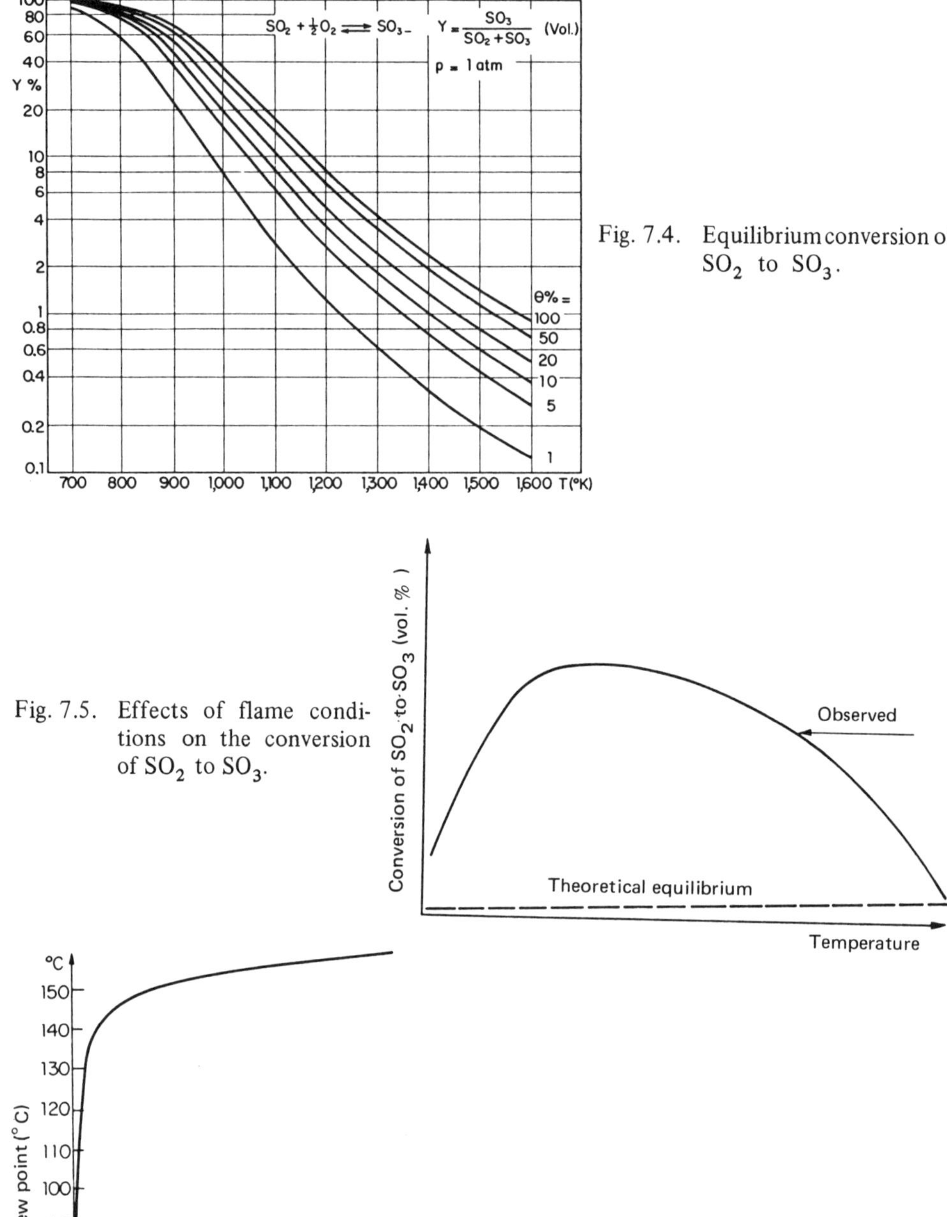

Fig. 7.4. Equilibrium conversion of SO₂ to SO₃.

Fig. 7.5. Effects of flame conditions on the conversion of SO₂ to SO₃.

Fig. 7.6. Temperatures of stack corrosion: the acid (H₂SO₄) dew point of flue gases as a function of their SO₃ concentration.

and H_2SO_4. These acids condense on walls whose temperature is lower than the acid dew point, which is higher than the dew point of water, depending on the acid concentration and hence on the concentration of SO_3. Figure 7.6 presents data from several authors on the relation between the acid dew point and the SO_3 concentration in the combustion gases.

It is noticed that a very low proportion of SO_3 in the flue gases raises the temperature of the acid dew point to more than $150°$ C. Beyond 40 ppm SO_3 this dew point temperature increases slowly.

Condensed sulfurous and sulfuric acids attack metals. The iron in steel is converted to greenish ferrous sulfate, $FeSO_4$, which in turn tends to be oxidized to ferric sulfate that is yellow, orange or brown, depending on the degree of hydration. If the temperature of surfaces in contact with flue gases is less than the dew point of steam, condensation becomes abundant and dissolved sulfur oxides participate in the corrosion, which can become virulent.

7.3.1. Controlling corrosion due to sulfur

There are three ways for fighting corrosion due to sulfur:
(a)   Prevent the formation of sulfur oxides, particularly SO_3.
(b)   Neutralize the acids.
(c)   Prevent acid condensation.

The formation of sulfur oxides can be avoided by burning a fuel with little or no sulfur. Since the desulfurization of heavy petroleum fuels is not economically feasible, the current option remains that of selecting fuels from crude oils that have little natural sulfur in them and of using such fuels for industrial operations in which corrosion due to sulfur is particularly harmful. Also, it is possible to limit the excess air for combustion, so that SO_2 is formed while the amount of SO_3 remains limited. To effectively limit SO_3 formation, however, it is necessary to hold the excess of air to 5% or less, and this is difficult because it requires both that the burner assures a correct mixture of fuel and air and that each one of several burners in a heater is equipped with a control and regulation system. Consequently, the cost of the necessary equipment restricts the limitation of excess air to large heaters in generating stations.

In some cases, pulverization of silica on the tubes of boilers or superheaters where harmful deposits occur can prevent the catalytic action of the deposits, thus reducing the local amount of SO_3. Frequent cleaning of the tubes with air or steam complete this action.

Neutralization of sulfurous and sulfuric acids in the combustion products can be achieved with injections of magnesium, magnesium carbonate, zinc oxide, dolomite, or ammonia, which in certain furnaces considerably reduces the corrosion of the air preheaters. However, such injections require great care to assure good distribution of the product and avoid plugging the flue gas passages, as well as to avoid carrying along solid particles into the atmosphere and prevent insufficient neutralization to, for example, ammonium bisulfate.

The third method of avoiding acid corrosion, i.e., preventing condensation, requires that all metallic surfaces in contact with the acid flue gases be maintained at a temperature above the acid dew point. In this case all the oxides of sulfur are found to be carried by the flue gases and contribute to atmospheric pollution, and the thermal efficiency of the heater is reduced by the amount of sensible heat carried out in the hot flue gases.

7.4 POLLUTION BY NITROGEN OXIDES

Industrial flames which are turbulent flames of premixed or diffused gases produce nitrogen oxides in their flue gases as the result of complex chemical and physical phenomena. The chemical phenomena determine different kinetic formation/destruction reaction mechanisms; the physical phenomena, principally diffusion and turbulence, determine conditions of stratification and fluctuation in the composition, as well as in local temperature. Depending on specific conditions, these related factors lead to great variation in speeds of formation or reduction of nitrogen oxides.

First, we will review the different kinetic reaction mechanisms, then, we will consider the effects on these mechanisms of the particular conditions existing in turbulent diffusion flames, and finally, we will describe methods of controlling nitrogen-oxides pollution in industrial fireboxes.

7.4.1.　Kinetic reactions involving nitrogen oxides

For convenience, experimental study of the kinetic mechanisms is done on premixed laminar flames, which are controlled by a single variable when possible.

The bond energy of the nitrogen atom in a nitrogen compound, on the one hand, and the composition of the reaction medium, on the other, determine three simultaneous reaction mechanisms of NO formation within the flame.

Bond energy

Among the possible bonds, the nitrogen triple bond $N\equiv N$ is characterized by a very high energy (945 kJ/mol) compared to the carbon-nitrogen triple bond, $C\equiv N$ (791 kJ/mol) and single bonds such as $N-C$ and $N-H$ which are on the order of 450 kJ/mol.

Composition

Depending on the state of combustion a number of co-reactants for the formation of NO will appear and disappear. In hydrocarbon flames two distinct reaction mediums occur (Fig. 7.7), each characterized by the nature of its free radicals. Thus:

(a) In the oxidation zone or flame front there is a predominance of hydrocarbon radicals, fragments of fuel molecules (Fig. 7.7 left).

(b) In the combustion products (right side of Fig. 7.7) there is a dominance of H, OH and O radicals, and hydrocarbon radicals disappear.

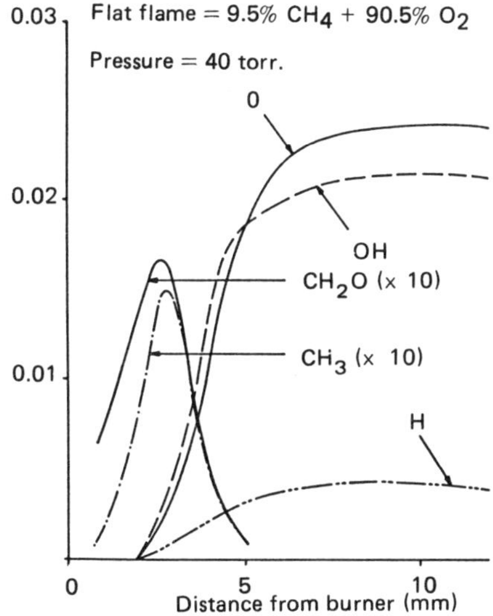

Fig. 7.7. Reaction mediums in a flame: the appearance and disappearance of reactive radicals as a function of distance from a burner.
(From: Peeters J. et Mahnen G. *Fourteenth Symposium International on Combustion*. The Combustion Institute, Pittsburgh, 1973, p. 133).

The combination of the bond energy of the existing nitrogen molecules with the composition of the reaction medium leads to three distinct mechanisms for NO formation termed "thermal, fuel-NO" and "prompt":

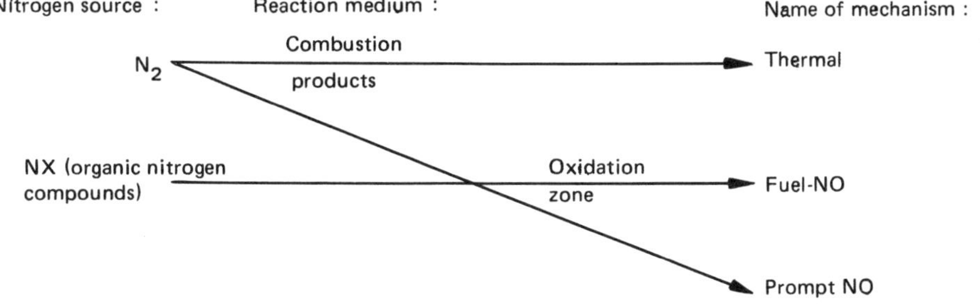

7.4.1.1. The thermal mechanism of NO formation

A large number of reactions take part in this type of reaction, the most important being (Refs. 7.16 and 7.17):

$$N_2 + O \underset{-1}{\overset{1}{\rightleftharpoons}} N + NO \quad (E_1 = 314\,\text{kJ/mol}) \quad (E_{-1} = 0)$$

$$NO + O \underset{-2}{\overset{2}{\rightleftharpoons}} N + O_2 \quad (E_2 = 163\,\text{kJ/mol}) \quad (E_{-2} = 29\,\text{kJ/mol})$$

$$N + OH \underset{-3}{\overset{3}{\rightleftharpoons}} NO + H \quad (E_3 = 0) \qquad (E_{-3} = 165\,\text{kJ/mol}?)$$

Except for very rich hydrocarbon/air mixtures, reactions numbered 3 and -3 can be ignored.

Rates of formation and decomposition of NO:

(a) Retaining only the four reactions numbered $1, -1, 2$ and -2.

(b) Assuming a constant concentration of the N atoms, the following kinetic equation is derived for the speed of formation of NO (Ref. 7.13):

$$\frac{d(NO)}{dt} = \frac{2k_1(N_2)(O)}{1 + \dfrac{k_{-1}(NO)}{k_{-2}(O_2)}} - \frac{2k_2(NO)(O)}{1 + \dfrac{k_{-2}(O_2)}{k_{-1}(NO)}} \tag{7.1}$$

where

k = the equilibrium constant,

$1, -1, 2, -2$ = subscripts denoting the specific reaction,

$(\ \ )$ = concentration of the molecule indicated.

The first (positive) group on the right-hand side of this equation represents the rate of NO formation, whereas the second (negative) group represents the rate of NO decomposition.

When (O_2) is considerably larger than (NO), which is usually the case, this equation can be simplified to:

$$\frac{d(NO)}{dt} = 2k_1(N_2)(O) - 2\frac{k_2 k_{-1}}{k_{-2}}(NO)^2(O_2)^{-1}(O) \tag{7.2}$$

In order to further simplify this relation, the concentration of nascent (O) might be defined as a function of the measurable molecular oxygen, (O_2). Two ways of doing that will be shown below. With the assumption that the equilibrium $O_2 \underset{-4}{\overset{4}{\rightleftharpoons}} 2O$ is achieved between O_2 and O, it is possible to write:

$$(O) = (O)_{\text{equil.}} = K_4^{1/2}(O_2)^{1/2} \tag{7.3}$$

Substituting Eq. (7.3) in Eq. (7.2) then gives:

$$\frac{d(NO)}{dt} = 2k_2 K_4^{1/2}(N_2)(O_2)^{1/2} - \frac{k_2 k_{-1} K_4^{1/2}}{k_{-2}}(NO)^2(O_2)^{-1/2} \tag{7.4}$$

However, the assumption of equilibrium between O and O_2 is valid only relatively far behind the flame front in the combustion products. Closer to the flame front the actual rate of formation of NO is several times greater than that calculated by the first term on the right of Eq. (7.4).

The second simplification is to assume equilibrium between the species H, O, OH, H_2, O_2 and H_2O. If the relatively slow tri-molecular reactions are disregarded, this equilibrium is controlled by the four following reactions (Ref. 7.18):

$$H + O_2 \underset{-5}{\overset{5}{\rightleftharpoons}} OH + O$$

$$O + H_2 \underset{-6}{\overset{6}{\rightleftharpoons}} OH + H$$

$$H_2 + OH \underset{-7}{\overset{7}{\rightleftharpoons}} H_2O + H$$

$$OH + OH \underset{-8}{\overset{8}{\rightleftharpoons}} H_2O + O$$

According to these, we have:

$$(O) = \frac{K_5 K_7 (H_2)(O_2)}{(H_2O)} \tag{7.5}$$

After substitution of Eq. (7.5) in Eq. (7.2), the kinetic equation for (NO) becomes:

$$\frac{d(NO)}{dt} = 2k_1 K_5 K_7 \frac{(N_2)(H_2)(O_2)}{(H_2O)} - 2\frac{k_2 k_{-1} K_5 K_7}{k_{-2}} \frac{(NO)^2 (H_2)}{(H_2O)} \tag{7.6}$$

Because reactions 5 to 8 are very rapid compared to reactions 1 and 2, this assumption appears more realistic than the preceding one. Also the rates of NO formation, as determined by experiment, approach to within a factor of about 2 to the rates calculated with the first term on the right of Eq. (7.6).

According to this mechanism the rate of NO formation depends greatly on temperature. Its overall energy of activation will be 503 kJ/mol at 1,000 K and 445 kJ/mol at 2,000 K: thus the name "thermal NO". This dependence on temperature, as well as oxygen concentration, explains the variation of NO emissions as a function of the fuel/oxidizer equivalence ratio of the flame. Figure 7.8 shows the dependence of the rate of formation of the thermal NO in relation to the temperature ($C_2H_4/O_2/N_2$ flames; equivalence ratio = 1.) Figure 7.9 shows the influence of equivalence ratio and nitrogen content on the thermal NO emission as measured in premixed flames.

The speeds of thermal decomposition given by the second terms of Eqs. (7.1), (7.2), (7.4) and (7.6) become measurable only when the NO concentrations are high. In hydrocarbon/air flames at atmospheric pressure this speed of decomposition can be disregarded.

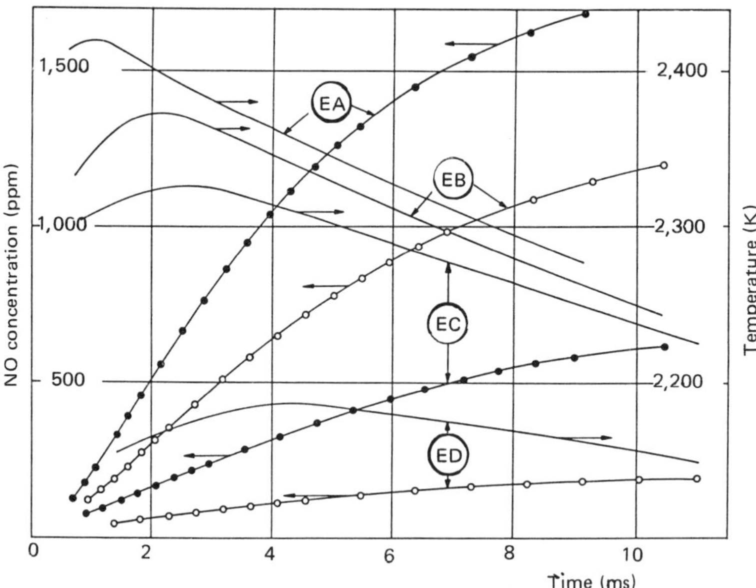

Fig. 7.8. Variations of NO in flat flames, as a function of residence
time. The flames were mixtures of $C_2H_4/O_2/N_2$ with an
equivalence of 1 and an N_2 concentration of 64%. Flame
temperatures were regulated by the rate of heat removal.
(Source : *IFP, Dept. of Documentation*).

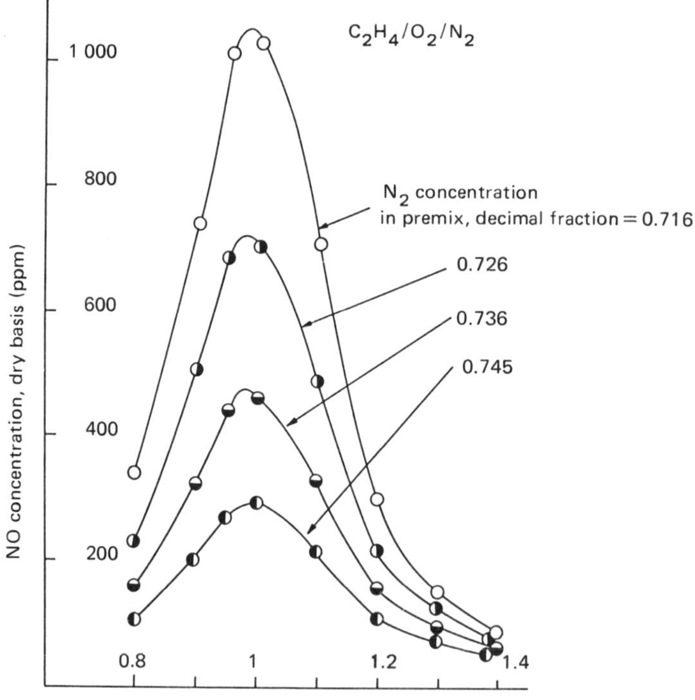

Fig. 7.9. Variations in NO in premixed flames, as a function of fuel/
oxidizer equivalence ratio and N_2 concentration.
(Source: *IFP, Dept. of Documentation*).

7.4.1.2.   The fuel-NO mechanism

Very early in a flame's oxidation zone, organic nitrogen compounds are transformed into intermediate carbon-nitrogen species by reaction with hydrocarbon radicals. The importance of the hydrocarbon radicals is shown by the rapid formation of cyanides (CN and HCN) in the flame front of a hydrocarbon/oxygen/argon mixture seeded with ammonia (see Fig. 7.10). The nitrogen intermediates are molecules or molecular fragments that are relatively small and characterized by the CN bond (such as HCN) or the NH bond (such as NH). The different intermediates are shown in Figs. 7.10, 7.11, 7.12 and 7.13, for flames doped by NH_3, C_2N_2, HCN or NO respectively, as the primary nitrogen.

These intermediate nitrogen species, (as well as nascent nitrogen) are further transformed by a double reaction path (Refs. 7.19 and 7.20):

1. Transformation of the intermediate nitrogen group into NO, through reaction with an oxygen source, principally molecular oxygen.

2. Transformation of the intermediate nitrogen group into N_2 by the reaction with a nitrogen source, principally NO itself.

The formation of nitrogen oxide and molecular nitrogen by the fuel-NO mechanism is shown on the left side of the drawing below. On the right is the thermal mechanism described in the preceding section. The competition of these two simultaneous reaction paths will finally determine the yield of NO from a fuel, as the fraction of the primary nitrogen compound that is transformed into nitrogen oxide.

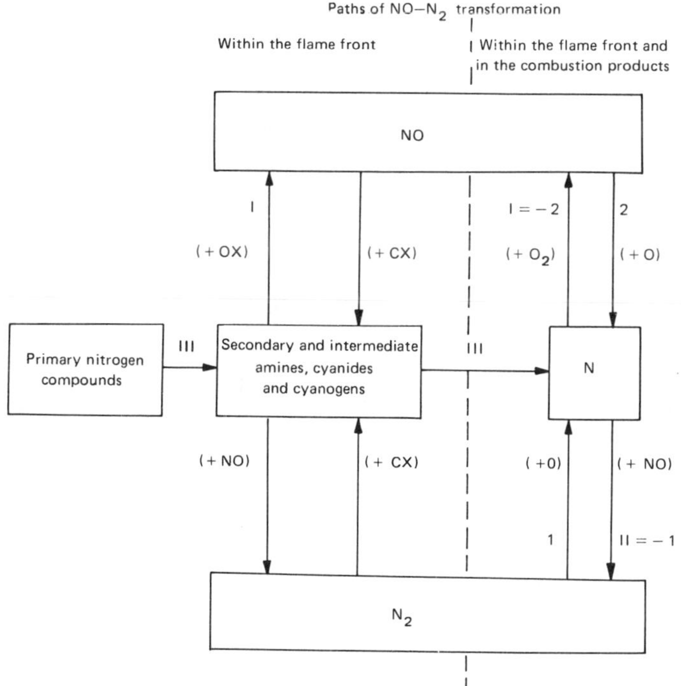

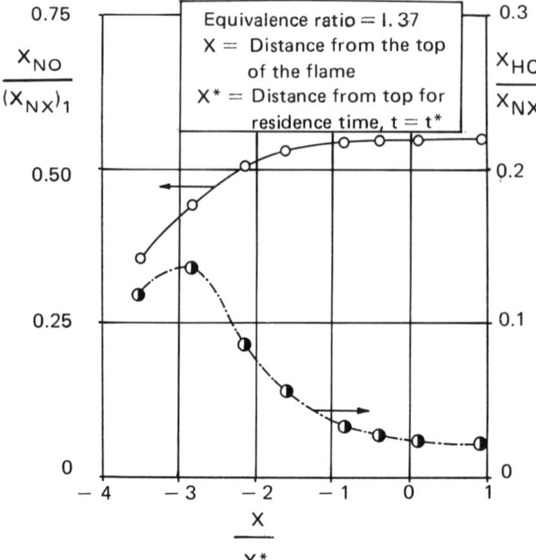

Fig. 7.10. Variation of NO and HCN concentration in a flat flame seeded with ammonia. The combustion premixture was (vol. %): 7.6 C_2H_4, 16.6 O_2, and 75.5 A with 1,630 ppm NH_3 added.
(Source: *IFP, Dept. of Documentation*).

Fig. 7.11. Variations in concentration of NO and nitro-carbons in a flat flame seeded with cyanogen. The combustion premixture was (vol. %): 7.5 C_2H_4 21.2 O_2, and 71.3 A, with 260 ppm C_2N_2 added.
(From: C. Meyer, G. De Soete. *Rev. Générale de Thermique* No. 142, 1973. p. 913).

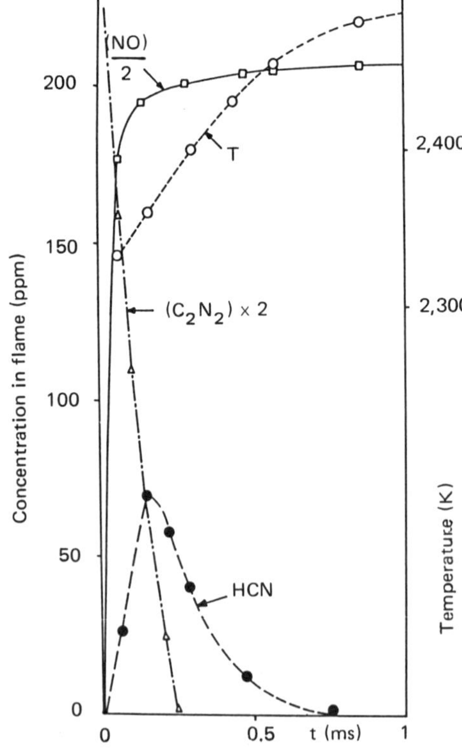

Fig. 7.12. Variations in concentration of nitrogen compounds in a flame seeded with hydrogen cyanide. The combustion premixture was methane, oxygen, and argon with a stoichiometric equivalence ratio of 1.61 and 1,025 ppm of HCN added.

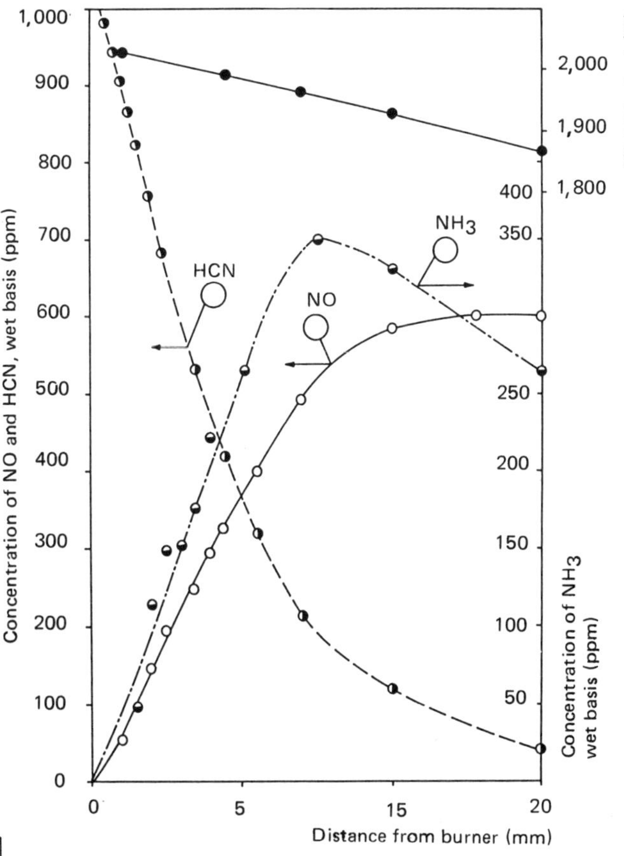

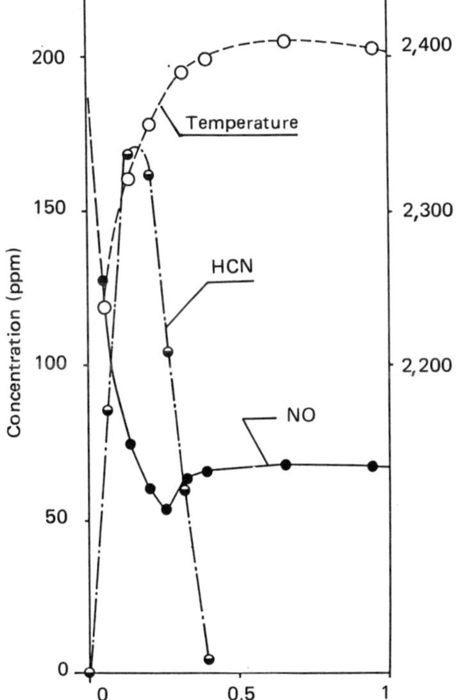

Fig. 7.13. Variations in concentration of nitrogen compounds in a flat flame seeded with NO. The combustion premix was (vol. %): 10.8 C_2H_4, 20.5 O_2, and 68.7 A, plus 290 ppm NO.
(Source: *IFP Dept. of Documentation*).

The speeds of NO formation and N_2 formation depend on the nature of the primary nitrogen compound. Table 7.1 gives some experimental equilibrium constants and activation energies.

TABLE 7.1

EQUILIBRIUM CONSTANTS AND ACTIVATION ENERGIES FOR FORMATION OF NO AND N_2 FROM COMMON NITROGEN SOURCES

Nitrogen Source	k_{A_0} (S^{-1})	E_A (kJ/mol)	k_{B_0} (S^{-1})	E_B (kJ/mol)
NH_3	$4 \cdot 10^6$	134	$1.8 \cdot 10^8$	113
$(CH)_2$	$4 \cdot 10^8$	167	$1.1 \cdot 10^{10}$	134
HCN	10^{10}	280	$3 \cdot 10^{12}$	250
N^-	$1.2 \cdot 10^7 \, X_{RC}$	250	—	—

In Table 7.1, X_i is the mole fraction of the species i, X_{N_-} designating that of the primary nitrogen compound. It is important to note that for a given primary nitrogen species, the difference, $E_A - E_B$ is always relatively small: only 20-40 kJ/mol.

As a function of the reaction rates of:

$$NO, \text{ formation, } V_A = k_{A_0} X_{N_-} X_{O_2}^b \exp\left(-\frac{E_A}{RT}\right) \qquad (7.7)$$

and

$$N_2, \text{ formation, } V_B = k_{B_0} X_{N_-} X_{NO} \exp\left(-\frac{E_B}{RT}\right) \qquad (7.8)$$

an integration up to a maximum of X_{NO} gives the approximate following equation for the yield R (Refs. 7.16, 7.17 and 7.18):

$$R \simeq \left[\frac{k_A X_{O_2}^b}{k_B (X_{N_-})_0}\right]^{1/2} = \exp\left(-\frac{E_A + E_B}{2RT}\right)\left[\frac{k_{A_0} X_{O_2}^b}{k_{B_0} (X_{N_-})_0}\right]^{1/2} \qquad (7.9)$$

According to this relation the yield for NO will:

(a) Decrease when the oxygen concentration decreases (see Fig. 7.14).

(b) Decrease when the primary nitrogen concentration increases (see Fig. 7.15).

(c) Vary only very slightly with temperature (see Fig. 7.16) according to a coefficient whose value is between 10 and 20 kJ/mol.

The rate of NO formation by the fuel-NO mechanism is incomparably higher than the rate determined by the thermal mechanism, despite the relatively low concentration of nitrogen reactant for the fuel-NO mechanism (see Fig. 7.17). Consequently, the reaction is practically complete at a short distance behind the flame front.

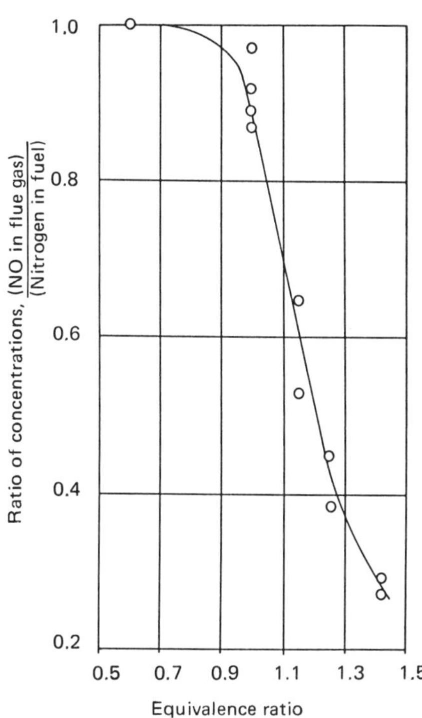

Fig. 7.14. Effects of stoichiometric equivalence ratio on the generation of NO in the combustion of a nitrogen-containing fuel mixture. The combustion premix was $C_2H_4/O_2/He$, seeded with 700-1,300 ppm of NH_3.
(From: G. De Soete. *Rev. Institut Franç. du Pétrole,* vol. 28, 1973, p. 95).

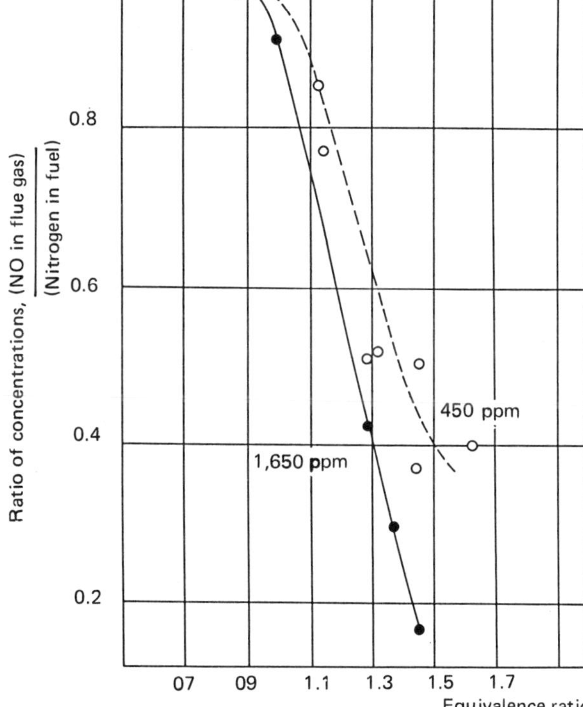

Fig. 7.15 Effect of initial nitrogen content on the generation of NO in the combustion of a nitrogen-containing fuel mixture. The combustion premix was $C_2H_4/O_2/A$, seeded with 450 ppm or 1,650 ppm of ethane diamine.
(From: G. De Soete. *Rev. Institut Franç. du Pétrole,* vol. 28, 1973. p. 95).

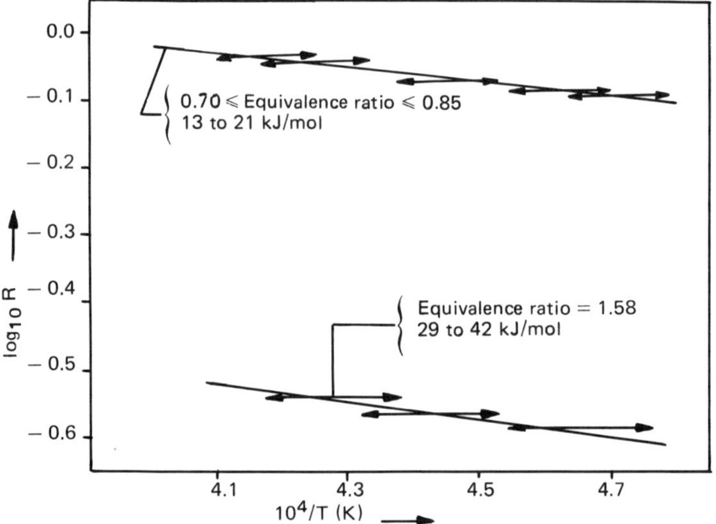

Fig. 7.16. Effects of combustion temperature on the generation of NO in the combustion of a nitrogen-containing fuel mixture. The fuel premix was $C_2H_4/O_2/A$, plus 260 ppm C_2N_2.
(From: G. De Soete. *Rev. Institut Franç. du Pétrole,* vol. 33, 1978, p. 747).

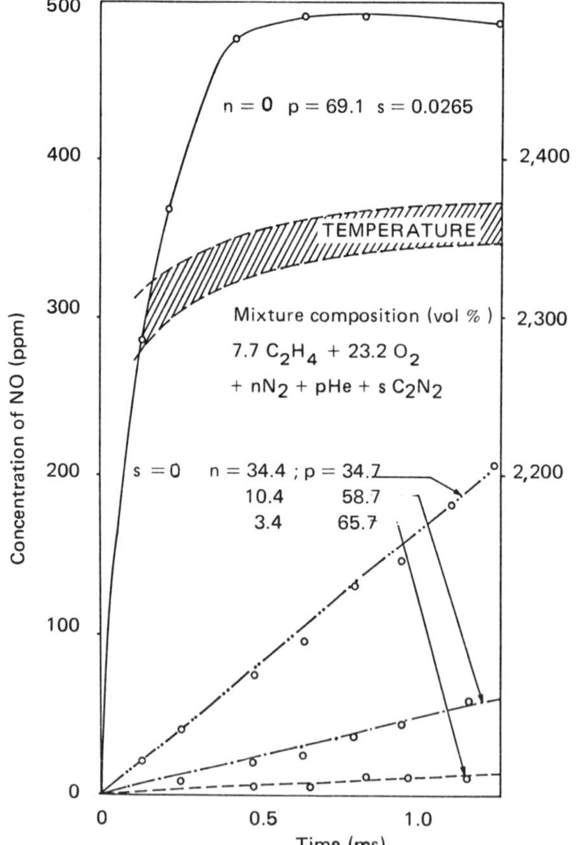

Fig. 7.17. Effects of nitrogen source on speed of NO formation during combustion. A fuel mixture seeded with cyanogen (solid line) is compared to mixtures containing various concentrations of N_2.
(From: G. De Soete. *Rev. Institut Franç. du Pétrole,* vol. 33, 1978, p. 747).

Nitrogen oxides also behave as primary nitrogen sources when they are added to an inflammable mixture (see Figs. 7.10, 7.18 and the diagram of fuel-NO versus thermal mechanisms). This similarity acquires particular interest for emission-reduction techniques to be discussed later, nitrogen oxide entering a flame front being partially transformed into molecular nitrogen by exactly this same fuel-NO mechanism.

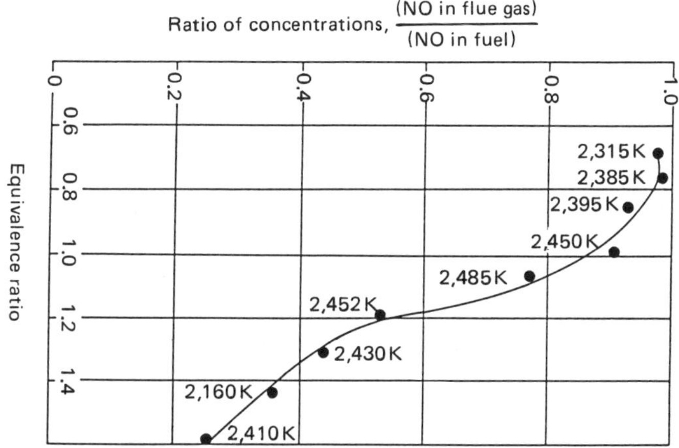

Fig. 7.18.   Effect of equivalence ratio on the persistence of NO present in a fuel mixture. Mixtures contained ethylene, oxygen and argon seeded with about 220 ppm of NO. Experimental temperatures are indicated by the data points.
(Source : *IFP Dept. of Documentation*).

7.4.1.3.   The mechanism of "prompt" NO

The relative slowness of NO formation by the thermal mechanism is due to the high endothermic heats (314 kJ/mol) required for reaction (1), which is the only reaction to rupture the triple bond of molecular N_2 in the zone following the flame. Because hydrocarbon radicals are present in the flame front, however, molecular N_2 bonds can be ruptured by reactions 9 and 10, which are only slightly endothermic and therefore more rapid (Ref. 7.19):

$$N_2 + CH \xrightarrow{9} N + HCN \qquad (\Delta H \cong +\; 8 \quad kJ/mol)$$

$$N_2 + C_2 \xrightarrow{10} 2\,CN \qquad (\Delta H \cong +\; 28 \quad kJ/mol)$$

By the reaction products they lead to, such reactions enter directly into the formation of nitrogen intermediates in the fuel-NO mechanism. Thus this mechanism of "prompt NO" is only a particular case of the fuel-NO mechanism, with N_2 behaving like an organic nitrogen compound in the flame front.

The overall rate of advance NO formation by this route can be approximated experimentally, and it is given in the last line of Table 7.1 (with $X_N = X_{N_2}$) and with X_{RC} = initial mole fraction in the fuel. This rate is definitely greater than the rate of the thermal mechanism as shown by the profiles of NO concentration measured behind premixed flames (Fig. 7.19). In the case of rich mixtures, these curves show a discontinuity close to the flame front. In Fig. 7.19 nitrogen intermediate, HCN, is also seen indicated; an extended analysis of the phenomenon shows this to be practically the only nitrogen intermediate in the mechanism of prompt NO (Refs. 7.21 and 7.23).

In Fig. 7.20 the amount of prompt NO formed is shown as a function of the equivalence ratio of the mixture with temperature as parameter. These curves exhibit a maximum, which is explained through the effects that hydrocarbon and oxygen concentrations have on the overall reaction rate, that is, when the equivalence ratio increases, the concentration X_{CX} increases at any given temperature and enhances the speeds of reactions 9 and 10 with a consequent increase in the amount of intermediate (HCN) formed. At the same time, however, the concentration of O_2 decreases, and this favors the preferential transformation of the intermediate HCN into N_2 (see fuel-NO mechanism above).

Except in the case of relatively rich mixtures, the contribution of prompt NO to the total emission of NO in the hydrocarbon/air flames is relatively small. As can be seen in the diagram of reaction mechanisms the thermal NO, fuel-NO and prompt NO are not independent each of the other but are interconnected by:

(a) The formation of nitrogen atoms from the fuel-NO mechanism, possibly as intermediates (for example, by reaction 9).

(b) The tendency of NO formed by the thermal path to behave as a fuel nitrogen when it re-enters in an oxidation zone in the presence of *CX* radicals.

Thermodynamically speaking, there should be practically no NO_2 present at the temperatures that predominate in the flames. Nevertheless, many researchers mention appreciable quantities of NO_2 found in the flue gases by analysis – at least in certain particular cases (Refs. 7.21 and 7.24) as for example, when samples taken in the flame front are analyzed. As long as this NO_2 is not a product formed during the sample taking, or in the analyzer, it will be a matter of a concentration of NO_2 that is higher than the thermodynamic equilibrium, formed from NO by the reaction:

$$NO + HO_2 \overset{11}{\rightarrow} NO_2 + OH$$

7.4.1.4. Reduction of nitrogen oxides on solid particles

The reduction of nitrogen oxide on the surface of solid catalysts has been known for a long time and is currently used in the treatment of exhaust gases from internal combustion engines. More recently, it has been discovered that

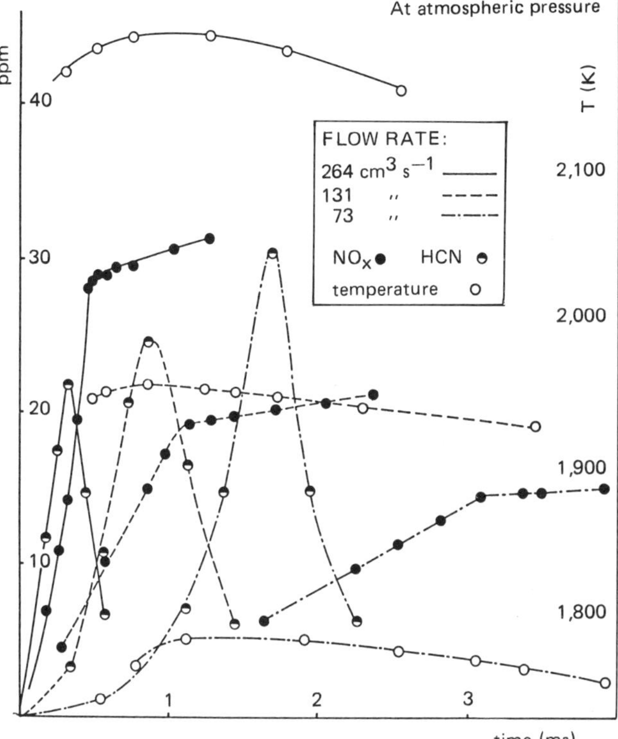

Fig. 7.19. Experimental demonstration of NO formation via carbo-nitrogen intermediates, or the "prompt NO" mechanism. The fuel premix contained (vol. %): C_2H_4 10.39, O_2 23.9 and N_2 65.8.

(From: G. De Soete. *Fifteenth Symposium International on Combustion*. The Combustion Institute, Pittsburgh, 1974, p. 1093).

Fig. 7.20. Effects of temperature and equivalence ratio on NO production via the "prompt NO" mechanism.
(From: G. De Soete. *Fifteenth Symposium International on Combustion*. The Combustion Institute, Pittsburgh, 1974, p. 1093).

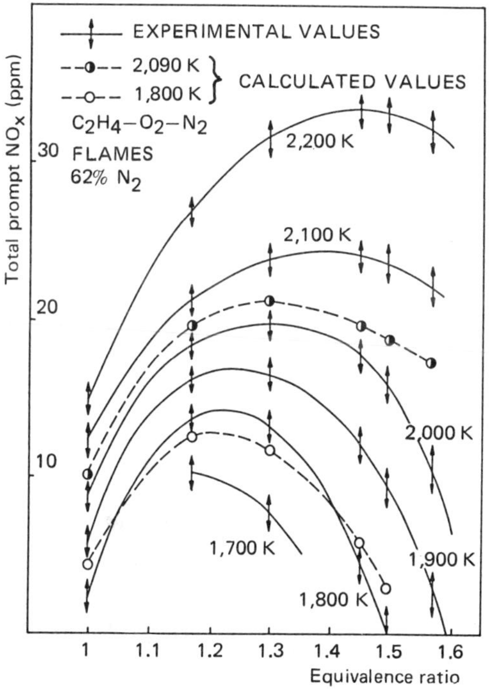

reactions transforming NO into molecular nitrogen also take place on the surfaces of the soot or carbon particles present in the flame and combustion products of industrial furnaces (Refs. 7.25 and 7.26). These reactions involve a relatively complex mechanism that passes through ammonia and hydrogen cyanide. Details of this mechanism are given in the references at the end of this chapter. It can be summarized by the following diagram where $(-\,C)$ and $(-\,H)$ represent solid-bound carbon and hydrogen atoms (Ref. 7.27).

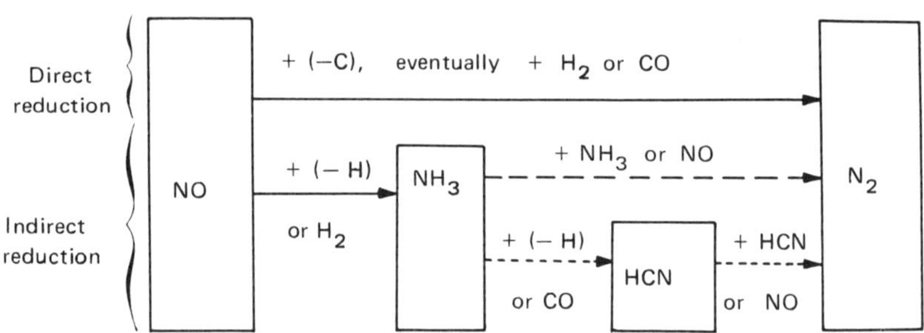

This diagram shows a direct reduction of nitrogen oxide into molecular nitrogen, as well as an indirect reduction that goes through the formation of ammonia and hydrogen cyanide. These reactions are not properly catalytic, because carbon and hydrogen atoms can participate as reactants. Normally, direct reduction dominates, principally at high temperatures: NO is reduced by the carbon atoms of the soot, charcoal or coke according to the following overall reaction:

$$NO + a\,(-\,C) \xrightarrow{12} 0.5\,N_2 + (2a - 1)\,CO + (1 - a)\,CO_2$$

The proportions of CO and CO_2 from this reaction depend on the temperature and kind of particle. The reaction takes place through a dissociative adsorption of a molecule of NO on two carbon sites; the oxygen atom being chemisorbed to form an oxy-carbon surface complex, while the nitrogen is weakly adsorbed:

$$\begin{vmatrix} -C \\ -C \end{vmatrix} + NO \;\rightarrow\; \begin{vmatrix} -C-O \\ -C---N \end{vmatrix}$$

By subsequent rapid surface diffusion, the nitrogen atoms recombine into molecular nitrogen:

$$2\,(-C---N) \;\rightarrow\; 2\,(-C) + N_2$$

The oxy-carbon surface complex decomposes and the oxygen is desorbed in the form of CO or CO_2:

$$\begin{vmatrix} -C-O \\ \end{vmatrix} \rightarrow \begin{vmatrix} \\ -C \end{vmatrix} + CO \quad or \quad \begin{vmatrix} -C-O \\ -C-O \end{vmatrix} \rightarrow \begin{vmatrix} \\ -C \end{vmatrix} + CO_2$$

The presence of gaseous reducing molecules, such as molecules of hydrogen or CO favors the NO reduction because they react with the surface complexes to form H_2O or CO_2.

Nevertheless, the presence of a hydrogen source, either as H_2, or as solid-bound hydrogen atoms, initiates the indirect reduction in parallel with direct reduction. The ammonia thus formed is ultimately transformed into molecular nitrogen either directly as by reacting with NO, or indirectly, as by passing through the intermediate formation of HCN, as shown in the drawing.

The importance of this mechanism of NO reduction on soot particles formed by industrial flames from heavy fuel is still difficult to estimate. Considering the relatively high density of concentration of solid particles (coke, fly-ash) it could play a significant role in flames from pulverized coal.

7.4.2.   The effects of diffusion and turbulence on nitrogen oxide formation

Any variation in concentration or temperature of reactants N_2, NX, O_2, with space and time should in principle have an influence on the formation and reduction of nitrogen oxides, hence on their concentration in the flue gases. Within a flame, space variations of concentration and temperature both depend on local gradients, which results in a stratification of the reaction medium. Time dependent variations depend on the amplitude and frequency of the local fluctuations of concentrations and temperature.

In a turbulent diffusion flame, the laws of turbulence control both phenomena (the distribution of the intensity of turbulence and the degree of turbulent diffusion). The complexity of these phenomena renders turbulent diffusion flames inaccessible to meaningful experiments leading to an understanding of the effect of stratification and fluctuation on the NO emissions. For this reason we approach this problem through analogic studies of laminar premix flames subjected to well defined stratifications and periodic fluctuations of compositions and temperature.

7.4.2.1.   Effects of diffusion and stratification

The oxidation zone of a diffusion flame is characterized by the existence of a natural staging phenomenon. A study of the effects of such steps on nitrogen oxides can be carried out on two-stage premix flames made up of two fractions with different equivalence ratio; and the results can be useful for understanding phenomena of diffusion flames (Ref. 7.25).

A premixed flame composed of i different homogeneous fractions is defined. Let f_i be the volume fraction of the gases burned in layer i, and let N_i be the ratio of molecules of combustion products to molecules of this layer. It is possible to distinguish two extreme cases: **case A**, in which combustion products from one layer do not enter the second layer; and **case B**, in which the combustion products of one layer do enter the second layer.

Case A: If the combustion products formed in one layer are not introduced into the combustion front of the other layer, the total emission of nitrogen oxides $(NO)_s$ from the stratified flame will approach the arithmetical average for the different individual fractions $(NO)_{hi}$:

$$(NO)_s = \frac{\sum_i f_i N_i (NO)_{hi}}{\sum_i f_i N_i} \tag{7.10}$$

The existence of this limiting case was partially verified by:

1.   Non-stabilized flames propagating across stratified inflammable mixtures, as occurs in stratified combustion engines; and

2.   Stabilized flames that are artificially stratified when the layers i are either both rich or both poor.

The case of non-stabilized flames is illustrated in Fig. 7.21. It will be noticed that when the total NO emissions of these stratified flames are calculated according to Eq. 7.10, they can be either greater or smaller than those of non-stratified flames with identical overall equivalence ratios.

Case B: The results are completely different when the nitrogen oxides along with the combustion products of one layer are introduced into the oxidation zone of another layer, because the introduced nitrogen oxides are subjected to the fuel-NO reaction mechanism and partially reduced to molecular nitrogen. This case can be interpreted in the light of results obtained with a bi-stratified flame stabilized so that all the combustion products including the nitrogen oxides from layer f_i do cross the combustion zone of second layer f_2. After mixing with the reactants already present in layer f_2, the introduced combustion products contribute to a reaction medium characterized by a mixed composition (C_m) which will be the locus of the second oxidation zone. Assume that the mixing time is negligible compared to the duration of the chemical reactions (a case difficult, if not impossible to achieve) and allow that there is no thermal NO (either because the thermal NO is low relative to that formed by the fuel-NO mechanism or because the reactants do not contain N_2). Then the combustion zone of the f_1 fraction is fed by concentration $(NX)_1$ of organic nitrogen and is characterized by fuel-NO yield R_1. The second layer f_2, is fed by a concentration $(NX)_2$ and is characterized by yield R_m, which takes into account the mixed composition C_m. The ratio of the rate of flows of reactants feeding the two layers, f_1/f_2 will be designated by ε. The overall yield of fuel-NO characterizing the bi-stratified flame is then given by :

$$R_s = \frac{R_1 R_m (NX)_1 + \varepsilon R_m (NX)_2}{(NX)_1 + \varepsilon (NX)_2} \tag{7.11}$$

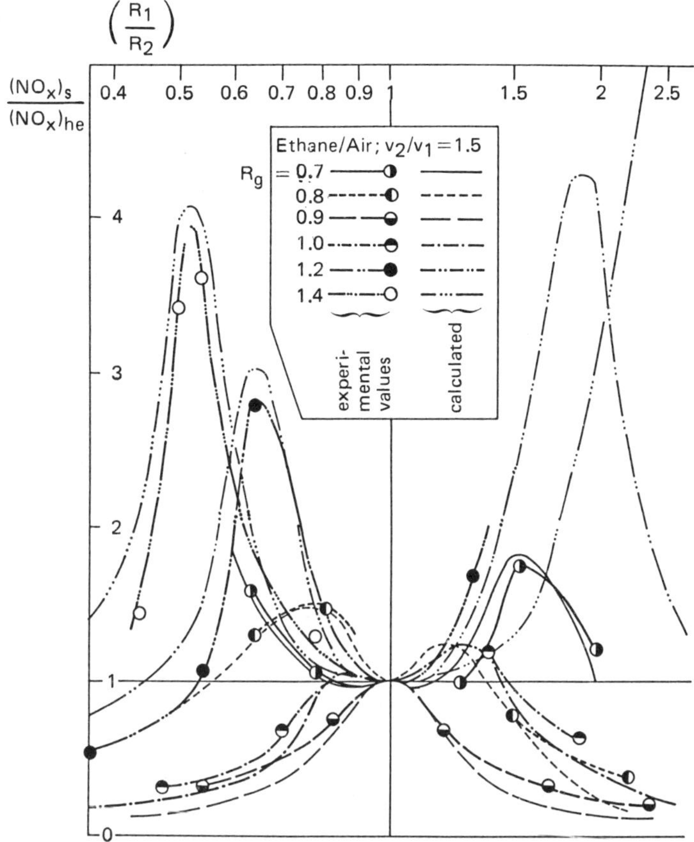

Fig. 7.21. A comparison of experimental and calculated concentrations of NO in non-stabilized flames.

The first term in the numerator of this expression represents the contribution of the first layer to the total yield, R_s, and is thus equal to the yield obtained in the combustion products of the first layer $(R_1 (NX)_1)$ multiplied by the yield R_m characteristic of the second layer. The second term of the numerator corresponds to the contribution of the second layer in the total yield R_s. The validity of Eq. 7.11 has been demonstrated experimentally, most particularly in the extreme cases where either $(NX)_1$ or $(NX)_2$ were equal to zero.

It is interesting to note that, according to Eq. 7.11, the total emission of nitrogen oxide $(NO)_s$ of these stratified flames is less than that of a non-stratified flame $(NO)_h$ of identical overall equivalence ratio as can be seen from Fig. 7.22. The curves given in this figure have been calculated according to Eq. 7.11 for an inflammable mixture ethylene/oxygen/argon doped with ammonia. It is interesting to see that the rich → lean stratification is more favorable from the point of view of emission reduction than is a lean → rich stratification.

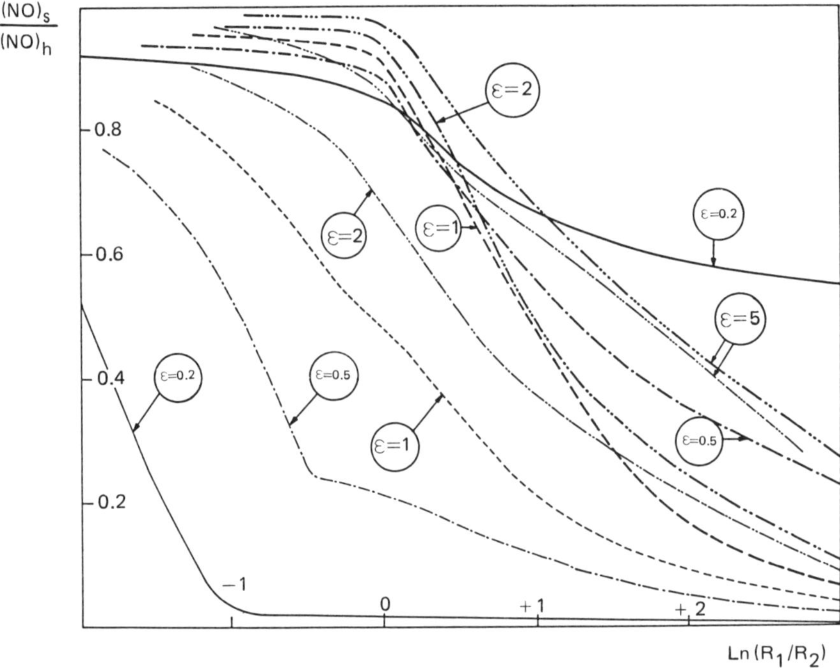

Fig. 7.22. Formation of NO in stratified and non-stabilized flames, as calculated from Eq. (7.11).

These fundamental results, as well as similar results obtained with the thermal mechanism, furnish quantitative as well as qualitative information on the possibilities of techniques for reducing nitrogen oxides by combustion in stages.

7.4.2.2. Effects of turbulence

Especially in the case of diffusion flames, turbulence provokes fluctuations in the local composition and hence in the combustion temperature. No matter what the mechanism for forming nitrogen oxides, its overall reaction-rate, V_F, is function of the instantaneous concentrations of oxygen (O_2), molecular nitrogen (N_2), or organic nitrogen compounds (NX), as well as temperature :

$$V_F = f\,[(O_2)\,(N_2) \quad \text{or} \quad (NX)\,(T)] \tag{7.12}$$

The time average of the reaction rate V_F, which will determine the NO concentration in the flue gases, can not be identified simply with a reaction speed expressed as a function of the average values of the concentration $(\overline{O_2})$, $(\overline{N_2})$ or $(\overline{NX})$ and the temperature, $\overline{T}$.

This problem was treated theoretically by several researches (Refs. 7.29 and 7.30). A verification of these theories by experimental measurements obtained with turbulent diffusion flames has not yet been supplied. Certain experimental studies carried out on laminar flames whose composition and temperature were subjected to periodic fluctuations, have nevertheless already shown that favorable as well as unfavorable effects should be expected from the point of view of NO_x emission (Ref. 7.31).

7.4.3.   Controlling pollution from nitrogen oxides

The chemistry described above indicates that measures to control pollution from nitrogen oxides differ depending on whether the source of nitrogen is the atmosphere or the fuel. Either way, the potential pollution is important because nitrogen oxides are poisonous.

7.4.3.1.   Atmospheric nitrogen

The points at which it is possible to intercept the formation of nitrogen oxides are first the amount of excess air, and second, the temperature of the flue gases. The residence-time at high temperature of the combustion products is also involved; however, this factor is difficult to determine exactly in large turbulent flames, since it depends on the geometry of the enclosure, the thermal charge, and the characteristics of the burners.

Excess of air

In most fireboxes, combustion takes place in the presence of an excess of air varying from 4-5 to 40-50%. An excess of air, if it does permit complete combustion of the fuel, tends to decrease the temperature of the flame. If the increase in oxygen concentration favors formation of nitrous oxide, the corresponding decrease in flame temperature goes in the opposite direction. Figure 7.23 indicates the nitrous oxide concentration in flue gases from an experimental furnace using natural gas which is a fuel that did not have combined nitrogen. Starting at an inverse equivalence ratio of air to fuel of around 8, the concentration of NO begins by increasing with the excess air, then after an inverse equivalence ratio of about 11, decreases as the temperature progressively declines with the increase in oxygen concentration. This concentration profile holds for industrial heaters burning heavy fuel oil; the influence of excess air varies depending on the type of heater, as is indicated for two heaters in Fig. 7.24, one heater equipped with front wall firing, the other with angular firing.

Temperature

The influence of maximum recorded flame temperature has appeared already in Fig. 7.23. Similar effects are observed when the temperature is raised by preheating the combustion air. Thus, the NO content in an experimental furnace

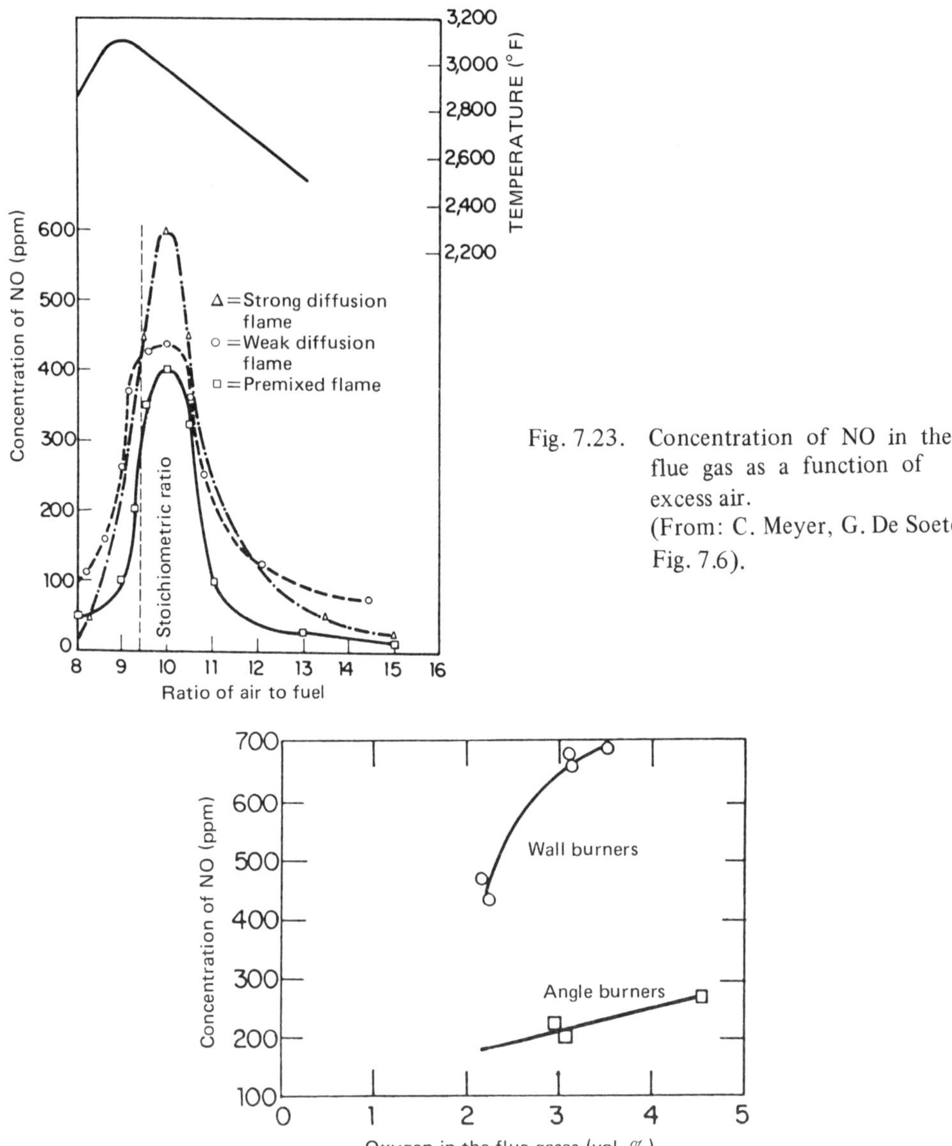

Fig. 7.23. Concentration of NO in the flue gas as a function of excess air.
(From: C. Meyer, G. De Soete, Fig. 7.6).

Fig. 7.24. Concentrations of NO in the flue gases, as a function of burner position.
(From : C. Meyer, G. De Soete, Fig. 7.7).

using natural gas goes from 100 ppm without air preheat to 700 ppm when the air is preheated to 600° C (Ref. 7.33). In a natural gas-fired boiler associated with a 250 MW turbo-alternator group, the NO content of the flue gas goes from 100 ppm without preheating to 200 ppm with air preheated to 250° C (Ref. 7.33).

7.4.3.2.　Nitrogen in the fuel

As indicated by the foregoing chemical equations for the fuel-NO mechanism (Table 7.1), organic nitrogen in the fuel is easily transformed into nitrous oxide, the energy of activation being relatively moderate. This recommends a search for a way to reduce the nitrogen content of the fuel by pre-treatment. We have seen that hydrotreating (i.e., catalytic hydrogenation) simultaneously reduces the sulfur content and nitrogen content of distilled petroleum fuels. But this process is economically applicable only to distilled products, and not to heavy residual fuels which have the most sulfur and nitrogen.

Certain burners with rotating air partially achieve combustion in two concentric streams of air and the delay in the mixing in the secondary air generally leads to a reduced NO content. The first stage occurs at high temperature but with a slight deficiency of air, while the second occurs at a lower temperature with complementary air necessary for combustion.

True two-stage combustion is achieved when a special injection device permits introducing secondary air at the desired point downstream of the primary combustion zone. An experimental study, which was carried out with such a device plus means for cooling the gas before injecting secondary air, has shown that the emission of NO is reduced as the air to primary combustion is reduced. The most favorable conditions were obtained when primary air was 70% of stoichiometric; results with two fuel oils having nitrogen contents are summarized in Table 7.2.

TABLE 7.2

EFFECTS OF TWO-STAGE COMBUSTION ON
PRODUCTION OF NO

primary air = 0.7　of stoichiometric
total air　　= 1.15 of stoichiometric

Fuel flow rate	5 gal/hr NO_x (ppm)	3.5 gal/hr NO_x (ppm)
Fuel A (0.1% nitrogen)		
Without interstage cooling	147 (320)	111 (494)
15 cm between stages	116 (251)	95 (256)
30 cm between stages	104 (270)	79 (274)
Fuel B (1% S, 0.85% N)		
Without interstage cooling	173 (315)	135 (416)
15 cm between stages	144 (271)	118 (341)
30 cm between stages	115 (253)	92 (288)

Numbers in parentheses indicate NO_x concentrations when the total 1.15 of stoichiometric air is introduced in the first stage.

When applied in approximately the same way in the furnaces of power generators, this technique achieved average reduction in NO_x of 50% using natural gas and of 38% using fuel oil or coal (Ref. 7.32). One might have expected better results for the fuel oil and coal, which have more combined nitrogen in them than the natural gas, and it is possible that the operating conditions at the various installations were not exactly comparable. In any event, any action that tends to reduce the formation of NO_x will apply to both atmospheric nitrogen and combined nitrogen even though it may act in different ways.

Recirculating flue gas

Part of the cooled flue gas can be recycled either to the firebox or to the burners. Recirculation to the firebox affords a means for controlling heat transfer in convection-section superheaters in large boilers, but this recirculation has no influence on the flames and it does not affect the NO content of the flue gas. Recirculation to the burners, on the other hand, dilutes the combustion air with inert gases, thus reducing the oxygen concentration and the temperature of the flame, effects that both favor lower NO emissions.

During experiments, a recirculation of 18% of the flue gas to the combustion air caused the maximum temperature of the flame to fall 130° C while the NO emission was reduced 75%. The fuel was a domestic fuel oil (Ref. 7.32). Analogous results (Refs. 7.34, 7.35, 7.36 and 7.37) are summarized in Figs. 7.25 and 7.26 for liquid fuels and natural gas, respectively.

Injecting water into the flame

Injecting water, either as atomized liquid or steam, affords another way to lower temperature levels in the flame. Since the presence of water also reduces formation of solid particles, this technique also permits reducing the excess air. In a sample case, water injection in a furnace for a 250 MW generator lowered the NO emissions from 330 ppm to 110 ppm. However, this method lowers thermal efficiency and enhances the risk of corrosion.

Fluid-bed combustion

This relatively recent innovation consists of burning the fuel across a fluidized bed of solid particles. It results in excellent heat transfer and causes a considerable lowering of the combustion temperature levels.

Table 7.3 shows the results obtained by different authors (Ref. 7.38). The nitrogen oxide concentration in the flue gas is still high for coal with a high combined nitrogen content, due perhaps to observed temperatures lower than those existing in the zone of NO formation.

Fluidized bed combustion also permits reducing emissions of sulfur oxides. For this purpose alkaline refractory particles such as dolomite or magnesium are

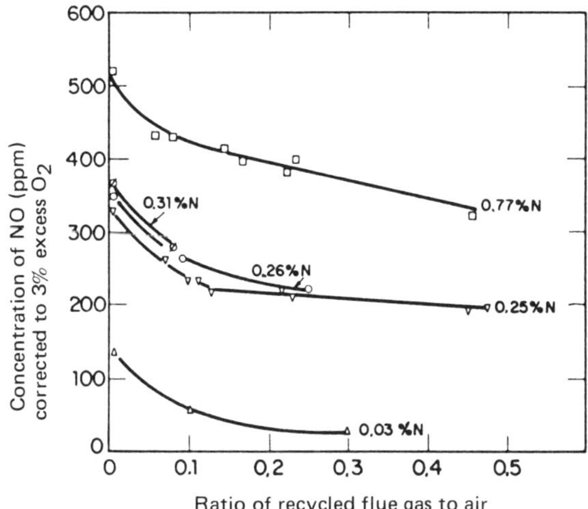

Fig. 7.25.   Effects of recycling flue gases on production of NO from fuel oil.
(From : C. Meyer, G. De Soete, Fig. 7.12).

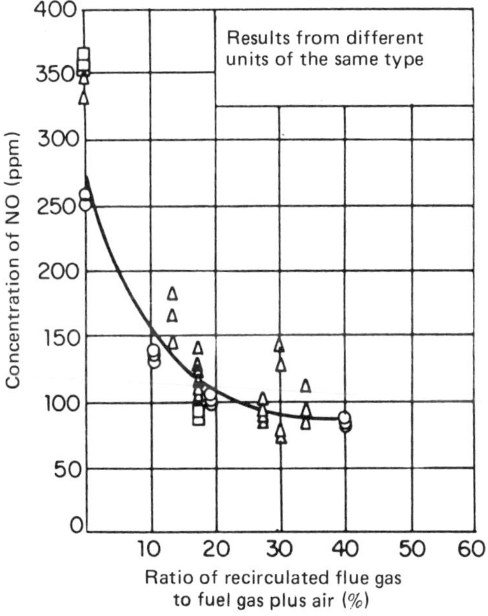

Fig. 7.26.   Effects of recycling flue gas on production of NO from natural gas.
(From: C. Meyer, G. De Soete, Fig. 7.13).

TABLE 7.3

NO$_x$ EMISSIONS FROM FLUIDIZED-BED COMBUSTION

Approximate maximum gas temperature ($^\circ$C)	Fuel	Nitrogen content of fuel (wt. %)	Type of unit	Composition of combustion air	NO concentration of dry flue gases (ppm)
870	Coal	1.4	Experimental fluid-bed furnace at pressure 1 atm.	O_2/N_2	580 ± 20
870	Coal	1.4		$O_2/$Argon	580 ± 20
870	Coal	1-1.5		Air	400 at 800
870	Natural gas	0		Air	60 at 100
840	Coal	1.4	Experimental fluid-bed furnace with two-stage combustion.	Air	70
800	Coal	1.9	Fluid-bed boiler at atm. pressure; 3 MW.	Air	400 at 500
800	Heavy fuel oil	0.3			70 at 160
850	Propane	0			15 at 20
800	Coal	1.9	Fluid-bed boiler for about 1.3 MW.	Air preheated to 240-300°C.	50 at 200

used, and the effect of these materials on the formation of NO remains controversial.

In sum, control of nitrogen oxide formation in combustion has principally to do with either manipulating the oxygen content in the combustion zone, or by reducing the temperatures reigning in this same zone. The choice of method depends on the type of firebox and the kind of fuel used.

REFERENCES

7.1   HUMERY, R. – *La lutte contre les fumées, poussières et gaz toxiques.* Dunod, Paris, 1933.

7.2   VAN TIGGELEN, A. – *Oxydations et combustions.* Chapter VIII. Editions Technip, Paris, 1968.

7.3   WAGNER, H. Gg. – "Soot formation in combustion". *17th International Symposium on Combustion. The Combustion Institute,* Pittsburgh, 1979, p. 3-19.

7.4   HAYNES, B.S., WAGNER, H. Gg. – "Soot formation". *Prog. Energy Combust. Sci.,* 1981, vol. 7, p. 229-273.

7.5 FENIMORE, C.P., JONES, G.W. – "Oxidation of soot by hydroxyl radicals". *J. Phy. Chem.*, 1967, **71**, p. 593-597.

7.6 HOWARD, J.B., NEOH, K.G. – "Particulate carbon-formation during combustion". *Proceedings of the General Motors Symposium*, 14-16 Oct. 1980, Warren, Michigan.

7.7 COTTON, D.J., FRISWELL, N.J., JENKINS, D.R. – "The suppression of soot emission from flames by metal additives". *Combustion and Flame*, 1971, **17**, p. 87-98.

7.8 FEUGIER, A. – "Effect of metal additives on the amount of soot emitted by premixed hydrocarbon flames". *Advances in Chemistry Series*, No. 166, *Evaporation – Combustion of Fuels*, 1978, Chapter 12.

7.9 NAGLE, J., STRICKLAND-CONSTABLE, R.F. – "Oxidation of carbon between 1,000-2,000° C". *Proceedings of the 5th Conference on Carbon, Pergamon Press*, Londres, 1962, p. 154-164.

7.10 FEUGIER, A. – "Soot oxidation in laminar hydrocarbon flames". *Combustion and Flame*, 1972, **19**, p. 249-256.

7.11 LEE, K.B., THRING, M.W., BEER J.M. – "On the rate of combustion of soot in a laminar soot flame". *Combustion and Flame*, 1962, **6**, p. 137-145.

7.12 MONNOT, G. – "Contribution à la mesure instantanée des températures de flammes éclairantes". *Rev. Inst. Franç. du Pétrole*, 1954, vol. IX, p. 11-12.

7.13 "Etude sur les émissions de poussières des installations thermiques fonctionnant au fuel oil lourd". *CPDP*, Paris 1977.

7.14 "L'évolution de l'utilisation des fuels lourds dans l'industrie". *Journée d'Etude de l'AFTP*, 8 Nov. 1979.

7.15 FEUGIER, A. – "Combustion des fuels lourds No. 2: Influence de la nature du fuel". *Internal Doc. IFP*, Ref. 28 989, March 1981.

7.16 DE SOETE, G.G. – "Overall kinetics of nitric oxide formation in flames". *La Revista dei Combustibili*, **29**, 1975, p. 35.

7.17 BOWMAN, C.T. – "Kinetics of pollutant formation and destruction in combustion". *Progress in Energy and Combustion Science*, vol. 1, No. 1, 1975, p. 33.

7.18 THOMPSON, D., BROWN, T.D., BEER, J.M. – "NO_x formation in combustion". *Combustion and Flame*, **19**, 1972, p. 69.

7.19 FENIMORE, C.P. – "Formation of nitric oxide from fuel nitrogen in ethylene flames". *Combustion and Flame*, **19**, 1972, p. 289.

7.20 DE SOETE, G.G. – "Overall reaction rates of NO and N_2 formation from fuel nitrogen". *Fifteenth Symposium International on Combustion. The Institute*, Pittsburgh, 1974, p. 1 093.

7.21 DE SOETE, G.G. – "La formation des oxydes d'azote dans la zone d'oxydation des flammes d'hydrocarbures". Rapport du Contrat No. 53-56. *Ministère de la Protection de la Nature et de l'Environnement*, June 1975 (Doc. *IFP*: Report No. 23 309).

7.22 FENIMORE, C.P. – "Formation of nitric oxide in premixed hydrocarbon flames". *Thirteenth Symposium International on Combustion. The Combustion Institute*, Pittsburgh, 1971, p. 373.

7.23 EBERIUS, K.H., JUST, Th., RITTWAGEN, H. – "NO formation in the primary reaction zone of low pressure flames". *Second European Symposium on Combustion. French Section of the Combustion Institute*, Orléans, 1975, p. 272.

7.24 MERRYMAN, E.L., LEVY, A. – "Nitrogen oxide formation in flames: the roles of NO_2 and fuel nitrogen". *Fifteenth Symposium International on Combustion. The Combustion Institute*, Pittsburgh, 1974, p. 1073.

7.25 DE SOETE, G.G., FAYARD, J. – "Reduction de l'oxyde d'azote par la suie dans les produits de combustion". *Rev. Inst. Franç. du Pétrole*, vol. 33, 1978, p. 947.

7.26 LEVY, J.M., CHAN, L.K., SAROFIM, A.F., BEER, J.M. – "NO/char reactions at pulverized coal flames conditions". *Eighteenth Symposium International on Combustion. The Combustion Institute*, Pittsburgh, 1981, p. 111.

7.27 DE SOETE, G.G. — "Heterogeneous reduction of nitric oxide on solid particles". *La Rivista dei Combustibili*, **35**, 1981, p. 1. — See also: "Reduction of nitric oxide by solid particles", Chapter 8 of "EPA Projects Decade Monograph on Coal Combustion", published by *US Environmental Protection Agency*, Research Triangle Park, N. Carolina (to be published). — "Aspects chimiques de la combustion du charbon pulvérisé". *Rev. Inst. Franç. du Pétrole*, vol. 37, 1982, No. 4, p. 503.

7.28 DE SOETE, G.G. — "Etude paramétrique des effets de là stratification de la flamme sur les émissions d'oxydes d'azote". *Rev. Inst. Franç. du Pétrole*, vol. 32, 1977, No. 3, p. 427 et No. 4, p. 595.

7.29 BORGHI, R. — *Adv. Geophys.*, **18B**, 1974, p. 349.

7.30 PRATT, D.T. — "Mixing and chemical reaction in continuous combustion". *Progress in Energy and Combustion Science*, **1**, 1976, p. 73.

7.31 DE SOETE, G.G. — "Effets des fluctuations basse fréquence de la vitesse, de la composition et de la température de flammes sur les émissions d'oxydes d'azote thermique". *Doc. IFP*, Report No. 26 189, 1978.

7.32 MEYER, C., DE SOETE, G. — "Formation des oxydes de l'azote dans les fours et les chaudières industriels". *RGT*, n° 142, Oct. 1973.

7.33 LEBLANC, B. — *Journée d'Etudes de la Sté des Ingénieurs Civils de France*, Paris, 1972.

7.34 MARTIN, G.B., BERKAU, E.E. — *14th Symposium on Combustion*, Penn. State University, 1972.

7.35 MEYER, C. MENEL, P. — Internal Report *IFP*, Paris, 1973.

7.36 MAUSS, F., MEYER, C., PERTHUIS, E. — *IFRF. Oil Commission*, June 1973.

7.37 BAGWELL, F.A. et *al.* — *JACA*, vol. 21, 1971, p. 702.

7.38 SHAW, J.T. — *J. of Inst. Fuel*, 1973. *Progress Review*, No. 64, p. 170.

8

heat exchange in the firebox

The preceding chapters have described combustion within fireboxes for the purpose of showing how flames may be adapted to different kinds of heating. To complete this description it has seemed worthwhile to offer calculation methods for determining the rate of heat transmission. This last chapter, consequently, describes the method of Hottel and Sarofim which is most widely accepted and which the authors have found satisfactory in actual applications.

8.1. REVIEW OF SOME BASICS

Radiant energy is exchanged through space without physical contact; it is made up of electromagnetic waves that can pass through a vacuum, as well as any gaseous, fluid or solid medium that does not absorb the waves and transform them into heat. To the extent that the electromagnetic waves of radiant energy are absorbed in any medium, they are converted into heat energy. In addition to passing through or being absorbed by a medium, radiant energy can be wholly or partially reflected or diffused from the surface of the medium.

8.1.1. Radiation from a solid surface, black body

A black body is understood as any body that completely absorbs incident radiation, no matter what its frequency and angle of incidence. When such a body is taken to absolute temperature θ, it radiates an energy at a rate, Q, identified by Stefan-Boltzmann as:

$$Q = S \sigma \theta^4 \tag{8.1}$$

with

Q = rate of energy emission,
θ = absolute temperature,
S = emissive surface,
σ = Stefan-Boltzmann constant.

The value of σ depends on the system of units that is chosen:

$$\sigma = 5.67 \quad 10^{-5} \ \text{erg/cm}^2.\text{s}.(\text{K})^4$$

$$\sigma = 4.93 \quad 10^{-8} \ \text{kcal/m}^2.\text{hr}.(\text{K})^4$$

$$\sigma = 0.173 \quad 10^{-8} \ \text{Btu/sq}.\text{ft}.\text{hr}.(^\circ\text{R})^4$$

This energy emitted by the black body is distributed among a spectrum of electromagnetic waves of different lengths λ according to Planck's law:

$$Q_\lambda = C_1 \lambda^{-5} \ \cfrac{1}{\exp\left(\dfrac{C_2}{\lambda\theta}\right) - 1} \tag{8.2}$$

C_1 and C_2 are two constants whose value depend on the chosen units:

$C_1 = 3.741 \quad 10^{-16} \ \text{W}.\text{m}^2$ (wave length expressed in meters),

$C_1 = 3.741 \quad 10^8 \quad \text{W}.\mu^4/\text{m}^2$ (wave length expressed in microns, surface in square meters),

$C_2 = 1.4388 \quad 10^{-2} \ \text{m (K)}$ (wave length expressed in meters, temperature in degrees Kelvin),

$C_2 = 14{,}388 \ \mu \ (\text{K})$ (wave length expressed in microns).

For sufficiently low temperatures ($\lambda\theta < 2{,}400\,\mu$ degrees) the -1 in the denominator can be neglected, and Planck's law can be expressed by the approximate formula, according to Wien:

$$Q_\lambda = C_1 \lambda^{-5} \ \exp\left(\frac{-C_2}{\lambda\theta}\right) \tag{8.3}$$

Formulas (8.2) and (8.3) show for a wave length λ_M a maximum energy emission whose value depends on the temperature θ:

$$\lambda_M \theta = 2{,}898 \ \mu \ (\text{K}) \tag{8.4}$$

8.1.2. Radiation from an object

Generally, bodies absorb only a part of the electromagnetic radiation that they intercept; the balance is transmitted or reflected.

For any body, absorption of radiation with wave length λ is related to that of a black body by a coefficient, α_λ, called the factor of monochromatic absorption. Necessarily:

$$0 < \alpha_\lambda < 1$$

From thermodynamic considerations, Kirchhoff showed that the ratio, ε_λ of energy of wave length λ, radiated by this body to that radiated by a black body at the same temperature was equal to the same factor of monochromatic absorption. If, ε_λ, is called the factor of monochromatic emission:

$$\alpha_\lambda = \varepsilon_\lambda \tag{8.5}$$

8.1.3.   Concepts of radiant energy transmission

8.1.3.1.   Energy radiated from a source

A source point naturally tends to radiate energy indiscriminately in all directions. The intensity of this emission accordingly does not involve area and is expressed only in terms of energy per unit time, or emissive power, as watts (W), kilowatts (kW) or kilocalories per hour (kcal/hr).

8.1.3.2.   Energy radiated in a given direction

From the indiscriminate radiation of a point source, there can be isolated a directional radiation defined by an infinitely small differential solid angle, $d\Omega$

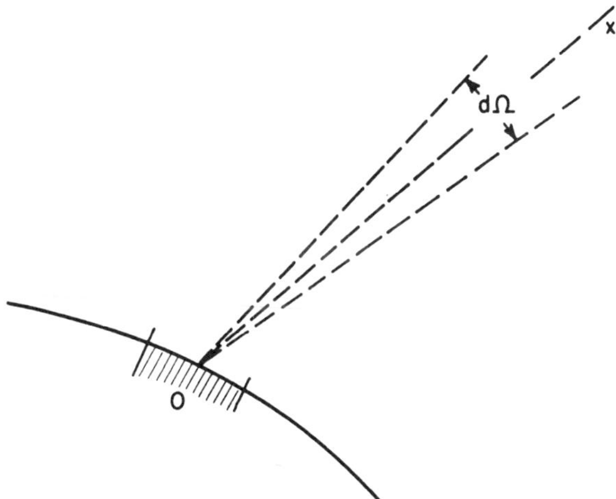

Fig. 8.1.   Intensity radiated from a source point.

(Fig. 8.1). The energy flowing in this direction is expressed as flux in terms of energy per unit time per unit angle, as watts per steradian (W/sr); and if Φ represents the total point radiation; the intensity I is the radiant flux in the direction or $d\Omega$, then:

$$I = \frac{d\Phi}{d\Omega}$$

(8.6)

8.1.3.3. Total emittance

If the theoretical source point is replaced by a radiant object, the emission of an elementary surface, dS, of this object radiates in all directions in front of it a flux $d\Phi$, of energy.

Total emittance, M, is written:

$$M = \frac{d\Phi}{dS}$$

(8.7)

It is expressed in such units as W/cm^2 or $kcal/hr.\,m^2$.

8.1.3.4. Luminance

The elementary surface dS radiates in all directions. In direction Ox making an angle β with the normal On to dS (see Fig. 8.2), considering the radiant flux $d\Phi$ in an elementary solid angle $d\Omega$, around Ox, luminance, L, is written:

$$L = \frac{\dfrac{d\Phi}{d\Omega}}{dA} = \frac{I}{dS \cos \beta}$$

(8.8)

with dA the projection of dS on normal plane to Ox at point O.

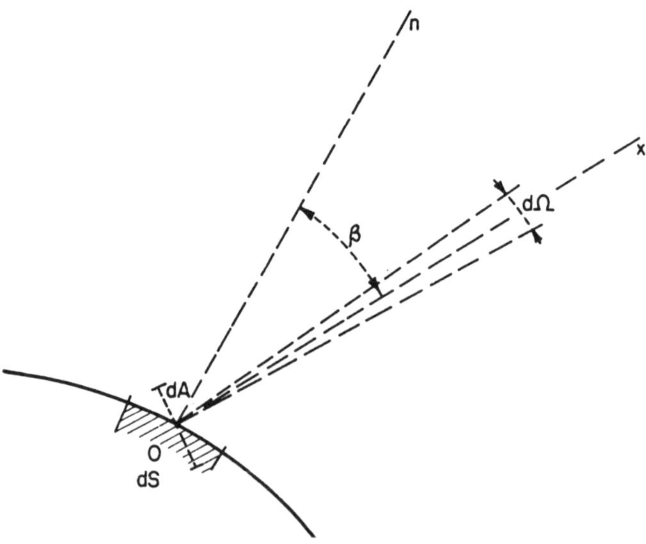

Fig. 8.2. Radiation in different angles from a source surface.

Because they are expressed in terms of geometry, rather than quality of the radiation, both luminance and emittance power apply equally to radiation at a specific wave length, λ, and to total radiation at all wave lengths. Convention identifies monochromatic radiation at a single wave length by the subscript λ. Thus:

(a) For monochromatic emittance power, M_λ . . .

$$M_\lambda = \left(\frac{\mathrm{d}M}{\mathrm{d}\lambda}\right)_\lambda \tag{8.9}$$

(b) For monochromatic luminance, L_λ . . .

$$L_\lambda = \left(\frac{\mathrm{d}L}{\mathrm{d}\lambda}\right)_\lambda \tag{8.10}$$

Also, convention identifies radiation by a black body by the superscript$^{\circ}$, as $L^{\circ}, M^{\circ}, L^{\circ}_\lambda, M^{\circ}_\lambda$.

8.1.3.5. Lambert's law

The foregoing derivation of luminance can be stated in terms known as "Lambert's Law": If the intensity of a radiation is independent of its direction, the energy radiated in any direction from a surface is proportional to the cosine of the angle of that direction from the normal to the surface. Furthermore, if the expression for luminance is integrated over an entire hemisphere, then the emissive power is related to luminance as:

$$M = \pi L \tag{8.11}$$

This relation is frequently used in calculations, even for bodies that follow Lambert's law imperfectly; and bodies following Lambert's law are called "gray" bodies.

8.1.4. Radiation between two solid surfaces

Imagine two differential surfaces $\mathrm{d}S_1$ and $\mathrm{d}S_2$ separated by distance r of non-absorbant and non-emissive medium (Fig. 8.3). Their respective temperatures are θ_1 and θ_2, and their emissivities, ε_1 and ε_2, are equal to their absorptivities, α_1 and α_2. The normal from each one of these surfaces makes an angle φ_1 or φ_2 with the line $O_1 O_2$ connecting the surfaces. Assume that the radiation of these surfaces follows Lambert's law.

From O_1, surface $\mathrm{d}S_1$ sees the surface $\mathrm{d}S_2$ at angle:

$$\frac{\mathrm{d}S_2 \, \cos \varphi_2}{r^2}$$

and surface $\mathrm{d}S_2$ sees the surface $\mathrm{d}S_1$ at angle:

$$\frac{\mathrm{d}S_1 \, \cos \varphi_1}{r^2}.$$

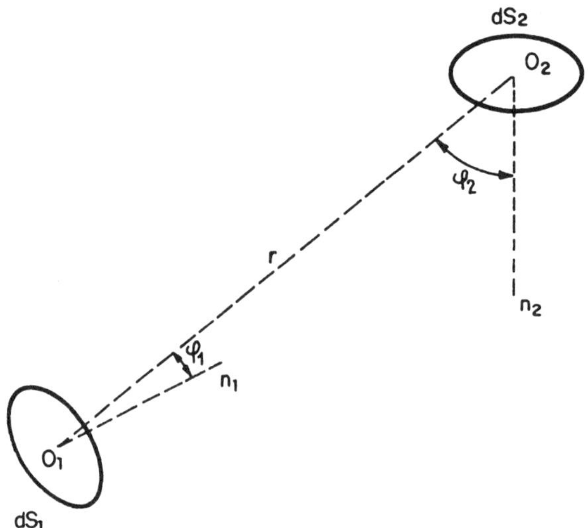

Fig. 8.3.   Radiation between two surfaces in a non-absorbent atmosphere.

The radiation along line $O_1 O_2$ that impinges on either of the two surfaces is
inversely proportional to the distance, r. From this, the energy passing from
dS_1 to dS_2 is defined as:

$$d\Phi_{1,2} = \varepsilon_1 \sigma \theta_1^4 \frac{dS_1 \cos \varphi_1}{\pi} \frac{dS_2 \cos \varphi_2}{r^2}$$

The surface dS_2 absorbs the fraction α_2 of the radiation $d\Phi_{1,2}$, and dS_2
emits toward dS_1 a flux $d\Phi_{2,1}$ of which dS_1 absorbs the fraction α_1.

Thus the radiant exchange between these two surfaces can be expressed by
the difference:

$$d\Phi = \alpha_2 \, d\Phi_{1,2} - \alpha_1 \, d\Phi_{2,1}$$

which can be further defined as:

$$d\Phi = \frac{dS_1 \cos \varphi_1 \cdot dS_2 \cos \varphi_2}{\pi r^2} \alpha_1 \alpha_2 \sigma (\theta_1^4 - \theta_2^4) \qquad (8.12)$$

The first term on the right-hand side of Eq. 8.12 involves only the geometry
of the surfaces, such as their respective distance and orientation: the last terms
involve the absorptivity of the two surfaces, α, the Stefan-Boltzmann constant,
σ, and temperature, θ. For each one of the surfaces, we have assumed that
factors α and ε are equal. These surfaces, therefore, do not have any selective
character vis-a-vis the wave length of radiation. If these surfaces also follow
Lambert's law, they are called gray. If the surfaces under consideration show

selective absorption as a function of the wave length, it is no longer valid to assume that the factors α and ε are equal, and it is necessary to calculate the exchange for each monochromatic wave length because α_λ does equal ε_λ in all circumstances. The total energy exchanged is the sum of the exchanges at every wave length.

8.1.5. Radiation between surfaces with finite dimensions

Formula 8.12 permits calculating the radiation between two surfaces of finite dimensions through a double integration along the surfaces S_1 and S_2. If these surfaces are gray ($\alpha_1 = \varepsilon_1$ and $\alpha_2 = \varepsilon_2$) and the temperatures uniform across the surfaces, the integration will result in a surface factor, SF, as follows:

$$SF = \iint_{S_1} \iint_{S_2} \frac{dS_1 \cos \varphi_1 \cdot dS_2 \cos \varphi_2}{\pi r^2} \tag{8.13}$$

SF depends only on the size of the surfaces, the distance between each of their elements, and their orientation. F is a geometric factor usually bearing subscripts identifying the surfaces between which the radiation takes place, as for example, $SF_{1,2}$ to indicate radiation between surfaces S_1 and S_2.

The energy directly exchanged is then expressed as:

$$\Phi = SF_{1,2} \, \alpha_1 \, \alpha_2 \, \sigma (\theta_1^4 - \theta_2^4) = SF_{1,2}(M_1^0 - M_2^0) \, \alpha_1 \, \alpha_2 \tag{8.14}$$

SF has the dimensions of a surface. To conveniently establish tables or graphs it is made dimensionless by dividing it by a surface such as S_1 or S_2. Two dimensionless factors are thus defined for characterizing the radiation between two surfaces:

$$F_{2,1} = \frac{SF}{S_2} \qquad \text{and} \qquad F_{1,2} = \frac{SF}{S_1} \tag{8.15}$$

with the emissivity equal to the absorptivity:

$$S_2 F_{2,1} = S_1 F_{1,2} \tag{8.15bis}$$

These factors are called factors of shape.

8.1.6. Absorption and emission of radiation by a gas

Although diatomic gases with symmetrical molecules, such as nitrogen and oxygen, are transparent throughout the whole visible and infrared spectrum, those gases with more complex molecules do not exhibit the same transparency. CO_2 and H_2O in particular show bands of absorption in the infrared; and according to Kirchhoff's law they must also radiate in these same wave bands. Other components (for example, carbon monoxide, methane, etc.) absorb and radiate in the infrared in the same way but their practical influence is slight compared to that of steam and CO_2.

Absorption of radiation by gases is on the molecular scale; thus it is proportional to the number of active molecules encountered by a given ray, i.e., proportional to the length of travel through the gas and the partial pressure of the gas.

Figure 8.4 represents a bundle of parallel monochromatic rays, passing through an absorbant gas while carrying radiant energy Φ_x to the cross-section at x. Between x and $x + dx$ a fraction $-\dfrac{d\Phi}{\Phi}$ of this energy will be absorbed proportional to the condition of the gas, as:

$$-\frac{d\Phi}{\Phi} = \alpha_\lambda \ dx \tag{8.16}$$

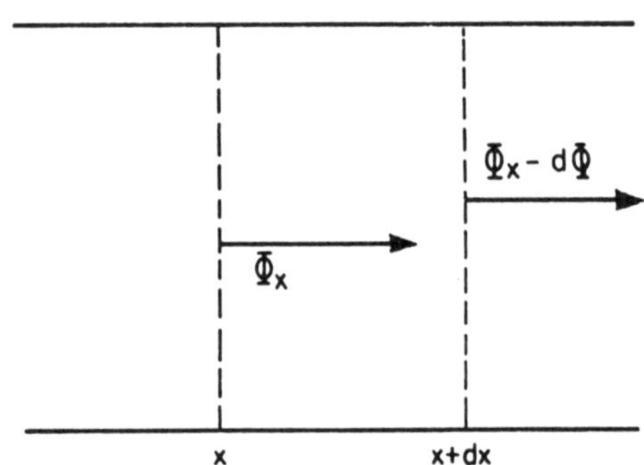

Fig. 8.4. The depreciation of radiant flux passing through an absorbent gas.

The coefficient α carries the subscript λ to show the relation refers to a given wave length. On integration, Eq. (8.16) becomes:

$$\Phi_{x,\lambda} = \Phi_{0,\lambda} \exp(-\alpha_\lambda x) \tag{8.17}$$

This expression is known as Beer's law. Since the number of molecules encountered are proportional to the path length, l, and the partial pressure, the expression can be written:

$$-\alpha_\lambda x = -\alpha_\lambda' \, pl \tag{8.18}$$

In technical literature, the monochromatic absorption factor α_λ is often written k_λ.

8.1.7.   Absorption spectrum of CO_2

This spectrum is made up of principally three bands (Fig. 8.5):

(a) Band 1, between 2.64 and 2.84 μ.
(b) Band 2, between 4.13 and 4.47 μ.
(c) Band 3, between 13 and 17 μ.

The coefficient of monochromatic absorption α_λ for each one of the bands is perceptively independent of the partial pressure in the mixture being considered.

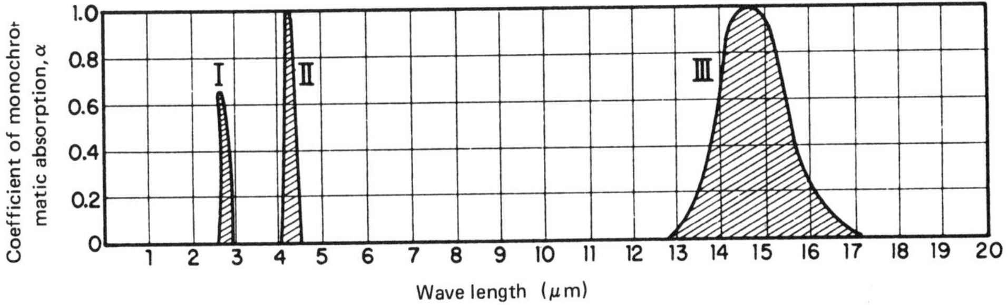

Fig. 8.5.   The absorption spectrum for a layer of carbon dioxide 6.3 cm thick at 20° C and a pressure of 1 atm.

8.1.8.   Absorption spectrum of steam

This spectrum is a little more complicated than that of CO_2. It can be approximated as 4 principal bands (Fig. 8.6) for which the peaks are: 1.36, 1.85, 2.70, 5.9 and 19.6 μ.

Unlike CO_2, steam presents for each band a monochromatic absorption factor α_λ that varies with the partial pressure of the steam. This is shown in Fig. 8.6 by a higher heavy line of pressure-times-path, pl, of 2.4 atm . m and the shaded areas for a pl of 0.24 atm : m.

8.1.9.   Effects of absorption/radiation by CO_2 and steam

Absorption by these two gases is not much influenced by temperature. For both of these gases, the principal absorbed wave lengths are well within the range infrared, on the order of 15 μ. As a result, these gases are more responsive to radiation at moderate temperature than to radiation at high temperature.

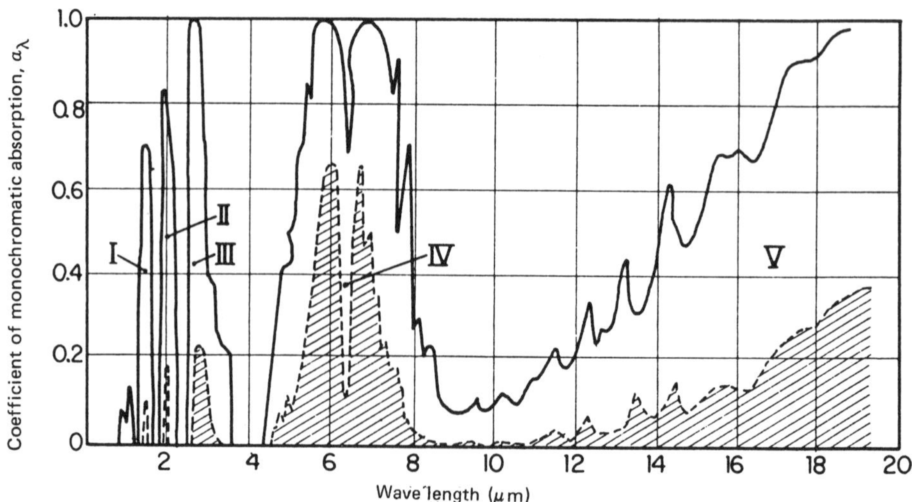

Fig. 8.6. Absorption spectrum for water vapor 1 m thick at 127° C and a pressure of 2.4 atm. The shaded portion corresponds to absorption for 0.24 atm/m.

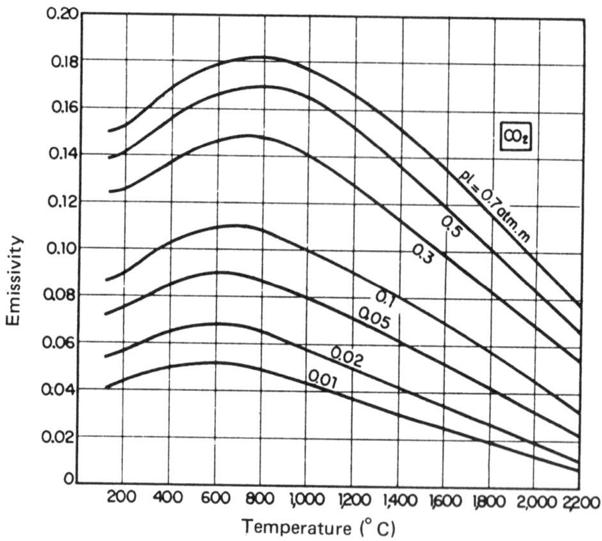

Fig. 8.7. The effects of temperature on the emissivity of carbon dioxide.

Conversely, their emissivity is higher at moderate temperatures than at high temperatures (Figs. 8.7 and 8.8). For CO_2, a maximum emissivity is observed for temperatures close to 800° C (Fig. 8.7) while the emissivity of steam drops from around 0.5 at ambient temperatures to around 0.20 at 2,000° C (Fig. 8.8).

The calculation of emissivity is described in conventional books. And there are nomographs, which permit calculating with an accuracy more than sufficient

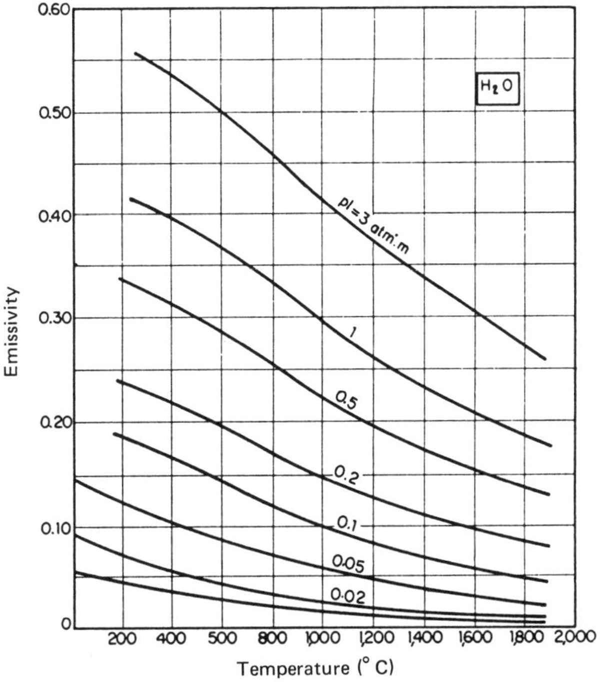

Fig. 8.8. The effects of temperature on the emissivity of water vapor.

for solving the usual problem of radiation, as well as polynomial equations that lend the calculation to hand-held computers.

Because of the superimposition of their absorption bands in certain ranges, a small interference appears between the absorption/radiation of CO_2 and steam, as they occur from the combustion of hydrocarbons. When this interference appears, the effective emissivity is slightly less than the sum of the emissivities of each of the constituents taken alone. Hottel and Sarofim ("Radiative transfer") furnish nomographs calculating the correction, which can often be omitted in a first approximation.

8.2. RADIATION IN A GRAY ENCLOSURE CONTAINING A GRAY GAS AT A UNIFORM TEMPERATURE

8.2.1. General

The enclosure comprises "gray" absorbing/radiating surfaces $S_1, S_2, \ldots, S_i, S_j, \ldots, S_n$, surfaces with zero-net-flux $S_R, S_S, \ldots$, and an absorbing/radiating gas filling the available volume. Any surfaces with zero-net-flux are adiabatic and return all of the radiant energy falling on them from the interior of the firebox. Furnace refractories are generally zero-net-flux surfaces, because the only heat passing through them is the relatively small heat loss to the atmosphere. Otherwise surfaces of refractory material are considered to be perfect black bodies.

For gray gases, Beer's law holds, and the radiation through the gas depends on a factor, $f = 1 - \alpha_g$, which varies with the exponential power of the distance traveled according to Eqs. (8.17) and (8.18). Thus, if the specific path of radiation 1,3 is three times the path 1,2, one will have $f_{1,3} = f_{1,2}^3$.

Introducing an exchange factor $\mathscr{F}$ which takes into account only the geometry of the enclosure and of the total emissivity of the radiator/absorbers the exchange between radiator, a, and an absorber, b, can be written:

$$\Phi_{a \rightleftarrows b} = S_a \, \mathscr{F}_{a,b} \; \sigma(\theta_a^4 - \theta_b^4) = S_b \, \mathscr{F}_{b,a} \; \sigma(\theta_a^4 - \theta_b^4) \qquad (8.19)$$

The factor $\mathscr{F}$ takes account of the fact that the net exchange $a \rightleftarrows b$ is made up of a direct exchange plus multiple exchanges with all surfaces, that is with $S_1, S_2, \ldots, S_R, S_S \ldots$ including the re-radiation from surfaces $S_R, S_S \ldots$, which have zero-net-flux.

Thus, for two surfaces S_i and S_j of the enclosure:

$$\Phi_{i \rightleftarrows j} = S_i \, \mathscr{F}_{i,j} \, \sigma(\theta_i^4 - \theta_j^4) \qquad (8.20)$$

This holds equally between surface S_i and the gas, as:

$$\Phi_{g \rightleftarrows i} = S_i \, \mathscr{F}_{i,g} \, \sigma(\theta_g^4 - \theta_i^4) \qquad (8.20 \text{ bis})$$

Consequently, the net quantity of heat lost by the hot gas can be written:

$$\Phi_{g \, net} = \sum_{i=1}^{i=n} \Phi_{g \rightleftarrows i} \qquad (8.21)$$

This last Eq. (8.21) defines the amount of radiant heat transmitted to **the heated surfaces**, including tubes, tube walls and charge, in the firebox of a furnace or fired heater. Accordingly, the problem of determining radiant heat transfer in a firebox resolves to an evaluation of the coefficients of net exchange of the type $\mathscr{F}_{i,g}$; and these are determined through constructing energy balances based on the absorbed flux and the radiated flux relative to each surface.

The complicated relations that might be expected from this approach are simplified a little through careful application of assumptions that simplify the relations without introducing an appreciable uncertainty to $\mathscr{F}$.

Thus starting from the assumption that the factor of net exchange, $\mathscr{F}$, is independent of the temperatures in the enclosure, Hottel (Ref. 8.1) assumes that all **but one** of the emitter-receivers of the system are at absolute zero. Under these conditions, the flux emitted by all but one of the emitter-receivers can be accounted entirely to the re-emittance of received radiation without the contribution of any radiation due to temperature. It remains to choose the exception, which becomes the source of energy.

No matter what — a given surface, or the gas — is chosen as the source emittance, this source forms the basis for an energy balance made with each one of the N surfaces $S_1, S_2, S_i, S_n, \ldots, S_R, S_S \ldots$ of the enclosure.

If a surface is chosen, the gas intervenes only by its transmission factors, f; if the gas is chosen as source emittance, it intervenes both through its transmission factors and its emittance M_g. **In the case of surface as chosen source,** mathematical analysis leads to a direct calculation of all the factors of the type $\mathcal{F}_{i,j}$ from which $\mathcal{F}_{1,g}$ is derived; and then the net exchange is arrived at by going from surface to surface and from gas to surface. **In the case of gas as chosen source,** the factor $\mathcal{F}_{i,g}$ is obtained directly, after which only the emittance from the gas to the surfaces is determined.

From this, it is apparent that the number of terms involved is fewer with the gas as emitter than with a surface as emitter. In either case, factor $\mathcal{F}_{i,g}$ will be expressed as a function of the emissivities $(\varepsilon_1, \varepsilon_2, \varepsilon_n, \varepsilon_g)$ of the different emitter-receivers and of the factors, F, accounting for the geometry of the enclosure. These last intervene, in any problem of radiant exchange (Section 8.1.5) between the possible surfaces.

Since each one of N surfaces of an enclosure may see itself, with a corresponding factor $F_{1,1}$, radiation calculations involve a total number of F factors $(F_{1,1}, F_{2,2}, \ldots, F_{i,i}, \ldots F_{n,n}, F_{R,R}, F_{S,S} \ldots)$. Taking into account the dynamic re-radiation between surfaces $(S_i F_{i,j} = S_j F_{j,i} \ldots)$ as expressed in Eq. (8.15 bis), as well as N relations of the type $\Sigma_N F_{i,N} = 1$ occurring within an enclosure, the total number of F factors is reduced from N^2 possibilities to $\dfrac{N(N-1)}{2}$ independent factors.

In calculating the exchanges between surfaces in the presence of a gas, the geometrical F factors act in terms of the type $S_i F_{i,j} f_{i,j} (= S_j F_{j,i} f_{j,i})$ in which the transmission factor, $f = 1 - \varepsilon_g$, for path $i \to j$ or $j \to i$ appears as $f_{i,j} = f_{j,i}$.

8.2.2. Equations for radiant heat transfer

8.2.2.1. With a surface as primary emittance

For the primary emitter (temperature not zero) any surface, S_i, is taken as S_1 at temperature θ_1 and emittance $\varepsilon_1 M_i$ (with $M_i = \sigma \theta_1^4$).

Since there is no heat transferred between S_1 and the zero-net-flux surfaces $S_R, S_S, \ldots$, the exchanges from surface to surface are going to involve only the coefficients $\mathcal{F}_{1,1}, \mathcal{F}_{1,2}, \ldots, \mathcal{F}_{1,N}$ in Eq. (8.20). Once these factors have been calculated, they are used to get $\mathcal{F}_{1,g}$ for Eq. (8.20 bis) by assuming that the fraction of primary radiation not absorbed by the N surfaces has been absorbed by the gas, that is:

$$S_1 \mathcal{F}_{1,g} = S_1 \varepsilon_1 - S_1 \mathcal{F}_{1,2} - \cdots - S_1 \mathcal{F}_{1,n} \tag{8.22}$$

Any radiation falling on a gray surface, S_i, is divided into an absorbed fraction, ε_i, and a reflected fraction $(1 - \varepsilon_i)$. so that the absorbed radiation represents a fraction, $\dfrac{\varepsilon_i}{1 - \varepsilon_i}$ of the reflected flux $(\Phi_r)_i$.

With the primary radiation coming from S_1 expressed in terms of flux density, φ_r, it is possible to write:

$$S_1 \mathcal{F}_{1,1} \, M_1^0 = \frac{\varepsilon_1}{1 - \varepsilon_1} S_1 (\varphi_r)_1$$

$$S_1 \mathcal{F}_{1,2} \, M_1^0 = \frac{\varepsilon_2}{1 - \varepsilon_2} S_2 (\varphi_r)_2 \tag{8.23}$$

$$S_1 \mathcal{F}_{1,n} \, M_1^0 = \frac{\varepsilon_n}{1 - \varepsilon_n} S_n (\varphi_r)_n$$

All the surfaces at zero temperature, $S_2, S_3, \ldots, S_n$, will exhibit radiation whose flux density is entirely due to reflection. However, the flux for S_1 will be composed of both reflected and emitted radiation as:

$$(\Phi_E)_1 = S_1 (\varepsilon_1 M_1^0 + (\varphi_r)_1) \tag{8.24}$$

In order to simplify the equations, Hottel (Ref. 8.1) relates the flux densities, φ_r, to black body emissivity M_1^0 of the primary source. Using the dimensionless term,

$$_1 \varphi_n = \frac{(\varphi_r)_n}{M_1^0}$$

the Eqs. of (8.23) become:

$$(\Phi_E)_1 = S_1 (\varepsilon_1 M_1^0 + (\varphi_r)_1)$$

$$S_1 \mathcal{F}_{1,1} = S_1 \frac{\varepsilon_1}{1 - \varepsilon_1} \, _1\varphi_1$$

$$S_1 \mathcal{F}_{1,2} = S_2 \frac{\varepsilon_2}{1 - \varepsilon_2} \, _1\varphi_2$$

$$S_1 \mathcal{F}_{1,n} = S_n \frac{\varepsilon_n}{1 - \varepsilon_n} \, _1\varphi_n \tag{8.25}$$

and Eq. (8.24) is then written as:

$$(\Phi_E)_1 = S_1 M_1^0 (\varepsilon_1 + {}_1\varphi_1) \tag{8.26}$$

The heat transfer problem is thus brought to the determination of ${}_1\varphi_1, {}_1\varphi_2, \ldots, {}_1\varphi_n$, which is achieved by writing a radiant energy balance for each of the surfaces making up the firebox: $S_1, S_2, \ldots, S_n, S_R, S_S \ldots$

For gray surfaces of type S_i having a coefficient of reflection $\rho_i = 1 - \varepsilon_i$, one writes that the reflected flux $(S_i \, {}_1\varphi_i \, M_1^0)$ is equal to ρ_i times the incident flux. For the surfaces (refractory) of type S_R with zero-net-flux, the radiant flux is equal to the incident flux, being a combination of reflection and re-emission. Incident flux on a surface is understood to include not only the sum of the flux coming from the other surfaces but also flux from the surface itself, when it can see itself.

The energy balance to and from S_1 is then written:

$$[S_1 F_{1,1} f_{1,1} M_1^0 (\varepsilon_1 + {}_1\varphi_1) + S_2 F_{2,1} f_{2,1} M_1^0 \, {}_1\varphi_2 + \ldots +$$

$$+ \ldots S_n F_{n,1} f_{n,1} M_1^0 \, {}_1\varphi_n + S_R F_{R,1} f_{R,1} M_1^0 \, {}_1\varphi_R + \ldots] \, \rho_1 = S_1 M_1^0 \, {}_1\varphi_1$$

Using the property of reversibility of the products SFf and grouping all the flux terms in the left-hand side, gives:

$$\left(S_1 F_{1,1} f_{1,1} - \frac{S_1}{\rho_1}\right) {}_1\varphi_1 + (S_1 F_{1,2} f_{1,2}) \, {}_1\varphi_2 + \ldots +$$

$$+ \ldots + (S_1 F_{1,n} f_{1,n}) \, {}_1\varphi_n + (S_1 F_{1,R} f_{1,R}) \, {}_1\varphi_R + \ldots = - S_1 F_{1,1} f_{1,1} \varepsilon_1$$

$$\tag{8.27}$$

The energy balance on S_2 is written:

$$[S_2 F_{2,2} f_{2,2} \, {}_1\varphi_2 + S_1 F_{1,2} f_{1,2} ({}_1\varphi_1 + \varepsilon_1) + \ldots +$$

$$+ \ldots + S_n F_{n,2} f_{n,2} \, {}_1\varphi_n + S_R F_{R,2} f_{R,2} \, {}_1\varphi_R + \ldots] \, \rho_2 = S_2 \, {}_1\varphi_2$$

that is:

$$(S_1 F_{1,2} f_{1,2}) \, {}_1\varphi_1 + \left(S_2 F_{2,2} f_{2,2} - \frac{S_2}{\rho_2}\right) {}_1\varphi_2 + \ldots +$$

$$+ \ldots + (S_n F_{2,n} f_{2,n}) \, {}_1\varphi_n + (S_2 F_{2,R} f_{2,R}) \, {}_1\varphi_R + \ldots = - S_1 F_{1,2} f_{1,2} \varepsilon_1$$

and so forth.

$$\tag{8.28}$$

The energy balance on S_R is written:

$$S_1 F_{1,R} f_{1,R} ({}_1\varphi_1 + \varepsilon_1) + S_2 F_{2,R} f_{2,R} \, {}_1\varphi_2 + \ldots +$$

$$+ \ldots + S_n F_{n,R} f_{n,R} \, {}_1\varphi_n + S_R F_{R,R} f_{R,R} \, {}_1\varphi_R + \ldots = S_R \, {}_1\varphi_R$$

that is:

$$(S_1 F_{1,R} f_{1,R})\, {}_1\varphi_1 + (S_2 F_{2,R} f_{2,R})\, {}_1\varphi_2 + \ldots +$$

$$+ \ldots + S_n F_{n,R} f_{n,R}\, {}_1\varphi_n + (S_R F_{R,R} f_{R,R} - S_R)\, {}_1\varphi_R + \ldots = - S_1 F_{1,R} f_{1,R} \varepsilon_1$$

$$(8.29)$$

and so forth.

Thus a system of N Eqs. (8.27), (8.28) ... (8.29) ... with N unknowns ${}_1\varphi_1, {}_1\varphi_2, \ldots, {}_1\varphi_n, {}_1\varphi_R, {}_1\varphi_S \ldots$

The resolution of such a system is usually done on a computer.

By following the conventional algebraic development for resolving linear systems, Cramer's rule indicates that the solution of the type, ${}_1\varphi_n$ is such that:

$$ {}_1\varphi_n = \frac{{}_1 D_n}{D} \tag{8.30}$$

D being the determinant of coefficients of ${}_1\varphi_1, {}_1\varphi_2 \ldots, {}_1\varphi_n, {}_1\varphi_R, {}_1\varphi_S \ldots$ and ${}_1 D_n$ the determinant obtained by replacing the D column relative to ${}_1\varphi_n$ by the second members of Eqs. (8.27), (8.28) ... (8.29).

Finally, according to Eq. (8.25) we have:

$$ S_1 \mathcal{F}_{1,n} = \frac{\varepsilon_n S_n}{\rho_n} \frac{{}_1 D_n}{D} \tag{8.31}$$

from which $S_1 \mathcal{F}_{1,g}$ by Eq. (8.22).

More generally, starting from an initial emitter, S_i, one is led to the terms of the type, $S_i \mathcal{F}_{i,n}$, by the same process and from which $S_i \mathcal{F}_{i,g}$.

Thus, this first method enables one to handle the transfers by radiation in the enclosure and obtain the exchanges from surface to surface as well as those from gas to surface.

The enclosure that does not contain an absorbant gas is a special case for which all the transmission terms f are equal to 1.

8.2.2.2. With the gas as primary emitter

The gas is taken as the initial emitter, all the other emitter-receivers being taken as at absolute zero.

This method then permits directly calculating $\mathcal{F}_{i,g}$ independently of $\mathcal{F}_{i,n}$, from the fact that the gas is taken as the only emitter of radiation in the system.

Saying that the flux absorbed by S_i represents a fraction $\dfrac{\varepsilon_i}{1 - \varepsilon_i}$ of the reflected flux, leads to:

$$S_1 \; \mathscr{F}_{1,g} \; M_g^0 = \frac{\varepsilon_1}{1 - \varepsilon_1} \, S_1 (\varphi_r)_1$$

$$S_2 \; \mathscr{F}_{2,g} \; M_g^0 = \frac{\varepsilon_2}{1 - \varepsilon_2} \, S_2 (\varphi_r)_2$$

$$S_n \; \mathscr{F}_{n,g} \; M_g^0 = \frac{\varepsilon_n}{1 - \varepsilon_n} \, S_n (\varphi_r)_n \tag{8.32}$$

Relating as before the density of the reflected flux to the black-body emittance of the primary emitter, by:

$$_g\varphi_n = \frac{(\varphi_r)_n}{M_g^0}$$

the Eqs. of (8.32) become:

$$S_1 \mathscr{F}_{1,g} = \frac{\varepsilon_1}{1 - \varepsilon_1} \, S_1 \, _g\varphi_1$$

$$S_2 \mathscr{F}_{2,g} = \frac{\varepsilon_2}{1 - \varepsilon_2} \, S_2 \, _g\varphi_2$$

$$S_n \mathscr{F}_{n,g} = \frac{\varepsilon_n}{1 - \varepsilon_n} \, S_n \, _g\varphi_n \tag{8.33}$$

The problem comes back therefore to the determination of $_g\varphi_1, \ldots, _g\varphi_n$. As before, a solution is obtained by writing an energy balance of radiation to and from **each one** of the N surfaces making up the firebox:

$$S_1, \ldots, S_n, S_R, S_S \ldots$$

But this time, the gas will get involved by its emittance as well as by its transmission factor. The flux reflected by **all** surfaces $S_1, \ldots, S_n$ not emitting radiation at zero temperature is equal to the flux emitted by the gas.

The energy balance is done the same as during the first method, by writing that:

(a) For a gray surface of type S_i having a coefficient of reflection ρ_i, the reflected flux $S_i \, _g\varphi_i \, M_g^0$ is equal to ρ_i times the incident flux.

(b) For a surface (refractory) of type S_R without flux, the escaping flux $_g\varphi_R \, M_g^0$ is equal to the incident flux.

But the incident flux on any surface, S_N, will be expressed by terms that take into account:

(a) The emission of all surfaces of which one part:

$$\Sigma_N SFf_g \varphi_N M_g^0$$

reaches the surface under consideration.

(b) The emission of the gas of which a part:

$$\Sigma_N SF\varepsilon_g M_g^0 = \Sigma_N SF(1-f) M_g^0$$

reaches this same surface.

The energy balance on S_1 is then written:

$$[S_1 F_{1,1} f_{1,1} \, {}_g\varphi_1 M_g^0 + S_1 F_{1,1} (1-f_{1,1}) M_g^0 + S_2 F_{2,1} f_{2,1} \, {}_g\varphi_2 M_g^0$$

$$+ S_2 F_{2,1} (1-f_{2,1}) M_g^0 + \ldots + S_n F_{n,1} f_{n,1} \, {}_g\varphi_n M_g^0 + S_n F_{n,1} (1-f_{n,1}) M_g^0$$

$$+ S_R F_{R,1} f_{R,1} \, {}_g\varphi_R M_g^0 + S_R F_{R,1} (1-f_{R,1}) M_g^0 + \ldots] \rho_1 = S_1 \, {}_g\varphi_1 M_g^0$$

By using the property of reversibility of the terms SFf and of f ($f_{2,1} = f_{1,2}\ldots$), on one hand, and by using the geometric property of the factors of shape $F_{1,1} + F_{1,2} + \ldots + F_{1,R} \ldots = 1$, on the other hand, all the terms of flux can be grouped on the left-hand side to obtain:

$$\left(S_1 F_{1,1} f_{1,1} - \frac{S_1}{\rho_1} \right) {}_g\varphi_1 + (S_1 F_{1,2} f_{1,2}) {}_g\varphi_2 + \ldots +$$

$$+ \ldots + (S_1 F_{1,n} f_{1,n}) {}_g\varphi_n + (S_1 F_{1,R} f_{1,R}) {}_g\varphi_R + \ldots$$

$$+ \ldots = - S_1 + (S_1 F_{1,1} f_{1,1} + S_1 F_{1,2} f_{1,2} + \ldots +$$

$$+ \ldots + S_1 F_{1,n} f_{1,n} + \ldots + S_1 F_{1,R} f_{1,R} + \ldots)$$

$$(8.34)$$

The energy balance on S_2 is written:

$$[S_2 F_{2,2} f_{2,2} \, {}_g\varphi_2 + S_2 F_{2,2} (1-f_{2,2}) + S_1 F_{1,2} f_{1,2} \, {}_g\varphi_1 + S_1 F_{1,2} (1-f_{1,2})$$

$$+ \ldots + S_n F_{n,2} f_{n,2} \, {}_g\varphi_n + S_n F_{n,2} (1-f_{n,2}) + S_R F_{R,2} f_{R,2} \, {}_g\varphi_r$$

$$+ S_R F_{R,2} (1-f_{R,2}) + \ldots] \rho_2 = S_2 \, {}_g\varphi_2$$

Finally, from the fact that $F_{2,2} + F_{2,1} + \ldots + F_{2,n} + F_{2,R} + \ldots = 1$:

$$S_1\,F_{1,2}\,f_{1,2}\,{_g\varphi_1} + \left(S_2\,F_{2,2}\,f_{2,2} - \frac{S_2}{\rho_2}\right){_g\varphi_2} + \ldots +$$

$$+ \ldots + (S_2\,F_{2,n}\,f_{2,n})\,{_g\varphi_n} + (S_2\,F_{2,R}\,f_{2,R})\,{_g\varphi_R} + \ldots$$

$$+ \ldots = -S_2 + (S_2\,F_{2,2}\,f_{2,2} + S_2\,F_{2,1}\,f_{2,1} + \ldots +$$

$$+ \ldots + S_2\,F_{2,n}\,f_{2,n} + S_2\,F_{2,R}\,f_{2,R} + \ldots$$

and so forth. (8.35)

The energy balance on S_R is written:

$$S_R\,F_{R,R}\,f_{R,R}\,{_g\varphi_R} + S_R\,F_{R,R}(1-f_{R,R}) + S_1\,F_{1,R}\,f_{1,R}\,{_g\varphi_1} +$$

$$+ S_1\,F_{1,R}(1-f_{1,R}) + \ldots + S_n\,F_{n,R}\,f_{n,R}\,{_g\varphi_n} + S_n\,F_{n,R}(1-f_{n,R}) + \ldots$$

$$+ \ldots = S_R\,{_g\varphi_R}$$

Finally, from the fact that $F_{R,R} + F_{R,1} + \ldots + F_{R,N} + \ldots = 1$:

$$(S_1\,F_{1,R}\,f_{1,R})\,{_g\varphi_1} + (S_2\,F_{2,R}\,f_{2,R})\,{_g\varphi_2} + \ldots + (S_n\,F_{n,R}\,f_{n,R})\,{_g\varphi_n} +$$

$$+ (S_R\,F_{R,R}\,f_{R,R} - S_R)\,{_g\varphi_R} + \ldots = -S_R + (S_R\,F_{R,R}\,f_{R,R} + S_R\,F_{R,1}\,f_{R,1} +$$

$$+ S_R\,F_{R,2}\,f_{R,2} + \ldots + S_R\,F_{R,n}\,f_{R,n} + \ldots)$$

(8.36)

and so forth.

Thus the coefficients of ${_g\varphi_1}, {_g\varphi_2}, \ldots, {_g\varphi_n}, {_g\varphi_R} \ldots$ of Eqs. (8.34), (8.35), $\ldots$, (8.36) $\ldots$ are identical to the coefficients of ${_1\varphi_1}, {_1\varphi_2}, \ldots, {_1\varphi_n}, {_1\varphi_R}$, of Eqs. (8.27), (8.28), $\ldots$ (8.29) $\ldots$ established for transfers with a surface as primary source.

Letting D stand for the determinant formed from these coefficients, we then have:

$$_g\varphi_n = \frac{_gD_n}{D} \tag{8.37}$$

$_gD_n$ being the determinant obtained by replacing the column D relative to $_g\varphi_n$ (that is, the same column of D) by the second members of Eqs. (8.34), (8.35) $\ldots$ (8.36) $\ldots$

These second members (with number N) can be put in the form $-S_N(1 - \Sigma_N\,Ff)$. Hottel remarks that $\Sigma_N\,Ff$ represents a weighted average of the transmission factors of the gas, these factors being relative to the different paths of radiation arriving at or going away from S.

$1 - \Sigma_N Ff$ will thus represent the corresponding absorption factor (thus of emission, since the gas is gray) such that the second members can be put in the form $S_N \varepsilon_{G,N}$. Thus, for a surface of nth order, for example, one will have for the second member of the energy equation:

$$- S_n \varepsilon_{G,n}, \text{ with } \varepsilon_{G,n} = 1 - (F_{n,1} f_{n,1} + F_{n,2} f_{n,2} + \ldots +$$

$$+ \ldots + F_{n,n} f_{n,n} + \ldots + F_{n,R} f_{n,R} + \ldots)$$

Therefore, in the expression Eq. (8.37), $_gD_n$ will be obtained by replacing the same column of D by:

$$- S_1 \varepsilon_{G,1} , - S_2 \varepsilon_{G,2}, \ldots, - S_n \varepsilon_{G,n} , - S_R \varepsilon_{G,R} , \ldots$$

Finally, we have, according to Eq. (8.33):

$$S_n \, \mathscr{F}_{n,g} = \frac{\varepsilon_n S_n}{\rho_n} \frac{_gD_n}{D} \tag{8.38}$$

8.2.3. Equilibrium temperature of surfaces with zero-net-flux

Starting from an initial emitter, a, (surface of type S_i, or gas) the above methods have arrived at the relative densities of reflected flux,

$$_a\varphi_1, {}_a\varphi_2, \ldots, {}_a\varphi_n, {}_a\varphi_R \ldots$$

from which the factors of exchange, $\mathscr{F}$, were derived. From this, the absolute value of the reflected flux, $_a\varphi$, due to the primary emitter will be of the form $S\varphi_r = SM_a^0 \, {}_a\varphi$ with $M_a^0 = \sigma\theta_a^4$.

Taking as initial emitter successively each of the N emitter-receiver surfaces of the enclosure and the gas, the actual flux reflected by any one of the N surfaces making up the enclosure will be the sum of the partial flux, $SM_a^0 \, {}_a\varphi$.

Thus for the surface of the nth order, we have:

$$(\Phi_r)_n = S_n [M_1^0 \, {}_1\varphi_n + M_2^0 \, {}_2\varphi_n + \ldots + M_n^0 \, {}_n\varphi_n + M_g^0 \, {}_g\varphi_n] \tag{8.39}$$

with:

$$M_1^0 = \sigma \theta_1^4$$

$$M_2^0 = \sigma \theta_2^4$$

$$M_g^0 = \sigma \theta_g^4$$

In particular, for a surface of the type S_R with zero-net-flux for which the flux $(\Phi_r)_R$ is equal to the flux emitted Φ_E, we have:

$$(\Phi_E)_R = S_R[M^0_1 \,{}_1\varphi_R + M^0_2 \,{}_2\varphi_R + \ldots + M^0_n \,{}_n\varphi_R + M^0_g \,{}_g\varphi_R] \qquad \text{(8.39 bis)}$$

Now, the laws of radiant exchange indicate that for such a surface with zero-net-flux, Φ_E is equal to black-body radiation. In a general way, the relation

$$\Phi_E = \frac{1 - \varepsilon}{\varepsilon} \Phi_{\text{net}} + \Phi^0$$

exists between the emitted flux, Φ_E, the black-body flux, Φ^0 and the earned flux Φ_{net}. Thus for a surface with zero-net-flux, $\Phi_{\text{net}} = 0$; $\Phi_E = \Phi^0$; and

$$\theta^4_R = (\theta^4_1 \,{}_1\varphi_R + \theta^4_2 \,{}_2\varphi_R + \ldots + \theta^4_n \,{}_n\varphi_R) + \theta^4_g \,{}_g\varphi_R \qquad \text{(8.40)}$$

Remember that, according to Hottel (Ref. 8.1):

$${}_1\varphi_R = \frac{{}_1D_R}{D} \,, \; {}_2\varphi_R = \frac{{}_2D_R}{D} \,, \; \ldots \,, \; {}_g\varphi_R = \frac{{}_gD_R}{D}$$

Consequently, the problems of radiant exchange from surface to surface or from gas to surface in the gray enclosures containing a gray gas should never at any time call upon the emission factor of surfaces with zero-net-flux. It does not matter in what way this type of surface sends back its incident flux, whether by reflection or by re-emission of the absorbed flux, since the gray gas does not at all modify the spectral distribution of the radiation that it transmits.

8.2.4. Applications and results of radiant transfer equations

A number of experimentally verified phenomena in actual fireboxes can be explained by the preceding development relative to a gray enclosure containing a gray gas with a uniform temperature. One can also predict the important influence that the position of the heat-receivers, as well as the emissive properties of the contents, will have on radiant heat transfer in a firebox.

We are going to treat some examples relative to enclosures with the geometry simplified for a first approximation. An additional simplification will consist in assuming the burning gases have an average emission factor, ε_G, characteristic of the average specific path of radiation in the firebox. A transmission factor $f = 1 - \varepsilon_G$ will correspond to ε_G.

8.2.4.1. Enclosures with one receiving surface and one refractory surface with zero-net-flux

For the moment, allow that all the receiving zones and all the refractory zones can be separated and grouped into two distinct zones:

S_1 = receiving surface.
S_R = refractory surface with zero-net-flux.

These two zones, of course, have uniform temperatures.

Let us apply the calculation method recommended by Hottel and developed in Section 8.2.2. (surface as source emittance).

According to Eq. (8.22):

$$S_1\, \mathscr{F}_{1,g} = S_1\, \varepsilon_1 - S_1\, \mathscr{F}_{1,1}$$

According to Eq. 8.25:

$$S_1\, \mathscr{F}_{1,1} = S_1\, \frac{\varepsilon_1}{\rho_1}\,_1\varphi_1$$

from which:

$$\mathscr{F}_{1,g} = \varepsilon_1 \left[1 - \frac{1}{\rho_1}\,_1\varphi_1 \right] \tag{8.41}$$

The energy balances carried out on S_1 and S_R are written according to Eqs. (8.27) and (8.29):

$$\left. \begin{aligned}
\left(S_1\, F_{1,1}\, f_{1,1} - \frac{S_1}{\rho_1} \right)\,_1\varphi_1 + \left(S_1\, F_{1,R}\, f_{1,R} \right)\,_1\varphi_R &= - S_1\, F_{1,1}\, f_{1,1}\, \varepsilon_1 \\[2ex]
\left(S_1\, F_{1,R}\, f_{1,R} \right)\,_1\varphi_1 + \left(S_R\, F_{R,R}\, f_{R,R} - S_R \right)\,_1\varphi_R &= - S_1\, F_{1,R}\, f_{1,R}\, \varepsilon_1
\end{aligned} \right\} \tag{8.42}$$

By putting:

$$S_1\, F_{1,1}\, f_{1,1} = \overline{11},$$

$$S_1\, F_{1,R}\, f_{1,R} = \overline{1R} \text{ and}$$

$$S_R\, F_{R,R}\, f_{R,R} = \overline{RR}$$

we have:

$$_1\varphi_1 = \frac{D_1}{D} = \frac{- \varepsilon_1 \begin{vmatrix} \overline{11} \,,\, \overline{1R} \\[1ex] \overline{1R} \,,\, \overline{RR} - S_R \end{vmatrix}}{\begin{vmatrix} \overline{11} - \dfrac{S_1}{\rho_1} \,,\, \overline{1R} \\[2ex] \overline{1R} \,,\, \overline{RR} - S_R \end{vmatrix}} = \frac{- \varepsilon_1 D'}{D}$$

we note that:

$$D = D' - \frac{S_1}{\rho_1} (\overline{RR} - S_R)$$

From which, according to Eq. (8.41):

$$\mathcal{F}_{1,g} = \varepsilon_1 \left[1 + \frac{\varepsilon_1}{\rho_1} \frac{D'}{D} \right]$$

Taking into account the relation existing between D and D', the exchange factor can be expressed in the form:

$$\frac{1}{S_1 \, \mathcal{F}_{1,g}} = \frac{1}{S_1} \left(\frac{1}{\varepsilon_1} - 1 \right) + \frac{1}{S_1 - \overline{11} + \dfrac{\overline{1R}^2}{\overline{RR} - S_R}} \qquad (8.42\,\text{bis})$$

Then put:

$$S_1 \overline{F}_{1,g} = S_1 - \overline{11} \frac{\overline{1R}^2}{\overline{RR} - S_R} \qquad (8.42\,\text{ter})$$

from which:

$$\frac{1}{\mathcal{F}_{1,g}} = \frac{1 - \varepsilon_1}{\varepsilon_1} + \frac{1}{\overline{F}_{1,g}} \qquad (8.43)$$

It thus appears that exchange factor, $\mathcal{F}_{1,g}$, takes the value, $\overline{F}_{1,g}$, when the surface, S_1, is black ($\varepsilon_1 = 1$).

It remains to clarify $\overline{F}_{1,g}$ by developing Eq. (8.42 ter). The enclosure having only two zones, S_1 and S_R, a single angle factor of type F, is independent: Hottel (Ref. 8.1) characterizes the enclosure by the direct angle factor $F_{R,1}$.

Thus we have:

$$\overline{1R} = S_1 F_{1,R} f_{1,R} = S_R F_{R,1} f_{1,R}$$

$$\overline{11} = S_1 F_{1,1} f_{1,1} = S_1 (1 - F_{1,R}) f_{1,1} = (S_1 - S_R F_{R,1}) f_{1,1}$$

$$\overline{RR} = S_R F_{R,R} f_{R,R} = S_R (1 - F_{R,1}) f_{R,R}$$

By putting these last values in Eq. (8.42 ter) the expression of $\overline{F}_{1,g}$ is finally obtained for the general case where a value characteristic of the path being considered ($f_{1,1}$, $f_{1,R}$, $f_{R,R}$) is assigned to the transmission factor of the gas.

The expression obtained is rather complicated and hard to handle; so it can be simplified if one takes: $f_{1,1} = f_{1,R} = f_{R,R} = f = 1 - \varepsilon_G$.

Under these conditions finally, one has the form that is usually adopted:

$$\bar{F}_{1,G} = \varepsilon_G \left\{ 1 + \frac{\dfrac{S_R}{S_1}}{1 + \dfrac{\varepsilon_G}{1 - \varepsilon_G} \cdot \dfrac{1}{F_{R,1}}} \right\} \tag{8.44}$$

$$\frac{1}{\mathscr{F}_{1,G}} = \frac{1 - \varepsilon_1}{\varepsilon_1} + \frac{1}{\bar{F}_{1,G}} \tag{8.43 bis}$$

$$\Phi_{G \rightleftharpoons 1} = S_1 \, \mathscr{F}_{1,G} \, \sigma(\theta_G^4 - \theta_1^4) \tag{8.43 ter}$$

From the fact that there is no mutual exchange between S_1 and S_R, $\Phi_{G \rightleftharpoons 1}$ represents in this particular case the net flux gained by the only receiving surface, S_1.

The hypothesis concerning the existence of two separate zones, S_1 and S_R, is rigorously viable only if the two kinds of receiving and refractory surfaces are intimately mixed in the same ratio at all points in the firebox: this is the checkered enclosure hypothesis of Hottel's.

Under these conditions:

$$F_{R,1} = \frac{S_1}{S_1 + S_R}$$

Taking $S_1 + S_R = S_t$ and $C = \dfrac{S_1}{S_t}$, Eq. (8.43 bis) becomes:

$$S_1 \mathscr{F}_{1,G} = \frac{S_t}{\dfrac{1}{\varepsilon_G} + \dfrac{1}{C\varepsilon_1} - 1} \tag{8.43 quater}$$

The hypothesis of the checkered enclosure applies rather well when the combustion chamber is made up of tubes placed in front of refractory, as in refinery tube stills, reforming furnaces, and furnaces of central power generators.

If, for purposes of **approximation**, Eq. (8.44) is used for an enclosure whose heated surface and refractory are separated, it is preferable to use the actual value of the angle factor $F_{R,1}$.

A certain number of practical applications, especially in fireboxes of tubular heaters that best approach the checkered enclosure, can be derived from the preceding relations.

a.   ***Influence of factors of emissivity.***

The increase of the factor of emissivity of the flame, ε_G, increases the transmission of heat, but not proportionally. The emissivity, ε_1, of the receiving walls affects the heat transfer as the emissivity of the flame increases. In particular, if the flame is very opaque ($\varepsilon_G \cong 1$) $F_{1,G} \cong \varepsilon_G \cong 1$, $\mathscr{F}_{1,G} \cong \varepsilon_1$ and the heat transfer by radiation becomes proportional to ε_1. If, on the contrary, the flame is very transparent, its emissivity has a dominant effect on the transfer and it becomes necessary to improve it in order to obtain better thermal efficiency (predominant influence of ε_G on $\mathscr{F}_{1,G}$, because of the large value taken by $\overline{F}_{1,G}$).

Of course, these observations are based on an assumption that the temperature θ_G is constant from one flame to the other; the possibilities offered by the choice of fuel, as well as by mixing with oxidizing air for modifying the temperature of the flame, even with constant excess air, is not taken into account here.

Figure 8.9 shows the values of $\mathscr{F}_{1,G}$ calculated by Godrige (Ref. 8.2) for different values of emission factors ε_G and ε_1, as a function of the ratio S_R/S_1 for a tubular firebox with refractory walls. The practical zone of use for these curves lies in the range of $S_R/S_1 < 1$. When $S_R/S_1 = 0.5$ and ε_1 goes from 0.9 to 0.7, for example, $\mathscr{F}_{1,G}$ decreases 21% if $\varepsilon_G = 0.9$ and only 15% if $\varepsilon_G = 0.5$.

b. *The contribution of refractory walls to radiant heat transfer.*

The net exchange between the flame and the refractory surface, such as is defined by Eq. (8.20 bis), is the result of a direct flame-to-surface transfer combined with a diffusion re-radiation from the refractory.

The coefficient $\mathscr{F}'_{1,g}$ characterizes the direct exchange, and by considering the gas as the only primary emitter in the enclosure, we have according to Eq. (8.33):

$$\mathscr{F}'_{1,g} = \frac{\varepsilon_1}{\rho_1}\, {}_g\varphi'_1$$

The energy balance made on S_1 is written, according to the reasoning described in the second method of Section 8.2.2:

$$\left[S_1\, F_{1,1}\, f_{1,1}\, {}_g\varphi'_1 + S_1\, F_{1,1}(1 - f_{1,1}) \right] \rho_1 = S_1\, {}_g\varphi'_1$$

that is:

$$\left(S_1\, F_{1,1}\, f_{1,1} - \frac{S_1}{\rho_1} \right) {}_g\varphi'_1 = -S_1\, F_{1,1}\,(1 - f_{1,1}) \tag{8.45}$$

Taking $f_{1,1} = f = (1 - \varepsilon_G)$ and combining Eq. (8.33) and Eq. (8.45) results in:

$$\mathscr{F}'_{1,G} = \varepsilon_1\, \frac{\varepsilon_G\, F_{1,1}}{1 - F_{1,1}\,(1 - \varepsilon_G)(1 - \varepsilon_1)}$$

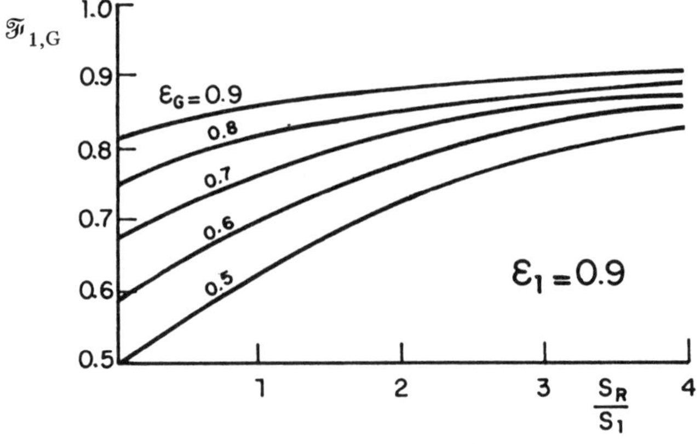

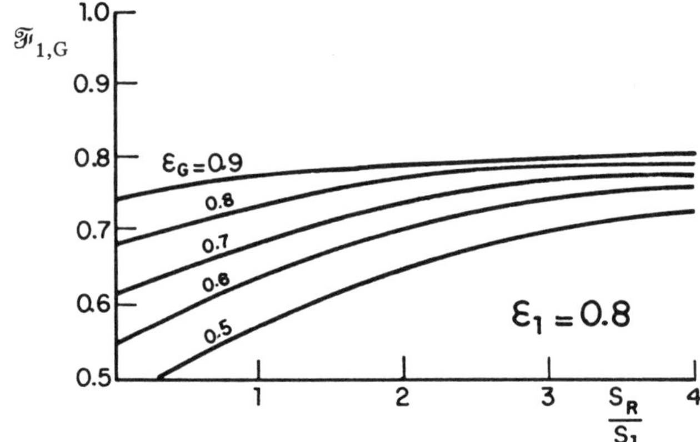

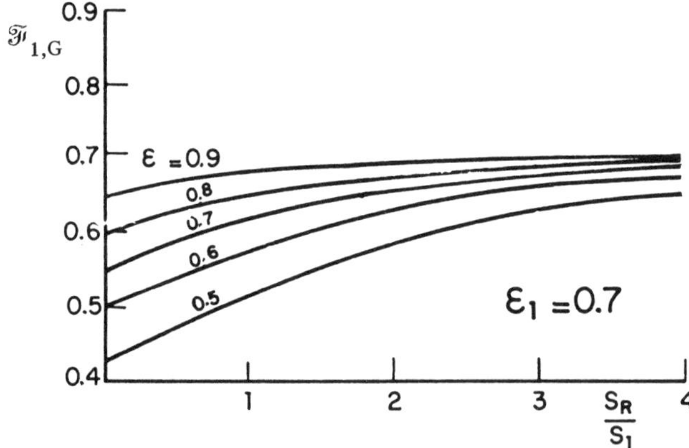

Fig. 8.9.   Gas-wall exchange factor, $\mathcal{F}_{1,G}$, as a function of emissivity, ε_1, and the ratio of zero-net-flux surface, S_R, to receiving surface S_1, as calculated by Godrige.

which can be put in the form:

$$\frac{1}{\mathcal{F}'_{1,G}} = \frac{1-\varepsilon_1}{\varepsilon_1} + \frac{1}{\varepsilon_G}\left\{1 + \frac{1}{\varepsilon_1}\left(\frac{1}{F_{1,1}} - 1\right)\right\} \tag{8.46}$$

As for the angle factor $F_{1,1}$ relative to the surface S_1, it becomes apparent that the same value as previously should be used, that is:

$$F_{1,1} = 1 - F_{1,R} = 1 - \frac{S_R}{S_1} F_{R,1}$$

To show what the contribution of the refractory can be in transmitting radiation, we will work with a firebox of the checkered-wall type in order to define the angle factors definitely. Under these conditions:

$$F_{R,1} = \frac{S_1}{S_1 + S_R} = F_{1,1}$$

Example. Imagine two values of the ratio $S_R/S_1 = 0.5$ and 1; and imagine two values for the emissivity of the flame: $\varepsilon_G = 0.75$ (a fuel oil flame) and $\varepsilon_G \doteqdot 0.25$ (a gas flame). Take $\varepsilon_1 = 0.9$, this factor corresponding rather closely to that for a surface either of oxidized steel or covered with soot. Then use the relations of Eq. (8.43 quater), in which:

$$C = \frac{S_1}{S_t} = \frac{1}{1 + \dfrac{S_R}{S_1}}$$

plus those of Eq. (8.46), in which:

$$F_{1,1} = F_{R,1} = C$$

The contribution of the refractory in the radiant transfer is given by the ratio:

$$\frac{(\mathcal{F}_{1,G} - \mathcal{F}'_{1,G})}{\mathcal{F}_{1,G}}.$$

The results obtained are shown in Table 8.1.

TABLE 8.1

$\dfrac{S_R}{S_1}$	ε_G	$\mathcal{F}_{1,G}$	$\mathcal{F}'_{1,G}$	Contribution of the refractory (%)
0.5	0,75	0.75	0.458	39
	0.25	0.322	0.158	51
1	0.75	0.783	0.342	56
	0.25	0.383	0.117	70

From this, the refractory enclosure comprises very effective support for radiation, especially for flames that are only slightly radiant.

It is conceivable that in a steel-industry furnace for heating ingots, melting, etc., the importance of the refractory material should be particularly great in radiant transfers to the charge because the walls of the firebox are surfaces with zero-net-flux. Nevertheless, because the charge is well separated from the refractory material, the preceding reasoning does not apply rigorously to heat-treating furnaces.

An interesting semi-quantitative approach was made by Jaegle and Leblanc on the experimental furnace of the *Groupement d'Etudes des Flammes de Gaz Naturel (GEFGN)* at Toulouse. The authors (Ref. 8.3) calculate the exchange $\Phi_{R \rightleftarrows 1}$ between refractory walls and a charge of cooled metallic blocks arranged in the bottom plane by assuming that the medium separating the two surfaces is transparent. To do this, they use the surface-temperature readings taken during the tests.

Relating $\Phi_{R \rightleftarrows 1}$ to the quantity of heat actually captured by the charge, Φ, they consider the evolution of the ratio $\Phi_{R \rightleftarrows 1}/\Phi$ as a function of the maximum measured emissive factor of the flame ε_G. The phenomenon is shown in Fig. 8.10. Even the extreme values of ε_G (0.25 with the gas and 0.9 with No. 2 heavy fuel oil) led to a variation of only 0.94 and 0.89 for the ratio. This shows that the radiant transfer from walls to charge is an essential part of the overall heat transfer.

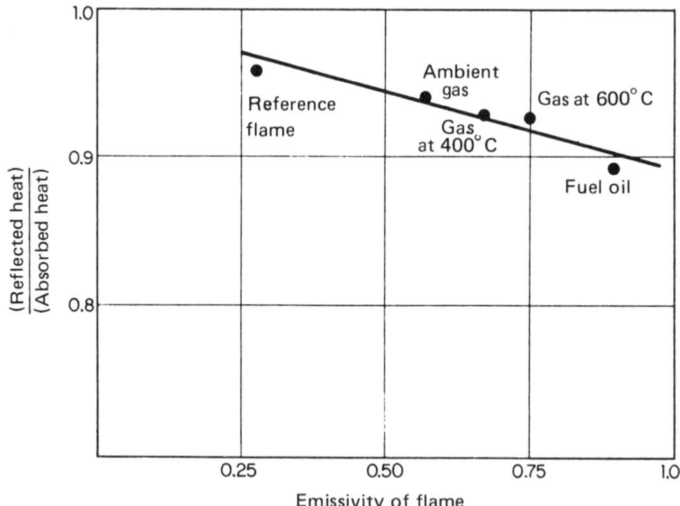

Fig. 8.10.　Importance of wall reradiation compared to direct flame radiation: the ratio of reradiated heat to heat duty varies only 0.94-0.89 as the flame's emissivity goes from 0.25 (for the gas) to 0.9 (for fuel oil).

Coming back to Eqs. (8.20 bis), (8.43), (8.44) and considering the ratio:

$$\frac{\Phi_{G \rightleftarrows 1}}{S_1} = \mathscr{F}_{1,G} \, \sigma(\theta_G^4 - \theta_1^4)$$

as representing the thermal load of the receiver, it becomes apparent that, other things being equal, this thermal load will increase with the ratio S_R/S_1 because of the increase in $\mathscr{F}_{1,G}$. For a fixed value of S_1, this increase in flux density will correspond to an increase in the quantity of heat captured by the receiver, and therefore, in an increase in the efficiency of the firebox. However, there is a chance that S_1 will vary when S_R increases to the detriment of S_1. This was the case of domestic boilers that, in a change-over from coal to fuel oil, were furnished with additional refractory bricks that hid a large fraction of the heat-capturing surface of the firebox.

Tests made at the *Institut Français du Pétrole (IFP)* on a commercial type heater thus showed that brickwork covering 35% of the initial surface of the firebox increased the thermal load of the surface offered to radiation by 50%, but actually reduced the heat captured by 5%.

8.2.4.2. Fireboxes without refractory

A firebox without refractory surface forms an unbroken absorber of radiation; as is the case with fireboxes with water curtains in which the tubes are joined, as well as fireboxes of domestic heaters without bricks and cylindrical fireboxes of heaters with flue gas tubes.

Take the simple case of different receiving surfaces at temperatures close enough to be grouped in one single zone, S_1, of uniform temperature. Under these conditions, S_1 is a reflection of itself. Then by saying:

$$S_R = 0 \text{ in Eq. (8.44)}$$
$$F_{1,1} = 1 \text{ in Eq. (8.46)}$$
$$\text{or } C = 1 \text{ in Eq. (8.43 ter)}$$

the gas-wall transfer is written:

$$\frac{1}{\mathscr{F}_{1,G}''} = \frac{1}{\varepsilon_1} + \frac{1}{\varepsilon_G} - 1 \tag{8.47}$$

and

$$\Phi_{G \rightleftarrows 1} = S_1 \, \mathscr{F}_{1,G}'' \, \sigma(\theta_G^4 - \theta_1^4) \tag{8.47 bis}$$

By comparing these expressions to Eq. (8.43) and Eq. (8.43 ter), it becomes apparent that, compared to a firebox with refractory, the use of all absorbent surfaces leads:

(a) To a lighter thermal load because of the reduction in $\mathscr{F}_{1,G}$.

(b) But to an increase in the amount of heat transmitted by radiation, because the relative increase in receiving surface is greater than the relative reduction in load.

Example. Take the example in the second part of Section 8.2.4.1. Although the receiving surface is S_1 in the checkered enclosure with refractory, in the firebox without refractory this surface becomes S_t, with the result that the amount of heat transferred varies by the ratio:

$$\frac{S_t \, \mathcal{F}''_{1,G}}{S_1 \, \mathcal{F}'_{1,G}} = \left(1 + \frac{S_R}{S_1}\right) \frac{\mathcal{F}''_{1,G}}{\mathcal{F}'_{1,G}}$$

while the flux densities are in the ratio:

$$\frac{\mathcal{F}''_{1,G}}{\mathcal{F}'_{1,G}}$$

For variations of this scope, one gets the values shown in Table 8.2.

TABLE 8.2

ε_G	$\dfrac{S_R}{S_1}$ initial	$\mathcal{F}'_{1,G}$	$\mathcal{F}''_{1,G}$	Transmitted flux		Flux density	
				With S_R	Integral receiver	With S_R	Integral receiver
0.75	0.5 1	0.75 0.783	0.692	1 1	1.38 1.77	1 1	0.92 0.88
0.25	0.5 1	0.322 0.383	0.244	1 1	1.14 1.28	1 1	0.76 0.64

To finish this section, we can do no better than to quote M. Veron of the *Conservatoire National des Arts et Métiers (CNAM)* (Ref. 8.4). Conclusions that seem obvious once these calculations are made are not self-evident, because the notion of water curtains was anticipated by an expectation that heat absorption over the receiving surface would be the same as before.

8.2.4.3. Schematic analysis of heat transfers during heating and melting by batch

A furnace for batch heating can be considered as two parts, a hearth holding the charge and a roof arching over it. The roof plays an important role in transmitting radiation because of its position; the thermal mass of the charge often determines the duration of the heating cycle.

In order to analyze the transfer of heat toward the charge, the hearth and roof are resolved into two generalized planes (1) and (2) that are exposed separately (Fig. 8.11).

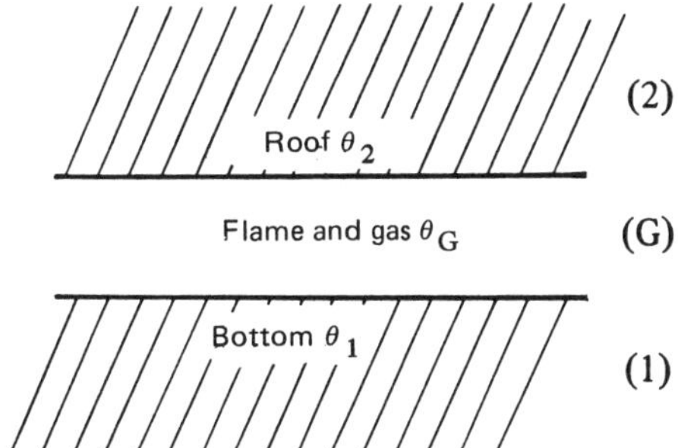

Fig. 8.11.　Exchange between hearth and roof in a furnace.

Under these conditions:

$$
\begin{aligned}
S_1 &= S_2 = S \\
F_{1,1} &= F_{2,2} = 0 \\
F_{1,2} &= F_{2,1} = 1
\end{aligned}
\qquad (8.48)
$$

In order to deal with this situation, the roof will not be treated as a surface with zero-net-flux.

Starting with the principle that $\theta_G > \theta_2 > \theta_1$, the density of the flux into the hearth is such that, according to Eq. (8.20) and Eq. (8.20 bis):

$$
\left(\frac{\Phi}{S}\right)_1 = \mathcal{F}_{1,2}\,\sigma(\theta_2^4 - \theta_1^4) + \mathcal{F}_{1,G}\,\sigma(\theta_G^4 - \theta_1^4)
\qquad (8.49)
$$

that is, by considering the black-body radiation of the three emitters and receivers:

$$
\left(\frac{\Phi}{S}\right)_1 = \mathcal{F}_{1,2}\,M_2^0 + \mathcal{F}_{1,G}\,M_G^0 - (\mathcal{F}_{1,2} + \mathcal{F}_{1,G})\,M_1^0
\qquad (8.49\,\text{bis})
$$

We can use the calculation method recommended by Hottel and described in Section 8.2.2 while taking into account the conditions of Eq. (8.48). Considering the hearth as initial emitter of the system, we have:

$$
\mathcal{F}_{1,G} = \varepsilon_1 - \mathcal{F}_{1,1} - \mathcal{F}_{1,2} \qquad \text{from (8.22)}
$$

$$
\left.
\begin{aligned}
\mathcal{F}_{1,1} &= \frac{\varepsilon_1}{\rho_1}\,{}_1\varphi_1 \\[2em]
\mathcal{F}_{1,2} &= \frac{\varepsilon_2}{\rho_2}\,{}_1\varphi_2
\end{aligned}
\right\} \qquad \text{from (8.25)}
$$

Energy balances around the hearth and the roof lead, according to Eqs. (8.27) and (8.28), to the system of equations:

$$\begin{cases} -\dfrac{1}{\rho_1}\,{}_1\varphi_1 + f\,{}_1\varphi_2 = 0 \\[2ex] f\,{}_1\varphi_1 - \dfrac{1}{\rho_2}\,{}_1\varphi_2 = -f\varepsilon_1 \end{cases}$$

with:

$$\begin{aligned} f &= 1 - \varepsilon_G \\ \rho_1 &= 1 - \varepsilon_1 \\ \rho_2 &= 1 - \varepsilon_2 \end{aligned}$$

from which:

$$_1\varphi_1 = \frac{{}_1 D_1}{D} = \frac{\begin{vmatrix} 0 & , & f \\[1ex] -f\varepsilon_1 & , & -\dfrac{1}{\rho_2} \end{vmatrix}}{\begin{vmatrix} -\dfrac{1}{\rho_1} & , & f \\[2ex] f & , & -\dfrac{1}{\rho_2} \end{vmatrix}} \quad \text{that is } _1\varphi_1 = \frac{\rho_1\,\rho_2\,\varepsilon_1\,f^2}{1-\rho_1\,\rho_2\,f^2}$$

$$_1\varphi_2 = \frac{{}_1 D_2}{D} = \frac{\begin{vmatrix} -\dfrac{1}{\rho_1} & , & 0 \\[2ex] f & , & -f\varepsilon_1 \end{vmatrix}}{\begin{vmatrix} -\dfrac{1}{\rho_1} & , & f \\[2ex] f & , & -\dfrac{1}{\rho_2} \end{vmatrix}} \quad \text{that is } _1\varphi_2 = \frac{\rho_2\,\varepsilon_1\,f}{1-\rho_1\,\rho_2\,f^2}$$

From which, according to Eq. (8.25):

$$\mathscr{I}_{1,1} = \frac{\rho_2\,\varepsilon_1^2\,f^2}{1-\rho_1\,\rho_2\,f^2} \tag{8.50}$$

$$\mathscr{I}_{1,2} = \frac{\varepsilon_1\,\varepsilon_2\,f}{1-\rho_1\,\rho_2\,f^2} \tag{8.51}$$

Going back to Eq. (8.22), one obtains:

$$\mathscr{I}_{1,G} = \varepsilon_1 - \frac{\rho_2\,\varepsilon_1^2\,f^2 + \varepsilon_1\,\varepsilon_2\,f}{1-\rho_1\,\rho_2\,f^2}$$

that is:

$$\mathcal{J}_{1,G} = \varepsilon_1 \frac{(1-f)(1+\rho_2 f)}{1-\rho_1 \rho_2 f^2} \tag{8.52}$$

Equations (8.49), (8.50), (8.51) and (8.52) then permit complete resolution of the problem of exchange by radiation.

M. Veron (Ref. 8.4) has analyzed three significant variations of the general condition.

First case: The roof temperature, θ_2, is almost the same as the gas temperature, θ_G. This is the general case of glass-melting furnaces.

If $\theta_2 = \theta_G$, we have $M_2^0 = M_G^0$ and Eq. (8.49) is written:

$$\left(\frac{\Phi}{S}\right) = (\mathcal{J}_{1,2} + \mathcal{J}_{1,G})(M_2^0 - M_1^0)$$

that is, according to Eq. (8.22):

$$\left(\frac{\Phi}{S}\right)_1 = (\varepsilon_1 - \mathcal{J}_{1,1})(M_2^0 - M_1^0)$$

By replacing $\mathcal{J}_{1,1}$ with its value taken from Eq. (8.50), one finally obtains:

$$\left(\frac{\Phi}{S}\right)_1 = \frac{\varepsilon_1[1-(1-\varepsilon_G)^2(1-\varepsilon_2)]}{1-(1-\varepsilon_1)(1-\varepsilon_2)(1-\varepsilon_G)^2}(M_2^0 - M_1^0) = \mathcal{J}(M_G^0 - M_1^0) \tag{8.53}$$

To illustrate the practical consequences resulting from the temperatures of the roof and of the flame being equal, some examples are treated here below. In these we assume $\varepsilon_1 = \varepsilon_2$ an hypothesis well verified for refractory surfaces.

From Table 8.3, it can be seen that the emissivity of the flame holds relatively minor importance for exchange coefficient $\mathcal{J}$, compared to the emissivity of the roof. With a black-body roof ($\varepsilon_2 = 1.0$), the flux density on the bottom is

TABLE 8.3

VALUES OF $\mathcal{J}$ WHEN THE ROOF TEMPERATURE
IS THE SAME AS THE GAS TEMPERATURE

$\varepsilon_1 = \varepsilon_2$ $\diagdown$ ε_G	0.2	0.8
0.5	0.405	0.495
0.9	0.848	0.897
1	1	1

theoretically independent of ε_G, and according to Eq. (8.53) this remains true if $\varepsilon_1 \neq \varepsilon_2$, since the absorption and the radiation of the gas is exactly compensated. This phenomenon is well known as the basis of the Kurlbaum (or auxiliary black-body) method for calculating the temperature of a radiating flame.

Second case: *The roof temperature, θ_2, is the same as the hearth temperature, θ_1.*

In this case, there is no longer mutual exchange between roof and hearth ($M_2^0 = M_1^0$) and Eq. (8.49) is written:

$$\left(\frac{\Phi}{S}\right)_1 = \mathscr{I}_{1,G}(M_G^0 - M_1^0)$$

Replacing $\mathscr{I}_{1,G}$ in this equation with its value from Eq. (8.52) leads to:

$$\left(\frac{\Phi}{S}\right)_1 = \frac{\varepsilon_1\,\varepsilon_G\,[1+(1-\varepsilon_2)(1-\varepsilon_G)]}{1-(1-\varepsilon_1)(1-\varepsilon_2)(1-\varepsilon_G)^2}(M_G^0 - M_1^0) = \mathscr{I}(M_G^0 - M_1^0) \quad (8.54)$$

From this, $\mathscr{I}$ can be calculated for different assumed values of emissivity, ε, as in Table 8.4.

TABLE 8.4

VALUES OF $\mathscr{I}$ WHEN THE ROOF TEMPERATURE
IS THE SAME AS THE HEARTH TEMPERATURE

$\varepsilon_1 = \varepsilon_2$ \ ε_G	0.2	0.8
0.5	0.167	0.445
0.9	0.196	0.735
1	0.2	0.8

Contrary to Table 8.3, Table 8.4 shows that the exchange coefficient varies a lot with the emissivity of the flame but is only slightly influenced by the emissivity of the roof with a transparent flame. Also, when exchange coefficients are compared for the same values of ε_2 and ε_G in Tables 8.3 and 8.4, it becomes apparent that the condition of roof temperatures equal to gas temperatures ($\theta_2 = \theta_G$) leads to higher heat transfer than when $\theta_2 = \theta_1$. Thus, it is important to keep the roof at a temperature as close as possible to the temperature of the flame.

Third case: *The roof has zero net radiant flux.*

This condition assumes that the amount of heat lost through the roof to the outside is balanced by convection heat transfer to the roof. This is about the case with open-hearth furnaces. The result is exchange by radiation only between the gases and the hearth, and Eq. (8.49) is written:

$$\left(\frac{\Phi}{S}\right)_1 = \mathcal{F}'_{1,G}(M^0_G - M^0_1)$$

$\mathcal{F}'_{1,G}$ being obtained by taking $\rho_2 = 1$ in Eq. (8.52).

Finally:

$$\left(\frac{\Phi}{S}\right)_1 = \frac{\varepsilon_1 \varepsilon_G(2 - \varepsilon_G)}{1 - (1 - \varepsilon_1)(1 - \varepsilon_G)^2}(M^0_G - M^0_1) = \mathcal{F}(M^0_G - M^0_1) \quad (8.55)$$

This third case was treated in a more general way in Section 8.2.4.1. The distinguishing feature of the present case lies in that $S_R/S_1 = 1$ and $F_{R,1} = 1$. Under these conditions Eq. (8.44) and Eq. (8.43) lead to:

$$\frac{1}{\mathcal{F}_{1,G}} = \frac{1 - \varepsilon_1}{\varepsilon_1} + \frac{1}{\varepsilon_G(2 - \varepsilon_G)} = \frac{1}{\mathcal{F}}$$

This expression gives the same value of exchange coefficient as obtained by Eq. (8.55).

Table 8.5 shows various exchange coefficients for assumed emissivities under this condition, as were shown for the other conditions of Table 8.3 and Table 8.4. It can be seen that the exchange coefficient is more affected by ε_1 when the flame is opaque than when the flame is transparent. This has already been pointed out in Section 8.2.4.1.

TABLE 8.5

VALUES OF $\mathcal{F}$ WHEN THE ROOF EXHIBITS ZERO-NET-FLUX

ε_1 \ ε_G	0.2	0.8
0.5	0.265	0.49
0.9	0.346	0.867
1	0.36	0.96

This third case leads to coefficients greatly dependent on ε_G, but intermediate between those obtained in the two preceding cases.

M. Veron points out that glass manufacturers usually operate according to the first case and metallurgists to the third case. Around 1936, there arose a difference of opinion concerning the value of radiant flames; the glass-makers denied any value, while the metallurgists were enthusiastic over it. Since that time, knowledge of the fundamentals plus experience have permitted better understanding of problems of radiant heat transfer, with a consequent decline of such extreme positions.

Since the optimum solution consists of operating with a roof temperature as near as possible to the flame temperature (first case), certain smelting furnaces with good control instruments operate almost to the limit at which the refractory begins to deform. Thus glass furnaces, as well as open-hearth furnaces treating special steels, operate with roof temperatures of 1,600-1,700° C.

The result is that the choice of fuels to give more or less transparent flames (heavy fuels or natural gas, for example) is governed more by economics than by heat transfer considerations. Nevertheless, glass manufacturers keep their preference for radiant flames, because the heat absorbed by a melted glass bath reaches a maximum at the wave lengths of radiation from carbon particles.

8.3. EXCHANGE BY RADIATION IN A GRAY ENCLOSURE CONTAINING A REAL GAS (NOT GRAY) AT UNIFORM TEMPERATURE

8.3.1. General

Products from the combustion of liquid or gaseous hydrocarbons contain molecules that are non-symmetric diatomic (CO, for example) or tri-atomic (principally CO_2 and H_2O), which through vibration and rotation permit absorption or emission of radiation in the infrared. Such absorption of a discontinuous segment of the energy of the spectrum is opposed to the hypothesis of gray gas.

Nevertheless, the hypothesis of the gray absorbent medium is confirmed within limits for the case of luminous flames filled with soot particles from cracked hydrocarbons. In these flames, the cloud of particles behaves statistically like a gray emitter although each elementary particle selects a part of the spectrum according to its composition and dimension. The work of the *IFRF* and the *GEFGN* is particularly directed toward fundamental questions relative to such flames.

If an industrial type radiating flame can be assumed as an overall gray body, the approximation becomes much less reliable with respect to the combustion gases at the exterior of the flame or when transparent flames are present. Under these conditions, reflected radiation passing through the flame from any point of the enclosure will have energy from a given region of the spectrum partially absorbed at each pass, until after a certain number of reflections and passages across the gas, the gas will behave practically as a transparent medium for spectral regions remaining.

The resulting change in the spectral distribution of transmitted energy prevents assigning a single value to the emissivity for multiple reflections from one surface to another within the enclosure. Unlike the case of a gray gas, where the emissivity of surfaces with zero-net-flux can be ignored, radiant transmission through a real gas must take into account spectral distribution of the energy reflected by the enclosure. Only a surface that reflects its incident flux by diffusion ($\rho_R = 1$) will exhibit zero-net-flux with the real gas in the same way that the theoretical surface of zero-net-flux behaves with a theoretical gray gas. Otherwise, a surface, S_R, with zero-net-flux should, strictly speaking, be treated like a receiver-emitter surface of type S_i, for which the temperature is determined by assuming that the sum of the exchanges of S_R with the other surfaces and the gas is nil.

Finally a supplementary complication from the failure of the gray gas hypothesis results from the non-equality of the total emissions and absorptions through real gases. The numerous theoretical and experimental works with which Hottel participated provide the authority in this matter. The numerical data and relations relative to ε_G and $\alpha_{1,G}$ for CO_2, H_2O and their mixture are widely used by heat transfer technicians (Ref. 8.1). $\alpha_{1,G}$ the absorption factor for a radiation from a black source (or a gray one) at θ_1, is equal to ε_G only if $\theta_1 = \theta_G$.

In the radiant exchange between a surface S_1 of the enclosure and the gas, for example, the exchange coefficient $\mathcal{F}_{G \to 1}$ involving ε_G should be distinguished from the exchange coefficient, $\mathcal{F}_{1 \to G}$, involving $\alpha_{1,G}$.

Thus we will have:

$$\Phi_{G \rightleftarrows 1} = S_1 \, \mathcal{F}_{G \to 1} \, M_G^0 - S_1 \, \mathcal{F}_{1 \to G} \, M_1^0 \tag{8.56}$$

instead of:

$$\Phi_{G \rightleftarrows i} = S_i \, \mathcal{F}_{i,G} \, (M_G^0 - M_1^0) \tag{8.20 bis}$$

in the case of the gray gas.

According to Hottel, a single value $\mathcal{F}_{1,G}$ can be adopted in practice with a satisfactory approximation only so long as θ_G does not go over twice θ_1: which can be the case in smelting furnaces but not in tubewall heaters.

Of course, not using the gray gas hypothesis increases the number and complexity of the calculations. Computer calculations permit the engineer to treat these problems numerically with enough rigor. Thus it has been possible to put the numerical data for the gases in polynomial form (Ref. 8.5). When algebraic solutions are desired, certain problems lend themselves to easy calculation by means of reasonable simplifications, such as neglecting the absorption by the real gas after a few passages.

8.3.2. Calculating radiant transfer with a real gas

One method used in practice consists of dividing the real gas into a number of gray gases that conform to the reasoning described in Section 8.2.2.

8.3.2.1. General method

a. *Formulating the emissivities of radiation and absorption*

In order to use the same types of exchange equations as those of Section 8.2.2., Hottel (Ref. 8.1) recommends treating the behavior of the real gas as similar to that of a collection of gray gases operating in the fractions $x, y, z \ldots$ of the energy spectrum.

The overall transmission factor f can then be represented to the desired degree of accuracy, through the equation:

$$f = xf_x + yf_y + zf_z + \ldots \tag{8.57}$$

with

$$x + y + z + \ldots = 1 \tag{8.58}$$

f thus appears as being a weighted sum of the partial transmission factors of the type:

$$f_x = \exp\left(- k_x pL\right) \tag{8.59}$$

k_x being the coefficient of absorption relative to the fraction x of the spectrum, p, the partial pressure of the gas and L, the specific path of the radiation. For a single value in any furnace or heater, L can be assumed about 3.4 times the average hydraulic diameter of the enclosure:

$$L = \frac{3.4\,\mathcal{V}}{S_p}$$

where S_p is the envelope of the volume $\mathcal{V}$.

According to Eqs. (8.57) and (8.58) we have:

$$1 - f = x(1 - f_x) + y(1 - f_y) + z(1 - f_z) + \ldots \tag{8.60}$$

Depending on whether $x, y, z \ldots$ and $f_x, f_y, f_z \ldots$ are evaluated from the emission factors of gas at θ_G or absorption factors of gas at θ_G for radiation coming from a black or gray surface at θ_i, $(1 - f)$ represents the total emissivity, ε_G, of the real gas or its total absorptivity, $\alpha_{i,G}$.

Equation (8.60) will then take the two following forms:

$$\varepsilon_G = \Sigma \, (x \, \varepsilon_x)_{\theta_G} \quad \text{or} \quad \alpha_{i,G} = \Sigma \, (x\alpha_x)_{\theta_G, \theta_i} \tag{8.61}$$

Using the most recent data, Hottel and Sarofim (Ref. 8.6) have put the emissivities of the mixtures CO_2, H_2O in a form well adapted for computer calculations. The authors limit themselves to three fractions of the spectrum, x, y, z chosen in such a way that the gas can be considered as transparent for the remaining fraction $u = 1 - (x + y + z)$.

From this we have:

$$f = x f_x + y f_y + z f_z + u \tag{8.57 bis}$$

Taking into account the definition Eq. (8.59) of the partial transmission coefficients, it is seen that u represents the asymptote of f when pL goes to infinity.

x, y and z are put in the polynomial form of third degree in θ. The coefficients of these polynomials, as well as the absorption coefficients, k, entering into the calculation of partial transmission factors Eq. (8.59), have been put into tables for an equimolar mixture of CO_2 and H_2O resulting from the combustion of a hydrocarbon of the type $(CH_2)_n$ with 15% excess air at a total pressure of 1 atm. and a partial pressure of CO_2 or H_2O of 0.115 atm.

This information permits calculating the partial emission and absorption factors and the corresponding energy fractions for the important variables encountered in heaters and furnaces:

$$L \quad \text{from} \quad 6 \text{ to } 13{,}700 \text{ mm}$$

$$\theta_G \quad \text{from} \quad 1{,}100 \text{ to } 2{,}200 \text{ K}$$

$$\theta_i \quad \text{from} \quad 550 \text{ to } 1{,}670 \text{ K}$$

b. Calculating the exchange coefficients

We have already pointed out in Section 8.3.1. that, by not following the gray gas hypothesis, the exchange between a gas or surface emitter and a receiving surface should be expressed in terms of both of two factors $\mathscr{F}$ of the two exchanges.

Thus between surface S_1 and the gas:

$$\frac{1}{S_1} \Phi_{G \rightleftarrows 1} = \mathscr{F}_{1 \leftarrow G}\, M_G^0 - \mathscr{F}_{1 \rightarrow G}\, M_1^0 \tag{8.56}$$

$\mathscr{F}_{1 \leftarrow G}$ refers to $\varepsilon_G\,(\theta_G)$ and $\mathscr{F}_{1 \rightarrow G}$ to $\alpha_{G,1}\,(\theta_G, \theta_1)$.

Between surfaces S_1 and S_2 we have:

$$\frac{1}{S_1} \Phi_{2 \rightleftarrows 1} = \mathscr{F}_{1 \leftarrow 2}\, M_1^0 - \mathscr{F}_{1 \rightarrow 2}\, M_1^0 \tag{8.56 bis}$$

$\mathscr{F}_{1 \leftarrow 2}$ refers to $\alpha_{G,2}\,(\theta_G, \theta_2)$ and $\mathscr{F}_{1 \rightarrow 2}$ to $\alpha_{G,1}\,(\theta_G, \theta_1)$.

For an enclosure made up of receiving-emitting surfaces, $S_1, S_2, \ldots, S_n$, as well as surfaces with zero-net-flux $S_R, S_S \ldots$ similar to perfect diffusers (when S_R, S_S are completely reflecting or white), it can be shown that the solution relative to $\mathscr{F}_{1,G}, \mathscr{F}_{1,2} \ldots$ equals the sum of a number of solutions obtained with gray gases, when each of these solutions is weighted according to the fraction of the energy spectrum it represents.

Introducing solution Eqs. (8.43, 8.44) in Eq. (8.65), thus:

$$\mathcal{F}_{1 \leftarrow G} = \cfrac{x}{\left(\dfrac{1}{\varepsilon_1} - 1\right) + \cfrac{1}{\dfrac{\varepsilon_G}{x}\left\{1 + \cfrac{S_R/S_1}{1 + \dfrac{\varepsilon_G/x}{1 - \varepsilon_G/x}\dfrac{1}{F_{R,1}}}\right\}}} \tag{8.69}$$

is obtained with:

$$x = \frac{\varepsilon_G^2}{2\,\varepsilon_G - \varepsilon_{2G}}$$

In order to calculate $\mathcal{F}_{1 \rightarrow G}$, in Eqs. (8.68) and (8.69) ε_G will be replaced by $\alpha_{G,1}$ and ε_{2G} by $\alpha_{2G,1}$.

Finally, the net flux accepted by S_1 will be:

$$\Phi_{G \rightleftarrows 1} = S_1\, \mathcal{F}_{G \rightarrow 1}\, M_G^0 - S_1\, \mathcal{F}_{1 \rightarrow G}\, M_1^0 \tag{8.56}$$

from the fact that there is no exchange between S_1 and S_R.

8.3.3.2. The entire enclosure is a radiation receiver with uniform temperature

This is the simplest case for a firebox without exposed refractory. It was treated by means of the gray-gas hypothesis in Section 8.2.4.2.

For a real gas, assuming that $S_R = 0$ in Eq. (8.69), one obtains:

$$\mathcal{F}_{1 \leftarrow G} = \frac{x}{\left(\dfrac{1}{\varepsilon_1} - 1\right) + \dfrac{x}{\varepsilon_G}}$$

and

$$\tag{8.70}$$

$$\mathcal{F}_{1 \rightarrow G} = \frac{x'}{\left(\dfrac{1}{\varepsilon_1} - 1\right) + \dfrac{x'}{\alpha_{G,1}}}$$

Equation (8.68) defines x or x' according to whether the basis is the factor of emission or absorption.

If the wall is "black", the exchange coefficients are written by saying $\varepsilon_1 = 1$ in Eq. (8.70):

$$\mathcal{F}_{1 \leftarrow G} = \varepsilon_G \quad \text{and} \quad \mathcal{F}_{1 \rightarrow G} = \alpha_{G,1}$$

Putting these values in Eq. (8.56) the conventional expression relative to exchange between a real gas and a black wall is found again:

$$\frac{1}{S_1}\,\Phi_{G \rightleftarrows 1} = \varepsilon_G\, M_G^0 - \alpha_{G,1}\, M_1^0 = \sigma\,(\varepsilon_G\, \theta_G^4 - \alpha_{G,1}\, \theta_1^4) \tag{8.71}$$

The exchange $\Phi_{G \rightleftarrows 1}$ relative to a gas and a gray wall can be put in a form somewhat like the preceding, by considering a half-gray gas as not modifying the spectral distribution of the radiation in the course of successive passages, but by maintaining the distinction between ε_G and $\alpha_{G,1}$.

Putting $x = x' = 1$ in Eq. (8.70), the gas-wall exchange can be written according to Eq. (8.56):

$$\frac{1}{S_1} \Phi_{G \rightleftarrows 1} = \left[\frac{\varepsilon_1 \varepsilon_G}{\varepsilon_1 + \varepsilon_G - \varepsilon_1 \varepsilon_G} \right] M_G^0 - \left[\frac{\varepsilon_1 \alpha_{G,1}}{\varepsilon_1 + \alpha_{G,1} - \varepsilon_1 \alpha_{G,1}} \right] M_1^0$$

The terms: (8.72)

$$\varepsilon_G - \varepsilon_1 \varepsilon_G = \varepsilon_G (1 - \varepsilon_1)$$

and

$$\alpha_{G,1} - \varepsilon_1 \alpha_{G,1} = \alpha_{G,1} (1 - \varepsilon_1)$$

being low relative to ε_1 in the usual case, they can be assigned a common value and we can write:

$$\frac{1}{S_1} \Phi_{G \rightleftarrows 1} = \left[\frac{\varepsilon_1}{\varepsilon_1 + \alpha_{G,1} - \varepsilon_1 \alpha_{G,1}} \right] (\varepsilon_G M_G^0 - \alpha_{G,1} M_1^0) \qquad (8.73)$$

According to Hottel, for $\varepsilon_1 > 0.8$, the term in the brackets can be replaced by:

$$\frac{1 + \varepsilon_1}{2}$$

from which:

$$\frac{1}{S_1} \Phi_{G \rightleftarrows 1} \simeq \left(\frac{1 + \varepsilon_1}{2} \right) (\varepsilon_G M_G^0 - \alpha_{G,1} M_1^0) \qquad (8.74)$$

From Eq. (8.73), H.N. Sharan (Ref. 8.5) established a method permitting rapid calculation of wall heating with the help of graphs or of a computer.

He verified that his method gave satisfactory results for the following parameters for heaters:

$\theta_G = 670\text{-}1{,}670\,\text{K};\ L[p_{H_2O}(p_{CO_2} + p_{H_2O})]^{0.5} = 0.02\text{-}2\,\text{atm/m}.$

$\theta_1 = 370\text{-}870\,\text{K}.$

$L = 0.1\text{-}5\,\text{m};\ p_{\text{box}} = 1\,\text{atm}.$

Equations (8.72), (8.73), (8.74) and the Sharan method do not take into account the modifications that occur in the spectral distribution of transmitted radiation during successive passages through the gas.

To justify this simplifying hypothesis, M. Veron (Ref. 8.4) proposes ignoring the successive reflections to the wall. Under these conditions, the gas-wall exchange can be written:

$$\frac{1}{S_1} \, \Phi_{G \rightleftarrows 1} = \varepsilon_G \, M_G^0 \, \varepsilon_1 - \varepsilon_1 \, M_1^0 \, \alpha_{G,1}$$

that is: (8.75)

$$\frac{1}{S_1} \, \Phi_{G \rightleftarrows 1} = \varepsilon_1 \, (\varepsilon_G \, M_G^0 - \alpha_{G,1} \, M_1^0)$$

The very simple structure of the enclosure envisaged here lends itself particularly well to the formulation of algebraic solutions based on successive simplifications.

There are close to ten methods at the disposal of the heat transfer technician for solving this problem of radiant exchange between a gas and its enclosure at uniform temperatures. Only experience of the engineer who has tried several cases can decide which is the better compromise to retain between the complexity of a calculation leading to a solution that is reputed to be accurate, or a more simple and more rapid formulation leading to an approximate result. To illustrate we give numerical examples characteristic of heaters.

a. *Comparing calculation methods for a transparent flame*

We assume two cylindrical fireboxes or heaters without bricks under the following conditions (Table 8.6).

Eight calculation methods have been used based on the emission and absorption factors obtained from Hottel's data in Chapter 4, Section 4.3, (Ref. 4.1). These methods can be identified in the following way:

I.	Hottel's simplified method (Section 8.3.2.2). One spectral band for the emission and one for absorption is considered. Eqs. (8.68) and (8.70) are applied.
II.	Method (I), in which $\mathcal{F}_{1 \rightarrow G} = \mathcal{F}_{1 \leftarrow G} = \mathcal{F}_{1,G}$.
III.	Use of Eq. (8.72).
IV.	Use of Eq. (8.73).
V.	Use of Eq. (8.74).
VI.	Sharan's method, Section 8.3.3.1, in which one uses the graph leading to an estimated radiation transfer coefficient of the type: $$\alpha_R = \frac{\Phi_R}{\theta_G - \theta_1}$$
VII.	Use of Eq. (8.75).
VIII.	The hypothesis of a gray gas, Section 8.2.4.2, for which Eq. (8.47) is applied.

TABLE 8.6

ASSUMED OPERATING CONDITIONS FOR COMPARING HEAT TRANSFER
CALCULATION METHODS

Characteristics Type of enclosure	Small firebox	Large firebox
Length/diameter ratio	3	
Diameter (m)	0.3	2
Receiving surface S_1 (m²)	0.99	44
Volume $\mathcal{V}$ (m³)	0.0637	18.84
Equivalent length of radiation (1) $$L = 3.8 \frac{\mathcal{V}}{S_1} \text{(m)}$$	0.245	1.63
Emissivity of wall ε_1	0.8	
Wall temperature θ_1 (K)	400 (hot water)	800 (steam)
Fuel	$(CH_2)_n$	
Excess air (%)	15	
Absolute pressure in firebox p_t (atm)	1	
Partial pressure of CO_2; p_{CO_2} (atm)	0.115	
Partial pressure of H_2O; p_{H_2O} (atm)	0.115	
Average temperature of the combustion gases, θ_G (K)	1,500	

(1) This relation for the equivalent length of radiation is viable for chambers neither too long nor too short (M. Veron).

Numerical constants for applying these various heat-transfer calculations are shown in Table 8.7. These numbers correspond to a high opacity for gases between the walls of a large firebox radiating at 400 K ($\alpha_{G,1} = 0.46$, etc). At this relatively low temperature, a black body emits radiation mostly in the deep infrared.

Sample results from calculations based on these data and methods are shown in Table 8.8. In addition to net flux, the gas-to-wall and wall-to-gas emissions are shown as:

(a) Gas-to-wall transfer $= \dfrac{1}{S_1}\, \Phi_{1 \leftarrow G}$

(b) Wall-to-gas transfer $= \dfrac{1}{S_1}\, \Phi_{1 \rightarrow G}$

(c) Net transfer to wall $= \dfrac{1}{S_1}\, \Phi_{G \rightleftarrows 1}$

TABLE 8.7

FACTORS OF EMISSIVITY AND ABSORPTION FOR USE IN EQ. 8.17
PLUS BLACK BODY EMISSIONS FOR CALCULATING THE OPERATIONS
ASSUMED IN TABLE 8.6

	ε_G	ε_{2G}	$\alpha_{G,1}$		$\alpha_{2G,1}$		Black-body emissions
			400 K	800 K	400 K	800 K	
Small firebox	0.09	0.11	0.20	0.16	0.28	0.22	1) Gas ($\theta_G = 1{,}500$ K): $M_G^0 = 248 \cdot 10^3 \dfrac{\text{kcal}}{\text{m}^2 \cdot \text{hr}}$ (28.8 W/cm^2)
Large firebox	0.2	0.28	0.46	0.35	0.61	0.48	2) Wall at θ_1:
These values used in method	I, II, III, IV, V, VII, VIII	I	I, III, IV, V, VII,		I		400 K: $M_1^0 = 1{,}255 \dfrac{\text{kcal}}{\text{m}^2 \cdot \text{hr}}$ (0.15 W/cm^2) 800 K: $M_1^0 = 20{,}100 \dfrac{\text{kcal}}{\text{m}^2 \cdot \text{hr}}$ (2.33 W/cm^2)

TABLE 8.8

CALCULATED HEAT FLUXES FOR THE OPERATIONS ASSUMED IN TABLE 8.6
(Fluxes are kcal/m^2 with W/cm^2 in parentheses. The flue gas is an
equimolar mixture of CO_2 and H_2O at 1,500 K)

Conditions	Exchange	I	II	III	IV	V	VI	VII	VIII
Small firebox at $\theta_1 = 400$ K	Wall ← gas	18,700	18,700	21,800	21,250	20,100		17,850	21,900
	Wall → gas	− 220	− 95	− 240	− 240	− 225		− 200	− 110
	Net	18,480 (2.15)	18,605	21,260	21,010	19,875	18,000	17,650	21,790
Small firebox at $\theta_1 = 800$ K	Wall ← gas	18,700	18,700	21,800	21,500	20,100		17,850	21,900
	Wall → gas	−2,780	−1,520	−3,100	−3,100	−2,890		−2,570	−1,770
	Net	15,920 (1.85)	17,180	18,700	18,400	17,210	16,300	15,280	20,130
Large firebox at $\theta_1 = 400$ K	Wall ← gas	43,200	43,200	47,300	44,500	44,700		39,700	47,250
	Wall → gas	− 495	− 220	− 520	− 520	− 520		− 460	− 240
	Net	42,705 (4.95)	42,980	46,780	43,980	44,180	44,300	39,240	47,010
Large firebox at $\theta_1 = 800$ K	Wall ← gas	43,200	43,200	47,300	45,700	44,700		39,700	47,250
	Wall → gas	−6,075	−3,500	−6,480	−6,480	−6,330		−5,630	−3,830
	Net	37,125 (4.3)	39,700	40,820	39,220	38,370	40,200	34,070	43,420

The calculated numbers show that for the relatively low temperature of 400 K, the wall-to-gas transfer is but a small part, on the order of 1-2% of the net heat transferred. These results correspond to the case of hot water heaters. For manual calculation, therefore, one might dispense with the evaluation of exchange-coefficient $\mathscr{F}_{1 \to G}$ and apply Method **II**, or even neglect the emissions of the wall entirely ($M_1^0 = 0$).

For higher temperatures on the order of 800 K, as occurs with steam generators, the wall-to-gas transfer amounts to 17% of the net exchange by Method **I** and to 9% by Method **VIII**. If Method **I**, which is recommended by Hottel, is taken as basis for comparisons, the net transfers obtained by the other methods can be given the numbers of relative values shown in Table 8.9.

TABLE 8.9

VARIATIONS BETWEEN THE RESULTS OF EIGHT CALCULATION METHODS
FOR THE OPERATIONS ASSUMED IN TABLE 8.6

Method →	I	II	III	IV	V	VI	VII	VIII
Small firebox, 400 K	1	1.01	1.165	1.135	1.075	0.975	0.955	1.18
Small firebox, 800 K	1	1.08	1.175	1.155	1.08	1.025	0.96	1.265
Large firebox, 400 K	1	1.01	1.095	1.03	1.035	1.04	0.92	1.10
Large firebox, 800 K	1	1.07	1.10	1.06	1.035	1.085	0.92	1.17

These comparisons indicated that the hypothesis of all-gray gas (Method **VIII**) leads to considerable over-estimation of the heat transferred, especially when the firebox is small. The comparison also indicates that Methods **V**, **VI** and **VII**, which are convenient and fast, lead to answers that are generally close enough to be satisfactory for this kind of problem.

b.   *Comparing calculation methods for a luminous flame*

The examples treated above were for combustion gases ($CO_2 + H_2O$) emitting only in the infrared. It is interesting to evaluate how much the presence of carbon particles permits approaching the hypothesis of a wholly-gray flame.

Using the subscripts *F, G* and *C* to designate the presence in a flame of CO_2, H_2O, and dispersed carbon particles, respectively, we have according to A. Gouffe (Ref. 8.7):

$$f_F = f_G \times f_C \tag{8.76}$$

We will limit ourselves to calculation of the flame-to-wall transfer and compare results obtained by Hottel's Method **I** (Ref. 8.1) to those obtained for the assumption of a wholly gray flame (Method **VIII**).

According to Eq. (8.76) we have:

$$\varepsilon_F = 1 - (1 - \varepsilon_G)(1 - \varepsilon_C) \tag{8.77}$$

Saying that the concentration of soot particles is such that:

(a) For the small firebox: $f_F = 0.82$ that is $\varepsilon_F = 0.18$.

(b) For the large firebox: $f_F = 0.27$ that is $\varepsilon_F = 0.73$.

Method VIII (Wholly gray flame): By using ε_F in Eq. (8.47) to calculate the exchange factor $\mathscr{I}_{1,F}$:

$$\frac{1}{\mathscr{I}_{1,F}} = \frac{1}{\varepsilon_1} + \frac{1}{\varepsilon_F} - 1$$

Method I (Gray gas in x fraction of the energy spectrum): The cloud of carbon particles is considered as emitting across the spectrum with an emissivity of ε_C, defined by Eq. (8.77).

In the x fraction of the spectrum defined by Eq. (8.68), the gaseous mixture $CO_2 + H_2O$ emits with an emissivity such that:

$$(\varepsilon_G)_x = \frac{\varepsilon_G}{x}$$

from which comes the emissivity of the flame defined by Eq. (8.77):

$$(\varepsilon_F)_x = 1 - \left[1 - (\varepsilon_G)_x\right]\left[1 - \varepsilon_C\right] \tag{8.77 bis}$$

In the remaining fraction of the spectrum, $y = 1 - x$ (where the gas is transparent), only the cloud of particles radiates, from which:

$$(\varepsilon_F)_y = \varepsilon_C$$

The result is that according to Eq. (8.62) the flame-to-wall transfer will be defined by:

$$\mathscr{I}_{1 \leftarrow F} = x(\mathscr{I}_{1,F})_{(\varepsilon_F)_x} + y(\mathscr{I}_{1,F})\varepsilon_G$$

where the partial coefficients correspond to "gray conditions" obtained by putting $(\varepsilon_F)_x$ and ε_G in Eq. (8.47).

Table 8.10 compares calculated results for the important intermediate stages. This table shows that the differences between the two methods are considerably reduced from the differences calculated for transparent flames in Table 8.9. Agreement is excellent for the case of the large firebox and within about 6.5% for the small firebox, whereas with the transparent flame (Table 8.9) the corresponding differences were respectively 9.5 and 17%. As might be expected, the assumption of gray flame corresponds more closely to a luminous flame than to a transparent flame.

It can be seen in these examples that the flame-to-wall heat transfer varies approximately as the ratio of the flame's emission factors, on the order of 1 to 2 for a small firebox and 1 to 3.5 for a large firebox, as conditions change from a transparent flame to a luminous flame.

The flux densities shown in Table 8.10 are representative of those for fireboxes of domestic hot water heaters, which are on the order of 40 th/m². hr as well as those encountered in very large generating stations, which are on the order of 150 th/m². hr.

TABLE 8.10

CALCULATED RADIANT HEAT TRANSFER FOR THE OPERATIONS
ASSUMED IN TABLE 8.6 WHEN THE FLUE GAS CONTAINS SOOT PARTICLES
$(\theta_F = 1{,}500 \text{ K})$

Conditions	Small firebox	Large firebox
ε_F	0.18	0.73
ε_G	0.09	0.20
ε_C	0.098	0.662
x	0.116	0.334
y	0.884	0.666
Method I		
$\mathcal{F}_{1 \to F}$	0.162	0.616
Transfer gas → wall: $\dfrac{\text{kcal}}{\text{m}^2 \cdot \text{hr}}, (\ ): \dfrac{\text{W}}{\text{cm}^2}$	40,200 (4.66)	152,500 (17.7)
Method VIII		
$\mathcal{F}_{1,F}$	0.172	0.618
Transfer gas → wall: $\dfrac{\text{kcal}}{\text{m}^2 \cdot \text{hr}}, (\ ): \dfrac{\text{W}}{\text{cm}^2}$	42,700 (4.95)	153,300 (17.8)

8.3.3.3. The enclosure comprises two receiving surfaces

This is a practical case allowing for algebraic calculations and representing a generalization of the problem (Section 8.2.4.3) of heat transmission between the hearth and the roof of a furnace heated by a gray gas.

The general case becomes significant when one of the two surfaces (S_2) is a refractory roof surface whose losses through conduction are such that it can no longer be considered as having zero-net-flux — or even when this surface has zero-net-flux, it reflects less than is absorbed. This last aspect of the problem becomes important when the assumption of gray gas is no longer used (see Section 8.3.1 and paragr. 8.3.2.1.b).

The mathematical development of the case has been done by Hottel (Ref. 8.1) who uses the method in Section 8.3.2.2 by saying, to simplify:

$$S_1 \mathcal{F}_{1 \leftarrow G} = S_1 \mathcal{F}_{1 \rightarrow G} = S_1 \mathcal{F}_{1,G}$$

$$S_2 \mathcal{F}_{2 \leftarrow G} = S_2 \mathcal{F}_{2 \rightarrow G} = S_2 \mathcal{F}_{2,G}$$

(The subscript G is used to designate the absorbing medium, whether gas or flame.)

According to Eqs. (8.65) and (8.66), the three factors solving the exchange problem by radiation are such that:

$$\mathcal{F}_{1,G} = x\,[\mathcal{F}_{1,G}]_{f_x} \qquad \mathcal{F}_{2,G} = x\,[\mathcal{F}_{2,G}]_{f_x} \tag{8.65}$$

and

$$\mathcal{F}_{1,2} = x\,[\mathcal{F}_{1,2}]_{f_x} + (1-x)\,[\mathcal{F}_{1,2}]_{f=1} \tag{8.65 bis}$$

The terms for part gray exchange $[\mathcal{F}_{1,G}]_{f_x}, \ldots, [\mathcal{F}_{1,2}]_{f=1}$ are calculated from the methods described in Section 8.2.2, but leaving out the S_R terms in the system of Eqs. (8.27), (8.28) or (8.34), (8.35) and by saying:

$$f_{1,1} = f_{1,2} = \ldots = f_x$$

We have:

$$S_1\,(\mathcal{F}_{1,G})_{f_x} = \frac{\varepsilon_1 S_1}{\rho_1}\,\frac{{}_G D_1}{D} \tag{8.38 bis}$$

According to Eqs. (8.34) and (8.35):

$$S_1(\mathcal{F}_{1,G})_{f_x} = \frac{\varepsilon_1 S_1}{\rho_1}\,\frac{\begin{vmatrix} \overline{11} + \overline{12} - A_1\,,\ \overline{12} & \dfrac{A_2}{} \\[2mm] \overline{21} + \overline{22} - A_2\,,\ \overline{22} - \dfrac{A_2}{\rho_2} \end{vmatrix}}{\begin{vmatrix} \overline{11} - \dfrac{A_1}{\rho_1}\,,\ \overline{12} \\[2mm] \overline{12}\,,\ \overline{22} - \dfrac{A_2}{\rho_2} \end{vmatrix}} \tag{8.78}$$

with:

$$\overline{11} = S_1 F_{1,1} f_x \qquad \overline{22} = S_2 F_{2,2} f_x$$

$$\overline{12} = S_1 F_{1,2} f_x \qquad \overline{21} = S_2 F_{2,1} f_x = S_1 F_{1,2} f_x = \overline{12} \tag{8.79}$$

For this type of enclosure with two surfaces ($N = 2$), the number of independent-angles factors is reduced to 1 (see Section 8.2.1); and by adopting $F_{1,2}$, the development of Eq. (8.78) leads to:

$$S_1\,(\mathcal{F}_{1,G})_{f_x} = (1 - f_x)\,\frac{\varepsilon_1}{\rho_1}\,\frac{1 + \dfrac{S_1}{S_2} + \left(\dfrac{1}{\rho_2 f_x} - 1\right)/F_{1,2}}{(B)_{f_x}} \tag{8.80}$$

with:

$$(B)_{f_x} = \frac{\left(\dfrac{1}{\rho_1} - f_x\right)}{S_2} + \frac{\left(\dfrac{1}{\rho_1} - f_x\right)\left(\dfrac{1}{\rho_2} - f_x\right)}{S_1 \, F_{1,2} \, f_x} + \frac{\left(\dfrac{1}{\rho_2} - f_x\right)}{S_1}$$

$S_2(\mathscr{F}_{2,G})_{f_x}$ is obtained by exchanging the subscripts 1 and 2 in the preceding equation:

$$S_1(\mathscr{F}_{1,2})_{f_x} = \frac{\varepsilon_2 S_2}{\rho_2} \, \frac{{}_1 D_2}{D} \tag{8.31 bis}$$

that is, according to Eqs. (8.27) and (8.28):

$$S_1(\mathscr{F}_{1,2})_{f_x} = \frac{\varepsilon_2 S_2}{\rho_2} \; \frac{\begin{vmatrix} \overline{11} - \dfrac{S_1}{\rho_1}, & -\overline{11}\ \varepsilon_1 \\[2mm] \overline{12} & , & -\overline{12}\ \varepsilon_1 \end{vmatrix}}{\begin{vmatrix} \overline{11} - \dfrac{S_1}{\rho_1}, & \overline{12} \\[2mm] \overline{12} & , & \overline{22} - \dfrac{S_2}{\rho_2} \end{vmatrix}} \tag{8.81}$$

Development of this leads to:

$$S_1(\mathscr{F}_{1,2})_{f_x} = \frac{\dfrac{\varepsilon_1 \varepsilon_2}{\rho_1 \rho_2}}{(B)_{f_x}} \tag{8.82}$$

As for $(\mathscr{F}_{1,2})_{f=1}$, it will be such that:

$$S_1(\mathscr{F}_{1,2})_{f=1} = \frac{\dfrac{\varepsilon_1 \varepsilon_2}{\rho_1 \rho_2}}{(B)_{f_x=1}} \tag{8.83}$$

Equations (7.65), (7.66), (8.80), (8.82), (8.83) thus cover treatment of the problem of exchange by radiation, with x and f_x defined by Eq. (8.68).

The transfer equations are written, **taking into account convection coefficient**, h:

$$\Phi_{G \rightleftarrows 1} = S_1 \, \mathscr{F}_{1,G} \, \sigma(\theta_G^4 - \theta_1^4) + h_1 S_1(\theta_G - \theta_1)$$

$$\Phi_{G \rightleftarrows 2} = S_2 \, \mathscr{F}_{2,G} \, \sigma(\theta_G^4 - \theta_2^4) + h_2 S_2(\theta_G - \theta_2) \tag{8.84}$$

$$\Phi_{2 \rightleftarrows 1} = S_1 \, \mathscr{F}_{1,2} \, \sigma(\theta_2^4 - \theta_1^4)$$

Now S_2 is considered as a surface of refractory material, which has temperature θ_2 somewhere between θ_1 and θ_G but **unknown a priori**, and which has thermal conductivity λ_2, as well as thickness a_2.

Designating θ_0 as the temperature outside the enclosure, a heat balance for S_2 can be written:

$$\text{heat earned} = \text{heat transmitted to the outside}$$

$$(\Phi_2) = S_2 \frac{\lambda_2}{a_2} (\theta_2 - \theta_0)$$

that is:

$$S_2 \frac{\lambda_2}{a_2} (\theta_2 - \theta_0) = \Phi_{G \rightleftarrows 2} + \Phi_{1 \rightleftarrows 2}$$

and according to Eq. (8.84):

$$S_2 \frac{\lambda_2}{a_2} (\theta_2 - \theta_0) = S_2 \mathfrak{F}_{2,G} \, \sigma(\theta_G^4 - \theta_2^4) + h_2 S_2 (\theta_G - \theta_2) -$$
$$- S_1 \mathfrak{F}_{1,2} \, \sigma(\theta_2^4 - \theta_1^4) \quad (8.85)$$

This last equation permits calculating θ_2, which can then be introduced in Eq. (8.84) to estimate the heat flux.

Special cases

Surface S_1 is the useful, or primary receiver, and all flux transmitted to the outside by surface S_2 corresponds to a loss. Thus two conditions may be assumed, depending on whether S_2 loses no heat or only that heat transmitted by convection:

1. **No heat loss:** this is the ideal case, in which transmission to the outside is zero; it occurs only when the refractory is an excellent insulator. Putting $\lambda_2 = 0$ in Eq. (8.85), we have:

$$S_1 \mathfrak{F}_{1,2} \, \sigma(\theta_2^4 - \theta_1^4) = S_2 \mathfrak{F}_{2,G} \, \sigma(\theta_G^4 - \theta_2^4) + h_2 S_2 (\theta_G - \theta_2) \quad (8.86)$$

that is:

$$\Phi_{2 \rightleftarrows 1} = \Phi_{G \rightleftarrows 2}$$

Because the amount of heat received by primary surface, S_1, is expressed as:

$$(\Phi_1)_{\text{net}} = \Phi_{2 \rightleftarrows 1} + \Phi_{G \rightleftarrows 1}$$

we have according to Eq. (8.86):

$$(\Phi_1)_{\text{net}} = \Phi_{G \rightleftarrows 1} + \Phi_{G \rightleftarrows 2}$$

This is the maximum flux that can be captured by S_1.

2. Heat lost by convection only: Eq. (8.85) is reduced to:

$$S_1 \mathcal{F}_{1,2}\, \sigma(\theta_2^4 - \theta_1^4) = S_2 \mathcal{F}_{2,G}\, \sigma(\theta_G^4 - \theta_2^4) \tag{8.87}$$

Because the balance of exchange by radiation between the inside surface S_2 of the refractory and the remainder of the enclosure is expressed by:

$$(\Phi_2)_{\text{radiation}} = S_2\, \mathcal{F}_{2,G}\, (\theta_G^4 - \theta_2^4) - S_2\, \mathcal{F}_{2,1}\, \sigma\, (\theta_2^4 - \theta_1^4)$$

Eq. (8.87) shows that $(\Phi_2)_{\text{radiation}}$ is zero.

This justifies describing the surface as having zero-net-radiant-flux, because the flux it loses to the outside is the same as that transmitted to it by convection. Such a condition closely resembles that of steel open-hearth furnaces where the arch is made of highly reflective but relatively conductive silica firebrick that is swept by the recirculating gases during melting-down (cf. Section 8.2.4.3, third case). The net flux gained **by radiation** by surface S_1:

$$(\Phi_1)^{\text{net}}_{\text{radiation}} = (\Phi_{G \rightleftarrows 1})_{\text{radiation}} + \Phi_{2 \rightleftarrows 1}$$

is written under these conditions as:

$$(\Phi_1)^{\text{net}}_{\text{radiation}} = (\Phi_{G \rightleftarrows 1})_{\text{radiation}} + (\Phi_{G \leftrightarrows 2})_{\text{radiation}}$$

because, according to Eq. (8.87) we have:

$$\Phi_{2 \rightleftarrows 1} = (\Phi_{G \rightleftarrows 2})_{\text{radiation}}$$

Taking θ_2^4 from Eq. (8.87) and placing it in Eq. (8.84), Hottel expresses the radiant flux gained by S_1 in the form:

$$(\Phi_1)^{\text{net received}}_{\text{radiation}} = (\Phi_G)^{\text{net lost}}_{\text{radiation}} =$$

$$\left[S_1\, \mathcal{F}_{1,G} + \cfrac{1}{\cfrac{1}{S_1\, \mathcal{F}_{1,2}} + \cfrac{1}{S_2\, \mathcal{F}_{2,G}}} \right] \sigma(\theta_G^4 - \theta_1^4) \tag{8.88}$$

This last Equation is viable when S_2 is a surface with zero-net-radiant-flux of type S_R. In the earlier derivations, S_R was simply designated as surface with "zero-net-flux" because convection was not considered. Accordingly, Eq. (8.88) takes into account the emission factor, ε_2 of the refractory material in Eqs. (8.65), (8.66), (8.80), (8.81), defining the exchange factors $\mathcal{F}_{1,G}, \mathcal{F}_{2,G}$ and $\mathcal{F}_{1,2}$.

It is easily shown that if $\rho_2 = 1$ ($\varepsilon_2 = 0$), the term between brackets in Eq. (8.88) is reduced to Eq. (8.69) of Section 8.3.3.1, which describes a refractory surface that is completely reflecting.

8.4   RADIANT EXCHANGE IN A GRAY ENCLOSURE WITH A HETEROGENEOUS ABSORBENT MEDIUM

8.4.1. General

Sections 8.3.2 and 8.3.3 were concerned with a uniformly emissive gas completely enveloped by the firebox. They permitted analyzing radiant transfers in which part of the enclosure acts on all the others and is acted on by them. The formulas thus obtained are mathematical analogies that conform to actual cases of heat exchange.

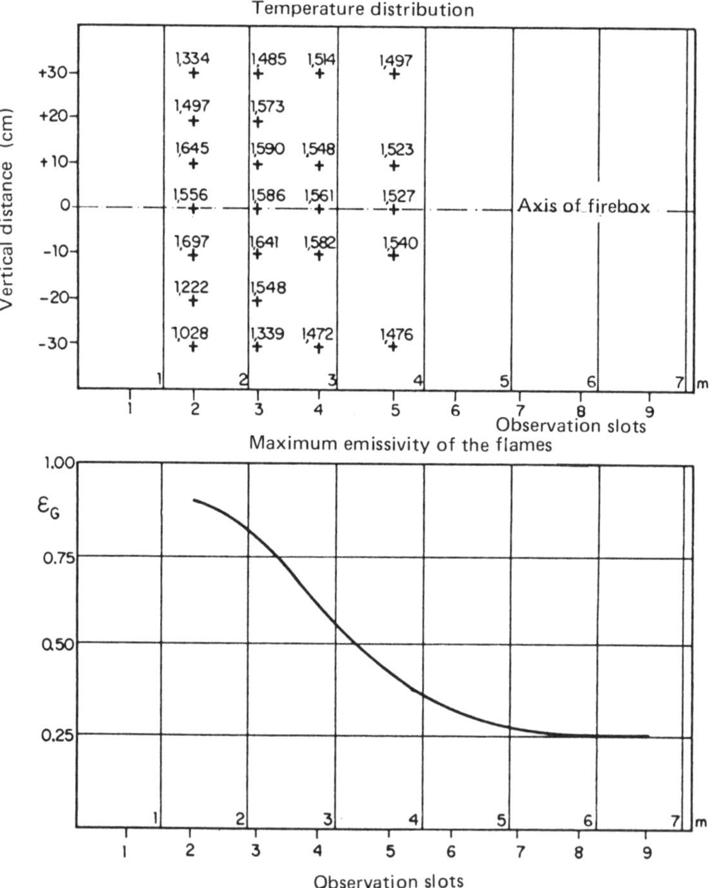

Fig. 8.12.   The distribution of temperatures and emissivities of combustion gases through the length of a firebox.

For example, measurement of the efficiency of a firebox and recordings of emissivity will permit assigning the gases an average temperature θ_G and an average emissivity from which the heat duty can be estimated for a surface not accessible to direct measurement; and from this, predictions can be made for modifying initial conditions.

When simplifying approximations are to be avoided, it is generally necessary to take into account the great variations within the absorbent medium for which Fig. 8.12 gives an example. Under these conditions, the complete treatment becomes very complicated; it requires subdividing the gaseous medium into differential layers, each of which is assigned homogeneous properties, after which the preceding methods are applied, all the while taking into account partial heat balances that define individual temperatures.

The innovators of these methods are Hottel and Cohen (Ref. 8.8), who in 1958 envisaged the case of rectangular fireboxes subdivided into rectangular layers. Again in 1965, Hottel and Sarofim used computers to study the exchange problems of cylindrical fireboxes subdivided into circular layers, taking into account the configuration of the flow (Ref. 8.6).

At this level of complexity, tests on experimental models supply a viable competitive method for treating the problem, although the criteria for correlations between models and actual units remain yet to be improved. The difficulty of the theoretical problem of any firebox can be reduced if legitimate simplifications based on experience can be found to aid in relatively rapid solutions.

8.4.2. Exchange calculations based on discrete envelopes of flame

The simplest model to account for the heterogeneity of a gaseous medium consists in defining a "flame" space and a "combustion-gas" space within the firebox, with each of the two zones having uniform emissive properties. This distinction gains validity from the fact that the emissive factors and temperatures encountered in the reacting medium are generally much higher than those of the corresponding combustion products.

8.4.2.1. The flame considered as the only source of radiation

For extremely radiant flames, it can be feasible to ignore emissions by the combustion products and consider only the radiation from the flame, whose envelope is S_F and whose average emission factor is ε_F. Taking into account that exchange factors of the type $\mathscr{F}_{i,G}$ (or $\mathscr{F}_{i,F}$) from the preceding sections were applied where an enclosure of total surface, S_t, was the envelope, the introduction of S_F will lead to defining new approximate gray solutions of the form:

$$(\mathcal{G}_{i,F})_{S_F} = \frac{S_F}{S_t} (\mathcal{G}_{i,F})_{S_F} \tag{8.89}$$

with $S_F < S_t$.

Equations of this type have sometimes been used to estimate the surface-envelope of flames in large combustion chambers, from readings of the duties at the walls and the emission factors. Thus, Godrige (Ref. 8.2) points out that for three types of heaters at generator stations, one was able to combine Eq. 8.89 and the mathematical model in Section 8.2.4.1 to determine that S_F represented a third of the surface of the water tubes. In fact, this flame is best considered as sufficiently opaque to be treated as a non-transparent radiant source ($f = 0$). One is then brought back to an interface exchange problem in non-absorbent medium. Also, this is the way the problem of exchange in fireboxes with grates for solid fuels is treated.

Thus, the enclosure in Section 8.2.4.1 has a receiving surface, S_1, plus a refractory surface with zero-net-flux, within which an opaque flame assumes a convex form with surface, S_F, generating a flame-wall such that:

$$\mathcal{G}'_{F,1} = \cfrac{1}{\cfrac{1}{\overline{F}_{F,1}} + \left(\cfrac{1}{\varepsilon_F} - 1\right) + \cfrac{S_F}{S_1}\left(\cfrac{1}{\varepsilon_1} - 1\right)}$$

with:

$$\overline{F}_{F,1} = \cfrac{1 - \cfrac{S_F}{S_1} F_{F,1}^2}{1 + \cfrac{S_F}{S_1}(1 - 2F_{F,1})}$$

$\overline{F}_{F,1}$ characterizing the position of the wall exposed to the flame.

The flame-wall exchange is then put in the form:

$$\Phi_{F\rightleftarrows 1} = S_F\, \mathcal{G}_{F,1}\, \sigma(\theta_F^4 - \theta_1^4) = S_1\, \mathcal{G}_{1,F}\, \sigma(\theta_F^4 - \theta_1^4)$$

In the case of the enclosure that is entirely receiver (i.e. $\overline{F}_{F,1} = 1$) and surface S_1 completely surrounds the flame) the result is:

$$\overline{F}_{F,1} = 1$$

and

$$\mathcal{G}'_{F,1} = \cfrac{1}{\cfrac{1}{\varepsilon_F} + \cfrac{S_F}{S_1}\left(\cfrac{1}{\varepsilon_1} - 1\right)}$$

Although the hypothesis of a completely opaque flame with uniform emissivity is not generally valid in a firebox, M.A. Field (Ref. 8.9) points out that useful results can nevertheless be arrived at with this concept.

8.4.2.2. Radiation from a flame versus radiation from burner gases

In its most simple form, the just-described mathematical model amounts to an assumption that the flame and the burned gases are similar to gray bodies with temperature θ_F and θ_G, and with emission factors ε_G and ε_F.

For a receiving wall of type S_i the exchange between flame-plus-gas and the wall is then written:

$$\Phi_{(F+G)\rightleftarrows i} = S_i\,\mathcal{F}_{i,G}\,\sigma(\theta_G^4 - \theta_i^4) + S_i\,\mathcal{F}_{i,F}\,\sigma(\theta_F^4 - \theta_i^4) \tag{8.90}$$

A. Hossard (Ref. 8.10) discusses an interesting practical example of this case, working with a set-up for experimentally studying transfers for different chamber pressures and various types of burners. The cylindrical firebox has a total surface, S_t, a receiving surface, S_1, and a refractory surface of zero-net-flux, $S_R = S_t - S_1$. This firebox is surrounded by a layer of water that makes it possible to take heat readings. With the experimental readings, the author establishes a mathematical model that permits him to finally characterize the different fuel oil flames that he has produced. This model involves the simplification of ignoring the successive reflections of the radiation on the walls.

Under these conditions, a factor of type $\mathcal{F}$ is written $\varepsilon_1\,\overline{F}, \overline{F}$, characterizing the exchange factor relative to a black receiving wall facing a refractory surface with zero-net-flux. Returning to Eq. 8.44 in Section 8.2.4.1, an equation in which for an approximation $F_{R,1} = \dfrac{S_1}{S_t} = C$ (according to the checkered surface theory) one obtains:

1. For gases enveloped by the enclosure:

$$\overline{F}_{1,G} = \frac{\varepsilon_G}{C + \varepsilon_G(1 - C)}$$

from which:

$$\mathcal{F}_{1,G} = \varepsilon_1\,\overline{F}_{1,G} = \frac{\varepsilon_1\,\varepsilon_G}{C + \varepsilon_G(1 - C)} \tag{8.91}$$

2. For a flame enveloped by surface, S_F:

$$\mathcal{F}_{1,F} = \frac{S_F}{S_t}\left[\frac{\varepsilon_1\,\varepsilon_F}{C + \varepsilon_F(1 - C)}\right](1 - \varepsilon_G) \tag{8.92}$$

This last equation derived from Eq. (8.89), takes into account the fact that the flame is surrounded by a gaseous medium whose transmission factor is

$(1 - \varepsilon_G)$. Thus according to Eq. (8.90) designating $(\Phi_1)_R$ as the net amount of heat received by the wall, leads to:

$$(\Phi_1)_R = S_1 \varepsilon_1 \sigma \left\{ \left[\frac{\varepsilon_G}{C + \varepsilon_G (1 - C)} \right] (\theta_G^4 - \theta_1^4) + \frac{S_F}{S_t} \times \right.$$

$$\left. \times \frac{\varepsilon_F}{C + \varepsilon_F (1 - C)} (1 - \varepsilon_G)(\theta_F^4 - \theta_1^4) \right\} \qquad (8.93)$$

On the other hand, θ_F and θ_G are related through an energy-balance of the heat introduced to the firebox. If the temperature of the gases leaving the firebox is equal to θ_G, and if H_0, H_F and H_G are the enthalpies of the reactants at θ_0, the combustion products at θ_F, and the flame at θ_G, respectively, we have successively:

1. From upstream to downstream of the firebox:

$$(H_0)_{\theta_0} = (\Phi_1)_R + (H_G)_{\theta_G} \qquad (8.94)$$

2. Inside the firebox:

$$(H_0)_{\theta_0} = \Phi_{F \rightleftarrows 1} + \Phi_{F \rightleftarrows G} + (H_F)_{\theta_F}$$

Thus, by clarifying the radiation terms:

$$(H_0)_{\theta_0} = \varepsilon_1 \sigma S_1 \frac{S_F}{S_t} \frac{\varepsilon_F}{C + \varepsilon_F (1 - C)} (1 - \varepsilon_G)(\theta_F^4 - \theta_1^4) +$$

$$+ \varepsilon_F \varepsilon_G \sigma S_F (\theta_F^4 - \theta_G^4) + (H_F)_{\theta_F} \qquad (8.95)$$

To complete his mathematical model, Hossard defines the dimensional characteristics of the flame and the emission factors of the gas-flame mediums for the equations as:

1. For the firebox gases:

$$\varepsilon_G = f_1 \quad \text{(pressure in the firebox, } L_G, \text{ excess air, } \theta_G)$$

$$\text{with } L_G \# 3.4 \frac{\text{gas volume}}{S_t + S_F} \qquad (8.96)$$

2. For the flame:

Volume $\mathcal{V}_F = f_3$ (H_0, excess air, pressure in the firebox, θ_F, burner geometry)   (8.97)

$$S_F = 4.84 \; \mathcal{V}_F^{2/3} \left. \right\} \text{ Shape of flame roughly} \qquad (8.98)$$

$$L_F = 0.705 \; \mathcal{V}_F^{1/3} \left. \right\} \qquad \text{spherical} \qquad (8.99)$$

$$\varepsilon_F = f_2 \text{ (pressure in the firebox, } L_F, \text{ excess air, } \theta_F) \qquad (8.100)$$

In this way eight equations, Eqs. (8.93) to (8.100), are developed for calculating the unknown quantities, $(\Phi_1)_R$, $\mathcal{V}_F$, S_F, L_F, θ_F, ε_F, θ_G, ε_G by using a computer.

This mathematical model, established industrially, is found to produce very good agreement between the pairs of values $(\theta_G, (\Phi_1)_R)$ measured and calculated on a combustion chamber operating:

(a) With two fuel oils of heavy and light specifications.
(b) With three types of burners.
(c) With pressures varying between 3 and 7 bars.

The development of Eqs. (8.96), (8.97) and (8.100) is necessarily specific to the conditions described; but the concept of this model provides an interesting approach for studying a transfer problem in an actual firebox.

8.4.3. Dividing the enclosure into elementary zones

The walls of the enclosure and the gaseous medium can be divided into isothermal zones.

In 1958, Hottel and Cohen published their work done with rectangular zones for studying exchange problems in rectangular fireboxes (Ref. 8.8).

In 1967, Hottel and Sarofim studied cylindrical fireboxes using the results of Erkku (Ref. 8.13) relative to circular zones.

The principle for solving this type of exchange problem is made up of two principal steps:

(a) A geometric part that determines the different exchange factors.
(b) A purely thermal part that simultaneously determines the temperatures and the heat flux for all the zones envisaged.

Details of these complicated calculations may be obtained from the references. We will discuss only the principles.

8.4.3.1. Exchange factors

There are three types of exchange factor: from surface to surface, from surface to gas, and from gas to gas. Hottel and his collaborators assigned these factors a surface area in order to be able to treat the gases and the walls in a consistent manner.

Thus by assigning symbol s to surfaces and symbol g to gases, the incident flux on a receiver of row j coming directly from an emitter of row i, without reflection to the walls, is written:

$$\Phi_{i \to j} = (\overline{s_i s_j})\, M^0_{s_i} \qquad\qquad \text{or } (\overline{g_i s_j})\, M^0_{g_i}$$
$$\text{or } (\overline{s_i g_j})\, M^0_{s_i} \qquad\qquad \text{or } (\overline{g_i g_j})\, M^0_{g_i} \tag{8.101}$$

where M_i^0 is the black-body emission from the body row i (surface or gas). The direct exchange factors, $\overline{ss}$, $\overline{sg}$, $\overline{gg}$ are the products of two terms:

1. A term with the dimensions of a surface area: Actual surface, S_i, for an emitting surface and equal to the product of volume, $\mathcal{V}_i$, times absorption coefficient K_i for a gas.

2. A dimensionless factor of direct reception, f.

This second term, f, has been tabulated as a function of the relative positions of the zones and the product KL, in which L represents a characteristic for the form, whether square, cubic or cylindrical. Thus f represents the fraction of a zone's radiation that is absorbed by another zone in a black enclosure and constitutes a generalization of the group Ff (or F in a non-absorbent medium) discussed in Section 8.1.

Hottel and Cohen present, as follows, the properties of the exchange factors and factors of direct reception. (Subscript 1 indicates the emitter):

1. **Properties of reversibility:**

$$\overline{s_1 s_2} = \overline{s_2 s_1}$$
$$\overline{g_1 g_2} = \overline{g_2 g_1} \, , \, \overline{g_1 s_2} = \overline{s_2 g_1}$$
$$4 K \mathcal{V}_{g_1} f_{g_1, s_2} = S_{s_2} f_{s_2, g_1} \tag{8.102}$$

where $K\mathcal{V}_{g_1}$ represents the equivalent of an emissive surface for the gas.

2. **Additive properties:**

$$\overline{s_1 s_1} + \overline{s_1 s_2} + \overline{s_1 s_3} + \ldots + \overline{s_1 g_1} + \overline{s_1 g_2} + \overline{s_1 g_3} + \ldots = S_{s_1} \left.\right\}$$
$$\overline{g_1 g_1} + \overline{g_1 g_2} + \overline{g_1 g_3} + \ldots + \overline{g_1 s_1} + \overline{g_1 s_2} + \overline{g_1 s_3} + \ldots = 4 K \mathcal{V}_{g_1} \left.\right\} \tag{8.103}$$

The overall factors for exchange by radiation, which take into account successive reflections to the walls, are represented by the symbols:

$\overline{SS}$ for exchange from surface to surface.
$\overline{SG}$ for exchange from surface to gas.
$\overline{GG}$ for the exchange between gaseous zones.

Since these have the dimension of a surface, they are the generalization of the groups of type $S\mathcal{F}$ described in Section 8.2.1. They are calculated from the direct exchange factors $\overline{ss}$, $\overline{sg}$ and $\overline{gg}$ defined previously, according to the methods derived from those in Section 8.2.2, as follows:

1. **The initial emitter is a surface of type S_i:**

S_1, for example, is an emitter with a temperature other than zero. The Eq. (8.25) is written:

$$\overline{S_1 S_n} = S_n \frac{\varepsilon_n}{1 - \varepsilon_n} \, _1\varphi_n$$

The equations that define $_1\varphi_1$, $_1\varphi_2$, $_1\varphi_n$ are the same kind as Eqs. (8.27), (8.28) ... in which $S_1\,F_{1,1}\,f_{1,1}$, $S_1\,F_{1,2}\,f_{1,2}$ ... are replaced by $s_1 s_1$, $s_1 s_2$ ...

2. The initial emitter is a gas of type g_r:

Taking the gas zone g_i as the only emitter of the system, Eq. (8.32) is now written:

$$\overline{G_i\,S}_n = \overline{S_n\,G}_i = \frac{\varepsilon_n}{1 - \varepsilon_n}\,S_n\,_{g_i}\varphi_n$$

The Equations (8.34), (8.35) ... , which define $_g\varphi_n$ become, for surfaces S_1 and S_2, for example:

$$(\overline{s_1 s}_1\,_{g_i}\varphi_1 + \overline{s_1 s}_2\,_{g_i}\varphi_2 + \ldots + \overline{g_i s}_1)\,\rho_1 = S_1\,_{g_i}\varphi_1$$

$$(\overline{s_1 s}_2\,_{g_i}\varphi_1 + \overline{s_2 s}_2\,_{g_i}\varphi_2 + \ldots + \overline{g_i s}_2)\,\rho_2 = S_2\,_{g_i}\varphi_2$$

The reflected flux densities thus calculated can be used to determine exchange factors of type $\overline{G_i G}_j$ relative to two gases i and j. Since the enclosure is composed of N surfaces, we have:

$$\overline{G_i G}_j = \overline{g_i g}_j + \Sigma_N (_{g_i}\varphi_n)\,(\overline{s_n g}_j)$$

This last relation accounts for the fact that the flux received by zone j is equal to the flux transmitted directly by zone i plus the sum of the fractions of the flux reflected by all the surfaces reaching j.

Finally the overall exchange factors have properties similar to those identified in Eq. (8.103) for the direct factors:

$$\overline{S_1 S}_1 + \overline{S_1 S}_2 + \overline{S_1 S}_3 + \ldots + \overline{S_1 G}_1 + \overline{S_1 G}_2 + \overline{S_1 G}_3 + \ldots = \varepsilon_1 S_1$$

$$(8.104)$$

$$\overline{G_1 G}_1 + \overline{G_1 G}_2 + \overline{G_1 G}_3 + \ldots + \overline{G_1 S}_1 + \overline{G_1 S}_2 + \overline{G_1 S}_3 + \ldots = 4K\,\mathcal{V}_{g_1}$$

$$(8.105)$$

When the gas medium can not be taken as entirely gray, the weighted coefficients described in Section 8.3.2 are applied to factors $\overline{SS}$, $\overline{GS}$ and $\overline{GG}$.

Thus for a real gas reduced to a gray component of the energy fraction x (according to the principle in Section 8.3.2.2) we have:

$$\overline{SS} = x(\overline{SS})_{f_x} + (1 - x)(\overline{SS})_{f=1}$$
$$\overline{GS} = x(\overline{GS})_{f_x}$$
$$\overline{GG} = x(\overline{GG})_{f_x}$$

Finally, the two directions that control a mutual exchange should be distinguished because of the difference between emission and absorption for a real gas (see Section 8.3.2.1). Thus between a surface S_i and a gas V_j, we will have:

$$\Phi_{V_j \rightleftarrows S_i} = \vec{G_j S_i} M_j^0 - \vec{S_i G_j} M_i^0 \quad (1)$$

$\vec{G_j S_i}$ being calculated from ε_j and $\vec{S_i G_j}$ from $\alpha_{i,j}$.

The symbol S in the second term of this expression can not be dissociated from G, since the two letters together represent the exchange factor. S should not be confused with S_i in the first term, which represents the active surface being considered.

Estimating the exchange factors requires knowing especially the local properties of the gaseous medium. Because of this, this first part of the problem is not totally independent of the second purely thermal part devoted to the simultaneous calculation of temperatures and flux in the enclosure.

8.4.3.2. Energy balances by zones

In its most complete form this second part of the problem takes into account exchange by convection as well as by radiation. To do the energy balances in an enclosure divided up into zones, it is necessary to know a priori the configuration of the flow and the field of internal concentrations. From this data, the net flux of sensible heat and the amount of heat liberated by combustion is derived in each one of the gas zones.

For an emissive medium similar to a group of gray components, the balance for a gray zone V_i is thus written:

$$\Sigma_j \vec{S_j G_i} M_{s_j}^0 + \Sigma_j \vec{G_j G_i} M_{g_j}^0 - 4\Sigma_n x k_x V_i M_{g_i}^0$$

+ net sensible flux in V_i across its frontiers
+ net convection flux on contiguous surfaces, if such occur
+ flux due to the combustion in zone $V_i = 0.$ (8.105)

The balance for a surface zone, S_i, is written:

$$\Sigma_j \vec{G_j S_i} M_{g_j}^0 + \Sigma_j \vec{S_j S_i} M_{s_j}^0 - S_i \varepsilon_i M_i^0 + h S_i (\theta_{g_k} - \theta_{s_i})$$

$$- \Phi_{\text{net gain}} \text{ (negligible term if } S_i \text{ has zero-net-flux like } S_R) = 0 \quad (8.106)$$

Subscript k designates the gaseous zone contiguous to the surface S_i and h is the convection coefficient.

With the system of equations established from equations of type Eqs. (8.105) and (8.106), the field of temperatures in the enclosure and the flux captured by the receiving surfaces can be determined.

The values obtained for the temperatures of gas zones can give new estimates of the exchange factors dependent on the emission and absorption properties of the medium, if those depart considerably from the original values.

These calculation methods are currently applied or foreseen in the field of large heaters where the design of a new enclosure requires knowledge extending beyond empirical data. To this end, the tests on models that permit predicting the fields of flow, of concentrations and even of temperatures, furnish the data necessary for exploratory calculations (Refs. 8.11 and 8.12).

8.5. CONCLUSION

The calculation of heat transfer within a fired enclosure is truly complex. Nevertheless, an examination of the methods established by Hottel and his school shows that the calculations for practical problems can often be simplified by means of assumptions based on experience.

It should be noted that the firebox's enclosure plays the roles not only of container but also of modifier and stabilizer for fired heat consumption. First of all, it isolates the fluids of combustion from the rest of the atmosphere and enhances internal recirculation currents for diluting the initial oxidizer and fuel with the combustion products. This modifies the chemical variables and the heat liberated at the burner outlet such that the type of enclosure has a primary influence in achieving the desired temperature and its exposure to the charge.

But this enclosure also plays a capital role in heat transfer by radiation, particularly when taken to a high temperature. The chemical energy available in the oxidizer-fuel mixture is irreversibly liberated in the flame in a very short time. This means that the heat energy is liberated within a small volume, which is, in fact, the surface of the flame multiplied by the thickness of the reaction zone. The result is that the firebox encloses considerable variations of temperature that must be accounted for by features of furnace design.

In spite of temperature variations, however, heat flows toward the charge much more steadily. This because the enclosure not only holds the reaction gases, but also reflects radiant energy coming from the hot gases and opposing walls so that this radiant energy travels in all directions throughout the enclosure at the speed of light thus becoming uniform through successive reflections, diffusions, absorptions and reemissions. The contained charge is thus bathed in a radiation of such complex origin that there is little variation from one point to another except when the hottest parts of the flame are purposely directed onto the charge as in steel melting furnaces.

This effect of smoothing temperature gradients through radiation is particularly marked in fireboxes such as melting furnaces, that contain charges heated to very high temperatures approaching those of the flame supplying the energy. Although convective heat transfer may not be active under these conditions, the transfer through radiation remains strong since it varies with the fourth power of the temperature.

The same holds true for a slight change in the operating conditions for a furnace or heater where the intensity of radiant transfer responds almost immediately to any change imposed by the controllers on the feed to the burners.

Finally, the Stefan-Boltzmann Equation relating radiation to the fourth power of the temperature shows that any operation permitting an increase in temperature of the flames will increase the rate of radiation inside the enclosure even when rapid combustion generates products with lower emissivities.

The application of all these effects to furnace design was still in the realm of art at the beginning of the 1950s. Since then, however, developments such as those to which this book has continually referred have enabled heating engineers to make more satisfactory calculations for their problems, even when those calculations do not provide rigorous models for a specific furnace or heater.

REFERENCES

8.1   HOTTEL, H.C. – Chapter IV in: Mc ADAMS, W.H. *Transmission de la chaleur*, 2nd ed. française. Dunod, Paris, 1961.

8.2   GODRIDGE, A.M. – *J. of the Institute of Fuel,*vol. 50, p. 300, July 1967.

8.3   JAEGLE, J. – Report GN-4 of the *GEFGN-DETN*, Nov. 1965.

8.4   VERON, M. – *Cours de Thermique Industrielle, 2e année : Transmission de la chaleur. CNAM* 1965.

8.5   SHARAN, H.N. – *J. of the Institute of Fuel*, vol. 44, p. 268, July 1963.

8.6   HOTTEL, C., SAROFIM, A.F. – *Int. J. Heat Mass Transfer.*, vol. 8, p. 1153, 1965.

8.7   GOUFFE, A. – *Rev. Générale de Thermique*, vol. 2, No. 19 et 20, July-August 1963.

8.8   HOTTEL, H.C., COHEN, E.S. – *AICHE, Journal*, vol. 4, No. 1, p. 3.

8.9   FIELS, M.A., GILL, C.W., MORGAN, B.B., HAWSKLEY, P.G.W. – *Combustion of pulverised fuel*, 3rd part. "Heat transfer". *BCURA,* 1967.

8.10 HOSSARD, A. – *Rev. Générale de Thermique*, No. 72, p. 1565, Dec. 1967.

8.11 LOISON, R. – Les modèles mathématiques de flammes. *VIIe Journée d'Etudes sur les Flammes*, Paris, April 1968.

8.12 MALHOUITRE, G. – "Modèles générateurs de vapeur". . . *VIIe Journée d'études sur les Flammes*, Paris, April 1968.

8.13 ERKKU, H. – Thesis of Chemical Engineering. *MIT.* Cambridge, March, 1959.

GENERAL REFERENCES TO RADIANT HEAT TRANSFER

GOUFFE, A. – *Transmission de la chaleur par rayonnement.* Eyrolles, Gauthier Villars. Paris, 1968.

HOTTEL, H.C., SAROFIM, A.F. – *Radiative transfer.* Mc Graw Hill Company. New York, 1967.

Mc ADAMS, W.H. – *Heat Transmission*, 3rd ed. Mc Graw Hill Company. New York, 1954. Published in French as: *Transmission de la chaleur,* 2nd ed. Dunod. Paris, 1961.

SCHACK, A. – *Der Industrielle Warmeübergang*, 3rd ed. Verlag Stahleisen mbH. Dusseldorf, 1948.

index

ACHEVE D'IMPRIMER
EN DECEMBRE 1984
PAR L'IMPRIMERIE LOUIS JEAN
05002 GAP
N° d'éditeur : 647
Dépôt légal : 516 . Octobre 1984

IMPRIME EN FRANCE